Waves and Oscillations

A Prelude to Quantum Mechanics

Walter Fox Smith

OXFORD

UNIVERSITY PRESS

2010

OXFORD
UNIVERSITY PRESS

Oxford University Press, Inc., publishes works that further
Oxford University's objective of excellence
in research, scholarship, and education.

Oxford   New York
Auckland   Cape Town   Dar es Salaam   Hong Kong   Karachi
Kuala Lumpur   Madrid   Melbourne   Mexico City   Nairobi
New Delhi   Shanghai   Taipei   Toronto

With offices in
Argentina   Austria   Brazil   Chile   Czech Republic   France   Greece
Guatemala   Hungary   Italy   Japan   Poland   Portugal   Singapore
South Korea   Switzerland   Thailand   Turkey   Ukraine   Vietnam

Published by Oxford University Press, Inc.
198 Madison Avenue, New York, New York 10016
www.oup.com

Library of Congress Cataloging-in-Publication Data
Smith, Walter Fox.
Waves and oscillations : a prelude to quantum mechanics /
Walter Fox Smith.
p. cm.
Includes index.
ISBN 978-0-19-539349-1
1. Wave equation. 2. Mathematical physics. I. Title.
QC174.26.W28S55 2010
530.12'4–dc22   2009028586

9   8   7   6   5   4   3

Printed in the United States of America
on acid-free paper

This book is dedicated to my mother, Barbara Leavell Smith, and to my wife, Marian McKenzie

Preface

To the student

I wrote this book because I was frustrated by the other textbooks on this subject. Waves and oscillations are enormously important for current research, yet other books don't stress these connections. The ideas and techniques that you will learn from this book are *exactly* what you need to be ready for a study of quantum mechanics. Every physics professor understands this linkage, and yet other books fail to emphasize it, and often use notations which are different from those used in quantum mechanics. Other books make little effort to keep you engaged. I can't teach you by myself, nor can your professor; you have to learn, and to do this you must be active. In this book, I've provided tools so that you can assess your learning as you go; these are described immediately after the table of contents. Use them. Read with paper and pencil handy. As a scientist, you know that only by understanding the assumptions made and the details of the derivations can you have your own logical sense of how it all fits together into a self-consistent whole. Visit this book's website. There, you will find links to current physics, chemistry, biology, and engineering research that is related to the topics in each chapter, as well as lots of other stuff, some purely fun and some purely educational (but most of it both). Hopefully, there will be a second edition of this book in the future; if you have suggestions for it, please e-mail me: wsmith@haverford.edu.

To the instructor

Please visit the website of this book. You'll find materials in the website that will make your life easier, including full solutions and important additional support materials for the end-of-chapter problems, lecture notes which complement the text (including additional conceptual questions, worked examples, applications to current research and everyday life, animations, and figures), as well as custom-developed interactive applets, video and audio recordings, and much more. The following sections can be omitted without affecting comprehension of later material: 1.10, 1.12, 2.3–2.6, 3.5–3.6, 4.5, 4.7–4.8, 6.6–6.7, 8.6–8.7, 9.9, 9.11, 10.8–10.9, and Appendix A. If necessary, one can skip all of chapter 6, except for the part of section 6.5 starting with the "Core example" through the end of the section; however omitting the rest of chapter 6 means that the students won't be exposed to any matrix math or to the idea of an eigenvalue equation. (They are exposed copiously to eigenvectors and eigenfunctions in other chapters, but the word "eigenvalue" is used only in chapter 6.) If you have questions or comments, please contact me: wsmith@haverford.edu.

Acknowledgments

This book builds on the enormous efforts of my predecessors. Like any textbook author, I have consulted many dozens of other works in developing my presentation. However, three stand out as particularly helpful: *Vibrations and Waves*, by A. P. French (Norton 1971), *The Physics of Vibrations and Waves, 6th Ed.*, by H. J. Pain (Wiley 2005), and *The Physics of Waves*, by H. Georgi (Prentice-Hall, 1993).

I am deeply grateful to my physics colleague Peter J. Love, who cheerfully answered endless questions from me, taught from draft versions of the book and gave me essential feedback, and made key suggestions for several sections. I am also most thankful to my other colleagues in physics who supported me in this effort and answered my many questions: Jerry P. Gollub, Suzanne Amador-Kane, Lyle D. Roelofs, and Stephon H. Alexander. I also received very valuable inputs from colleagues in math, particularly Robert S. Manning, and chemistry, including Casey H. Londergan, Alexander Norquist, and Joshua A. Schrier. I also thank Jeff Urbach of Georgetown University and Juan R. Burciaga of Lafayette College who used draft versions of the text in their courses, and provided helpful feedback.

I am profoundly thankful for the proof-reading efforts, and suggested edits and end-of-chapter problems from Megan E. Bedell, Martin A. Blood-Forsythe, Alexander D. Cahill, Wesley W. Chu, Donato R. Cianci, Eleanor M. Huber, Anna M. Klales, Anna K. Pancoast, Daphne H. Paparis, Annie K. Preston, and Katherine L. Van Aken. Special thanks are due to Andrew P. Sturner for his tireless efforts and suggestions, right up to the last minute.

Finally, I am most deeply grateful to my family, for their support and encouragement throughout the writing of this book. My children Grace, Charlie, and Tom checked up on my progress every day, and suggested things in everyday life connected to waves and oscillations. My good friend Michael K. McCutchan gave deep proofreading and editing help, and support of all kinds throughout. Finally, words cannot express my gratitude for the efforts of my wife, Marian McKenzie, who did almost all the computerizing of figures, helped with editing, and provided the much-needed emotional support. This book would never have been published without her encouragement.

Learning Tools Used in This Book

Throughout this text you will find a number of special tools which are designed to help you understand the material more quickly and deeply. Please spend a few moments to read about them now.

Concept test

This checks your understanding of the ideas in the preceding material.

Self-test

Similar to a concept test, but more quantitative. It will require a little work with pencil and paper.

Core example

Unlike an ordinary example, these are not simply applications of the material just presented, but rather are an integral part of the main presentation. There are some topics that are much easier to understand when presented in terms of a specific example, rather than in more abstract general terms.

Your turn

In these sections, you are asked to work through an important part of the main presentation. Be sure to complete this work before reading further.

Concept and skill inventory

At the end of each chapter, you'll find a list of the key ideas that you should understand after reading the chapter, and also a list of the specific skills you should be ready to practice.

Contents

Waves and Oscillations

1 Simple Harmonic Motion

All around us, sinusoidal waves astound us!
From "The Waves and Oscillations Syllabus Song," by Walter F. Smith

1.1 Sinusoidal oscillations are everywhere

You are sitting on a chair, or a couch, or a bed, something that is more or less solid. Therefore, every atom within it has a well-defined position. However, if you could look very closely, you'd see that *every one* of those atoms right now is vibrating relative to this assigned position. The hotter your chair the more violent the vibration, but even if your chair were at absolute zero, every atom would *still* be vibrating! Of course, the same is true for every atom in every solid object throughout the universe—right now, each one of them is vibrating relative to its assigned or "equilibrium" position within the solid.

The vibration of a particular one of these atoms might follow the pattern shown in the top part of figure 1.1.1. The pattern appears complicated, but we will show in the course of this book that it is really just a summation of simple sinusoids (as shown in the lower part of the figure), each of which is associated with a "normal mode" of the solid that contains the atoms. (Over the next several chapters, we'll explore what the term "normal mode" means.)

The complexity shown in the top part of the figure arises because the solid has many "degrees of freedom"; every one of the atoms in the solid can move in three dimensions, and each atom is affected by the motion of its neighbors. The approach of physics, and it has been enormously successful in an astonishing variety of situations, is to build up an understanding of complex systems through a thorough understanding of simplified versions. For example, when studying trajectories, we begin with objects falling straight down in a vacuum, and gradually build up to an understanding of three-dimensional trajectories, including effects of air resistance and perhaps tumbling of the object.

So, to understand the motion of the atom, we begin with systems that have only one degree of freedom, that is, systems that can only move in one direction and moreover don't have neighbors that move. A good example is a tree branch. If you pull it straight up and then let go, the resulting motion looks roughly as shown in figure 1.1.2. Again, we see a sinusoidal motion, although in this case it is "damped," meaning that over

1

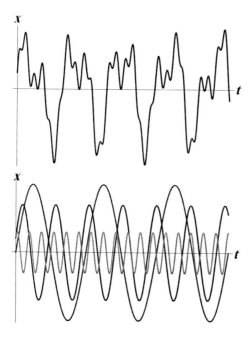

Figure 1.1.1 Top: motion of an atom in a solid. Bottom: Sine waves that, when added together, create the waveform shown in the top part.

time the motion decays away. Hold a pen or a pencil loosely at one end with your thumb and forefinger, with the rest of the pencil hanging below. Push the bottom of the pencil to one side, and then let go—the resulting motion looks similar to figure 1.1.2, though this time the quantity being plotted is the angle of the pencil relative to vertical.

In fact, if you take *any* object that is in an equilibrium position, displace it from equilibrium, and then let go, you'll get this same type of damped sinusoidal response, as we will show quite easily in section 1.2. This type of oscillation is enormously important, not only in the macroscopic motion of objects, machine parts, and so on but also, perhaps surprisingly, in the performance of many electronic circuits, as well as in

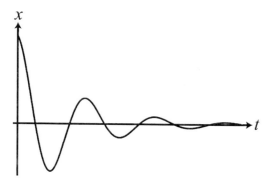

Figure 1.1.2 Motion of a tree branch when pulled up and then released.

the detailed understanding of the motions of atoms and molecules, and their interaction with light.

So, sinusoidal motion really is all around us, and something which any scientist must understand deeply. However, there is another perhaps even more important reason to study oscillations and waves: the mathematical tools and intuition you will develop during this study are *exactly* what you need for quantum mechanics! This is not surprising, since much of quantum mechanics deals with the study of the "wave function" which describes the wave nature of objects such as the electron. However, the connection of the field of waves and oscillations to that of quantum mechanics is much deeper, as you'll appreciate later. For now, rest assured that you are laying a very solid foundation for your later study of quantum mechanics, which is the most important and exciting realm of current physics research and application.

1.2   The physics and mathematics behind simple sinusoidal motion

To start our quantitative study, we follow the approach of physics and consider the simplest possible system: one with no damping. This means that all the forces acting on the object are conservative and so can be associated with a potential energy.

A body in stable equilibrium is, by definition, at a local minimum of the potential energy *versus* position curve, as shown in figure 1.2.1. For convenience, we choose $x = 0$ at the equilibrium position. Except in pathological cases, the potential energy function $U(x)$ near $x = 0$ can be approximated by a parabola, as shown. We write this parabolic or "harmonic" approximation in the form $U(x) \approx \frac{1}{2}kx^2 + \text{const.}$ for reasons that will become apparent in the next sentence.

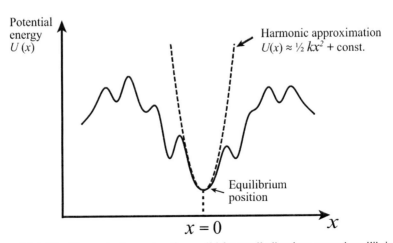

Figure 1.2.1 The Harmonic Approximation, valid for small vibrations around equilibrium.

The force acting on the body can then be found using $F = -\dfrac{dU}{dx} = -kx$. The relation

$$\boxed{F = -kx}$$

(1.2.1)

is known as "Hooke's Law," after its discoverer Robert Hooke (1635–1703).[1] The quantity k is called the "spring constant." To find the position of the body as a function of time, $x(t)$, we will follow a three-step procedure. We'll use the same procedure throughout the book, for progressively more complex systems. To save space, we simply write x remembering that this is shorthand for the function $x(t)$.

1. Write down Newton's second law for each of the bodies involved.
In this case, there is only one body, so we have

$$\left.\begin{aligned} F = ma = m\frac{d^2x}{dt^2} \\ F = -kx \end{aligned}\right\} \Rightarrow m\frac{d^2x}{dt^2} = -kx.$$

(1.2.2)

This is a "differential equation" or DEQ which simply means that it is an equation that involves a derivative. (If you haven't had a course in DEQs, don't worry; we'll go through everything you need to know for this course and for a first course in quantum mechanics.) This is called a "second order DEQ," because it contains a second derivative. The "solution" for this equation is a function $x(t)$ for which the equation holds true—in this case, a function for which, when you take two time derivatives and multiply by m (as indicated on the left side of the equation), then you get back the same function times $-k$ (as indicated on the right side of the equation). This is the solution that we are trying to find, since it tells us the position of the object at all times. One important thing to know right away is that there is *no general recipe* for finding the solution that works for all second-order DEQs. However, for many of the most important such equations in physics, we can *guess* a solution based on our intuition and then *check* to determine whether our guess is really right, as shown in the following steps.

---

1. Some scholars feel that Robert Hooke is one of the most underappreciated figures in science. He was the founder of microscopic biology (he coined the word "cell"), he discovered the red spot on Jupiter and observed its rotation, he was the first to observe Brownian motion (150 years before Brown), and discovered Uranus 108 years before the more-publicized discovery by Herschel. Unfortunately, it seems that Hooke spread himself too thin, and never got around to publishing many of his results. Hooke and Newton, though originally on friendly terms, later became fierce rivals. It appears that Hooke conceptualized the inverse square law of gravity and the elliptical motion of planets before Newton, and discussed this idea briefly with Newton. Newton (unlike Hooke) was able to show quantitatively how the inverse square law predicts elliptical orbits, and felt that Hooke was pushing for more recognition than he deserved in this very important discovery. Some scholars feel that, when Newton became the president of the Royal Society (the leading scientific organization of the time in England), he may intentionally have "buried" the work of Hooke, but there is no hard evidence to support this.

To save space, we write $\dfrac{d^2x}{dt^2}$ as $\ddot{x}$. (Each dot represents a time derivative,[2] so that $\dot{x}$ represents $\dfrac{dx}{dt}$.) We rearrange equation (1.2.2) slightly to give

$$\ddot{x} = -\frac{k}{m}x. \qquad (1.2.3)$$

This is called the "equation of motion."

2. Using physical intuition, guess a possible solution.

Observation of a mass bouncing on a spring suggests that its motion may be sinusoidal. The most general possible sinusoid can be expressed as

$$x = A\cos(\omega t + \varphi) \qquad (1.2.4)$$

The values of the "adjustable constants" A and φ depend on the initial conditions, as we will discuss later.

3. Plug the guess back into the system of DEQs to see if it is actually a solution, and to determine whether there are any restrictions on the parameters that appear in the guess.

In this case, the "system of DEQs" is the single equation (1.2.3). Before you look at the next paragraph, plug the guess (1.2.4) into (1.2.3), verify that it is indeed a solution, and find what the "parameter" ω must be in terms of k and m.

You should have found that

$$\omega = \sqrt{k/m} \qquad (1.2.5)$$

So, we see that sinusoidal vibration, also known as "simple harmonic motion" or SHM, is universally observed for vibrations that are small enough to use the Harmonic Approximation shown in figure 1.2.1.

As described in section 1.3, ω equals 2π times the frequency of the motion and is called the "angular frequency."

1.3 Important parameters and adjustable constants of simple harmonic motion

Figure 1.3.1 shows a graph of the SHM represented by equation (1.2.4). Any such sinusoidal motion can be described with three quantities:

1. The amplitude A. As shown, the maximum value of x is A, and the minimum value is $-A$.

---

2. The dot notation was invented by Isaac Newton. It is very convenient for us, because we have to deal with time derivatives so frequently. However, it is generally felt that, because historical English mathematicians continued to use this notation so long, they were held back relative to their German counterparts, who used Gottfried Leibniz's d/dt notation instead. (Leibniz's notation is more flexible, and we will use it where convenient.)

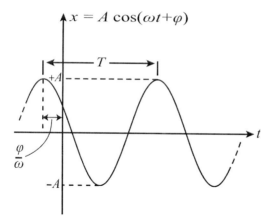

Figure 1.3.1 Simple harmonic motion of period T and amplitude A.

2. The period T. This is the time between successive maxima, or equivalently between successive minima. The period is the time needed for one complete cycle, so that when the time t changes by T, the argument of the cosine in $x = A \cos(\omega t + \varphi)$ must change by 2π. Therefore,

$$\omega(t + T) + \varphi = \omega t + \varphi + 2\pi,$$

so that

$$\boxed{\boxed{T = 2\pi/\omega}} \tag{1.3.1}$$

(This equation is shown with a double border because we'll be referring to it so frequently. Equations shown this way are so very important that you will find it helpful to begin memorizing them right away.) The frequency f is given by $1/T$, so that

$$\boxed{\boxed{\omega = 2\pi f}} \tag{1.3.2}$$

For this reason, ω is called the "angular frequency." We will use it *continually* for the rest of the text, so get accustomed to it now! We will encounter various different angular frequencies later, so we give the special name ω_0 to the angular frequency of simple harmonic motion, that is,[3]

$$\boxed{\boxed{\omega_0 \equiv \sqrt{k/m}}} \tag{1.3.3}$$

(Note: the "0" subscript here does not indicate a connection to $t = 0$, but it is universally used.)
3. The "initial phase" φ. The position at $t = 0$ is determined by a combination of A and φ. It is easy to find the relation between these two "adjustable constants"

---

3. Physicists use the symbol "$\equiv$" to mean "is defined to be."

on one hand and the initial position x_0 and the initial velocity v_0 on the other. From equation (1.2.4): $x = A \cos(\omega t + \varphi)$ we obtain:

$$x_0 = A \cos \varphi \quad \text{and} \quad v_0 = \left. \frac{dx}{dt} \right|_{t=0} = -\omega_0 A \sin \varphi$$

Your turn: From these, you should now show that

$$A = \sqrt{x_0^2 + \left(\frac{v_0}{\omega_0} \right)^2} \quad \text{(1.3.4a)} \quad \text{and} \quad \varphi = \tan^{-1} \left(\frac{-v_0}{\omega_0 x_0} \right). \text{ (1.3.4b)}$$

(We use the term "parameter" to refer to a quantity determined by the physical properties of a system, such as mass, spring constant, or viscosity. Thus, ω_0 is a parameter. In contrast, we use "adjustable constant" to designate a quantity that is determined by initial conditions. Thus, A and φ are adjustable constants.)

As mentioned earlier, the equation of motion (1.2.3) is a second-order DEQ, because the highest derivative is of second order. It can be shown that the most general solution to a second-order DEQ contains two (and no more than two) adjustable constants.[4] (We know that this must be true for our case, since we need to be able to take into account (1) the initial position and (2) the initial velocity when writing out a particular solution, therefore we need to be able to adjust two constants.) So, we can be confident that equation (1.2.4): $x = A \cos(\omega t + \varphi)$ is the *general* solution to equation (1.2.3): $\ddot{x} = -\frac{k}{m} x$. An example of a nongeneral solution would be $x = A \sin \omega_0 t$; you should verify that this satisfies equation (1.2.3). But this is the same as equation (1.2.4), with the particular choice $\varphi = -\pi/2$.

Look again at equation (1.3.3): $\omega_0 = \sqrt{k/m}$. There is something about it that is absolutely astonishing. *The angular frequency depends only on the spring constant and the mass – it doesn't depend on the amplitude!* It would be very reasonable to expect that, for a larger amplitude, it would take longer for the system to complete a cycle, since the mass has to move through a larger distance. However, at larger amplitudes the restoring force is larger and this provides exactly enough additional acceleration to make the period (and so ω) constant. The fact that the frequency is independent of amplitude is critical to many applications of oscillators, from grandfather clocks to radios to microwave ovens to computers. Most of these do not actually have separate masses and springs inside them, but instead have combinations of components which are described by exactly analogous DEQs, and so exhibit exactly analogous behavior. We'll explore many of these in chapter 2, but we start now with the two most basic, and most important, examples.

---

4. For the special case of a "linear" (meaning no terms such as x^2 or $x\dot{x}$), "homogeneous" (meaning no constant term) DEQ, such as equation (1.2.3), this theorem is often phrased in the alternate form, "The general solution of a linear, homogeneous second-order DEQ is the sum of two independent solutions." An example for our case would be $x = A_1 \cos \omega_0 t + A_2 \sin \omega_0 t$. However, you can easily show (see problem 1.7) that this can be expressed in the form $x = A \cos(\omega_0 t + \varphi)$, with $A = \sqrt{A_1^2 + A_2^2}$ and $\varphi = \tan^{-1}(-A_2/A_1)$.

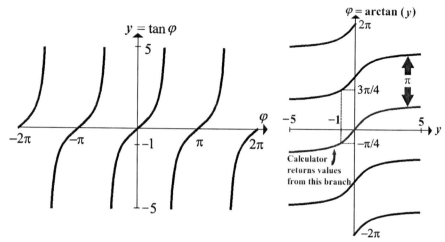

Figure 1.3.2 Left: the tangent function. Right: Because the arctan function is multivalued, you can add π to the result your calculator returns (shown by the curve which passes through the origin), and sometimes you need to do this to get the physically correct answer.

Aside: The arctangent function

The arctan function, which appears in equation (1.3.4b), is a slippery devil, because it's multivalued, that is, $\tan^{-1}x$ is only defined up to an additive factor of π. For example, $\tan^{-1}(-1)$ can equal either $-\pi/4$ or $3\pi/4$, as shown in figure 1.3.2b.

Your calculator is programmed always to return the value between $-\pi/2$ and $\pi/2$, but this is not always the correct answer for the particular situation. For example, consider a case with $A = 5$ m and $\omega_0 = 7$ rad/s, with $v_0 = -24.75$ m/s and $x_0 = -3.536$ m. If you use equation (1.3.4b) and plug in the numbers on your calculator, it will return $\varphi = -\pi/4$, but this is wrong, because $x = A\cos(\omega_0 t - \pi/4)$ would mean $x_0 = A\cos(-\pi/4) > 0$ and $\dot{x}_0 = -A\omega_0 \sin(-\pi/4) > 0$. To get the correct signs for x_0 and $\dot{x}_0$ you must add π to the result from your calculator, giving $\varphi = 3\pi/4$. So, every time you use your calculator to find $\tan^{-1}$, you must think carefully about the result, and use other information from the problem to determine whether you should add π to it to get the truly correct answer. See problem 1.10.

1.4   Mass on a spring

Any system described by a DEQ of the form (1.2.2), $m\ddot{x} = -kx$, has a time evolution of the form (1.2.4), $x = A\cos(\omega t + \varphi)$. The very simplest example is a mass that feels only one force, from an attached ideal spring. It is difficult to eliminate the force of gravity, so instead we often counteract it with a frictionless supporting surface, as shown in figure 1.4.1a. The spring has an equilibrium length ℓ. However, if we measure the position of the mass relative to its equilibrium position, as shown, then the force exerted by the spring has a very simple form:

$$F = -kx. \tag{1.4.1}$$

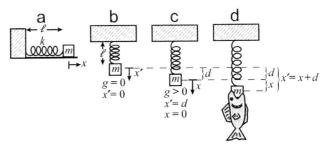

Figure 1.4.1 a: Mass on a frictionless surface. Important: The vertical line and horizontal arrow marked "x" at the bottom of the figure show the definition of x: it is zero at the position of the vertical line, and becomes positive in the direction of the arrow. In this example, this means that when the mass moves to the right of the position shown, x is positive, whereas if the mass moves to the left of the position shown then x is negative. We will use this combination of line and arrow to define the displacement x throughout the book. b–d: Mass on a vertical spring. The direction of positive x is downward.

As mentioned earlier, this is called Hooke's law. It simply states that, when the mass is to the right of its equilibrium position, so that $x > 0$, and the spring is stretched, the spring pulls back to the left, that is, in the $-x$ direction. If instead the mass is to the left of its equilibrium position ($x < 0$) and the spring is compressed, then (as predicted both by equation (1.4.1) and common sense), the spring pushes to the right, that is, in the positive x direction.

Often, we happen not to have any frictionless surfaces handy, so it is more convenient to suspend the mass vertically, as shown in figure 1.4.1 b–d. As a thought experiment, we consider what would happen in the absence of gravity, as shown in figure 1.4.1b. As before, we measure the position of the mass relative to its equilibrium position (in the absence of gravity); we'll call this x', as a reminder that this is before gravity is turned on. As shown, we define x' to be positive downward. The force of the spring is just the same as before:

$$F = -kx'. \tag{1.4.2}$$

Now, we turn on gravity, as shown in figure 1.4.1c. This causes the spring to stretch out by an additional distance d, so that $x' = d$, and the spring force is $F = -kx' = -kd$. (The spring force is negative, which means that it is upward.) At the new equilibrium position, the net force on the mass must be zero, that is, the spring force must cancel the force of gravity. Since we have defined the down direction to be positive, the force of gravity is positive, so

$$-kd + mg = 0 \Leftrightarrow d = \frac{mg}{k}. \tag{1.4.3}$$

We now measure the position x of the mass relative to its new equilibrium position, as shown in figure 1.4.1c. In figure 1.4.1d, an additional downward force is applied, stretching the spring further and so creating a positive x. We see that

$$x' = x + d$$

The total force on the mass is

$$F_{\text{Tot}} = -kx' + mg = -k(x+d) + mg = -kx - kd + mg.$$

Substituting for d from equation (1.4.3) gives

$$F_{\text{Tot}} = -kx - k\frac{mg}{k} + mg = -kx.$$

Thus, as long as we measure relative to the new equilibrium position, *the combined effects of the spring and gravity give a total force which follows Hooke's Law!*

So, for either the situation of figure 1.4.1a or 1.4.1c and d, we have a total force $F_{\text{Tot}} = -kx$, so we can use the result of equation (1.2.2), that is,

$$\left.\begin{array}{r} F = ma = m\ddot{x} \\ F = -kx \end{array}\right\} \Rightarrow m\ddot{x} = -kx$$

with the solution we found in section 1.2, $x = A\cos(\omega_0 t + \varphi)$.

1.5 Electrical oscillators

Consider the circuit shown in figure 1.5.1. The capacitor, designated C, stores electrical charge and potential energy, in much the same way that a spring can store potential energy. The capacitor always has equal and opposite charge q on its two plates. For example, at some instant in time it might have a charge $+1.2$ nC on the top plate and -1.2 nC on the bottom plate. At this instant, $q = +1.2$ nC. The capacitance C is defined as the ratio of the charge to the voltage across the capacitor:

$$C \equiv \frac{q}{V_c} \Leftrightarrow V_c = \frac{q}{C}. \tag{1.5.1}$$

The inductor, designated L, consists of a number of loops of wire. As you'll recall from a previous course, when electrical current I flows through the loops, it creates a magnetic field B, with associated magnetic flux ϕ_B linking through the loops. The inductance is defined as

$$L \equiv \frac{\phi_B}{I}. \tag{1.5.2}$$

Faraday's law tells us that there is an emf across the inductor given by

$$\varepsilon = -\dot{\phi}_B = -L\dot{I}. \tag{1.5.3}$$

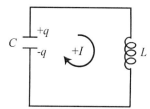

Figure 1.5.1 Electrical oscillator.

Recall that the current is defined to be a time rate of change of charge. We define a positive current to be one that flows clockwise in the circuit, as shown in figure 1.5.1. We also define q to be positive when the upper plate is positive, as shown. Current is the time derivative of charge, but, with our sign definitions, a positive I decreases the charge on the capacitor. Therefore,

$$I = -\dot{q}. \tag{1.5.4}$$

Combining this with equation (1.5.3) gives

$$\varepsilon = +L\ddot{q}. \tag{1.5.5}$$

This is the voltage across the inductor, so

$$V_{\mathrm{L}} = L\ddot{q}. \tag{1.5.6}$$

Next, we will apply Kirchhoff's loop rule, which says that when you go around the loop, the voltage changes must add up to zero:

$$V_{\mathrm{c}} + V_{\mathrm{L}} = 0 \Rightarrow \frac{q}{C} + L\ddot{q} = 0 \Leftrightarrow$$

$$L\ddot{q} = -\frac{1}{C}q \tag{1.5.7}$$

This is isomorphic to equation (1.2.2), $m\ddot{x} = -kx$, meaning that it is exactly the same, except with different symbols. Right away, then, we know that the solution, which must be isomorphic to $x = A\cos(\omega_0 t + \varphi)$, is $q = A\cos(\omega_0 t + \varphi)$. The isomorphism is summarized in table 1.5.1.

> **Your turn (answer below[5])**: Using the isomorphism, deduce what the angular frequency ω_0 is for the electrical oscllator.

Electrical oscillators are tremendously important in electronic circuits, from radio tuners to the clocks that regulate the speed of computers.

Table 1.5.1. Isomorphism between mechanical and electrical oscillators

Mass and spring	Electrical oscillator
Position relative to equilibrium x	Charge q on capacitor
Mass m	Inductance L
Spring constant k	Inverse capacitance $1/C$

---

5. Answer to self-test: Comparing equations (1.2.2) and (1.5.7), we see that m gets replaced by L, while k gets replaced by $1/C$. Therefore, $\omega_0 = \sqrt{\dfrac{k}{m}}$ becomes $\omega_0 = \sqrt{\dfrac{1}{LC}}$.

1.6   Review of Taylor series approximations

To move forward efficiently, we must take a little time now to go over two important mathematical techniques. Later in this chapter, we'll show that oscillatory motion can be expressed in a more elegant way by using complex exponential functions. However, to develop those, we'll need to use Taylor series, which we review in this section.

Much of the creative effort in physics is devoted to making reasonable approximations so that we can study the most important behaviors of complex systems without getting bogged down in a morass of hundreds of complex equations. The most important approximation tool is the Taylor series approximation.

The goal is to find the value of a function $f(x)$ at the position $x = x_0 + a$, if we are given complete information about the function at the nearby point x_0. The simplest approximation, shown by the dot labeled "zeroth order approximation" in figure (1.6.1), is simply to say that $f(x_0 + a) \approx f(x_0)$. We can get a better approximation (shown in gray) by using our knowledge of the slope of $f(x)$ at the point x_0. We write this slope as $\dfrac{df}{dx}\Big|_{x_0}$, which is read as "the derivative of f with respect to x, evaluated at x_0." The slope equals the "rise" over the "run," so by multiplying it by the run (i.e., by a), we get the rise, and by adding this to the initial value $f(x_0)$, we get a closer approximation to the true value $f(x_0 + a)$; in doing so, we approximate the function as a straight line. We can do even better by approximating $f(x)$ as a parabola, as shown by the dashed curve. If we wanted to get even more accurate, we could use a third-order approximation:

$$f\left(x_0 + a\right) \cong f\left(x_0\right) + a \left.\frac{df}{dx}\right|_{x_0} + \frac{a^2}{2!}\left.\frac{d^2f}{dx^2}\right|_{x_0} + \frac{a^3}{3!}\left.\frac{d^3f}{dx^3}\right|_{x_0}. \tag{1.6.1}$$

You can see the pattern. Assuming a is small, each additional correction term gets smaller and smaller, so that usually we don't need to go beyond a second-order approximation. (In fact, most often a first-order approximation will suffice.) The complete version would be

$$f\left(x_0 + a\right) = f\left(x_0\right) + a \left.\frac{df}{dx}\right|_{x_0} + \cdots + \frac{a^n}{n!}\left.\frac{d^nf}{dx^n}\right|_{x_0} + \cdots = \sum_{n=1}^{\infty} \frac{a^n}{n!}\left.\frac{d^nf}{dx^n}\right|_{x_0}. \tag{1.6.2}$$

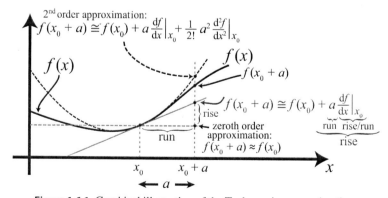

Figure 1.6.1 Graphical illustration of the Taylor series approximation.

As an example, let's find the Taylor series for the sine function:

$$f(\theta) \equiv \sin\theta \Rightarrow \frac{df}{d\theta} = \cos\theta, \ \frac{d^2f}{d\theta^2} = -\sin\theta, \ \frac{d^3f}{d\theta^3} = -\cos\theta, \ \frac{d^4f}{d\theta^4} = \sin\theta, \dots$$

Plugging this into equation (1.6.2), using θ as the variable instead of x, and expanding around $\theta_0 = 0$ gives

$$\sin\theta = \sin 0 + \theta\cos 0 + \frac{\theta^2}{2!}(-\sin 0) + \frac{\theta^3}{3!}(-\cos 0) + \frac{\theta^4}{4!}(\sin 0) + \frac{\theta^5}{5!}(\cos 0) + \cdots$$

$$\Rightarrow \sin\theta = \theta - \frac{\theta^3}{3!} + \frac{\theta^5}{5!} - \cdots \tag{1.6.3}$$

From this, we can see why the approximation

$$\sin\theta \cong \theta \ (\theta \text{ in radians}) \tag{1.6.4}$$

works so well for small θ: there is no second-order correction term – the next correction term is third order. (In fact, as you can show yourself on your calculator, this approximation works pretty well up to about $\theta = 0.4$ radians $= 23°$.)

> **Your turn:** Show that
> $$\cos\theta = 1 - \frac{\theta^2}{2!} + \frac{\theta^4}{4!} - \cdots \tag{1.6.5}$$

1.7   Euler's equation

We will see in section 1.9 that there is a different way of expressing the solution for simple harmonic motion, $x = A\cos(\omega t + \varphi)$, one which will become *much* more convenient when we begin treating more complicated systems. We will make use of Euler's equation:

$$e^{i\theta} = \cos\theta + i\sin\theta. \tag{1.7.1}$$

Here, $i \equiv \sqrt{-1}$. The proof of this statement, and also the understanding of what it means to have a complex number as an exponent, comes through consideration of series expansions. Using the Taylor expansions we just derived for cos and sin, we can express the right side of this as

$$\cos\theta + i\sin\theta = 1 + i\theta - \frac{\theta^2}{2!} - i\frac{\theta^3}{3!} + \frac{\theta^4}{4!} + \cdots \tag{1.7.2}$$

Now, we express the left side of equation (1.7.1) using a Taylor series, again expanding around $\theta_0 = 0$:

$$e^{i\theta} = e^{i0} + \theta i e^{i0} + \frac{\theta^2}{2!}i^2 e^{i0} + \frac{\theta^3}{3!}i^3 e^{i0} + \frac{\theta^4}{4!}i^4 e^{i0} + \cdots$$

$$= 1 + i\theta - \frac{\theta^2}{2!} - i\frac{\theta^3}{3!} + \frac{\theta^4}{4!} + \cdots$$

This is just the same as equation (1.7.2), which proves equation (1.7.1). This was first demonstrated by Leonhard Euler[6] in 1748. You will use Euler's equation every day for the rest of your life☺, so you are encouraged to commit it to memory.

1.8 Review of complex numbers

Let us briefly review complex numbers. It is helpful to use the "complex plane," as shown in figure 1.8.1a in which the vertical axis is used for the imaginary part of a number and the horizontal axis for the real part. A complex number z can be represented

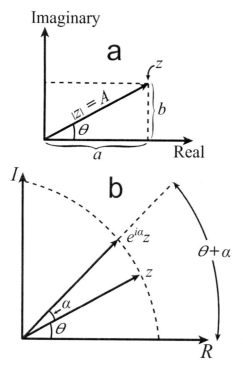

Figure 1.8.1 a: The complex plane. b: Multiplication by $e^{i\alpha}$ is equivalent to rotating counterclockwise by α in the complex plane.

---

6. Euler earned his Master's degree from the University of Basel at the age of 16. During his life, he published over 900 works. He is responsible for many of our mathematical notations, including $f(x)$ to denote a function, Δx to denote a difference, and e for the base of the natural logarithm.

as a vector in this plane, and we can write the "Cartesian representation" for the number as $z = a + ib$. We see from the diagram that the length or magnitude of the vector is given by $A = |z| = \sqrt{a^2 + b^2}$. Simple trigonometry then provides that

$$a = A \cos \theta \text{ and } b = A \sin \theta, \text{ so that}$$

$$z = a + ib = A \cos \theta + i(A \sin \theta) = Ae^{i\theta}$$

(using Euler's equation) and also

$$\theta = \tan^{-1}(b/a).$$

Your turn: If you've not already done so, read the aside about the arctan function in section 1.3. Then, explain why a more complete version of the above equation is

$$\theta = \tan^{-1}(b/a) + \begin{cases} 0 & \text{if } a > 0 \\ \pi & \text{if } a < 0 \end{cases}$$

Let us collect all these useful relations into a single box:

$z = a + ib$ "Cartesian representation"

– or –

$z = Ae^{i\theta}$ "Polar representation,"

where

$A = |z| = \sqrt{a^2 + b^2}$ and $\theta = \tan^{-1}(b/a) + \begin{cases} 0 & \text{if } a > 0 \\ \pi & \text{if } a < 0 \end{cases}$

What happens in the complex plane when we multiply z by $e^{i\alpha}$?

$$e^{i\alpha}z = e^{i\alpha} Ae^{i\theta} = Ae^{i(\theta + \alpha)}.$$

This is a vector in the complex plane which still has length A, but has been rotated counterclockwise by the angle α so that it now points in the direction given by $\theta + \alpha$, as shown in figure 1.8.1b. Thus,

Multiplying a number by $e^{i\alpha}$ is equivalent to rotating its vector counterclockwise by α in the complex plane.

Often we need to take the "complex conjugate" of a complex number:

To form the complex conjugate, simply replace every instance of i by −i.

For example, the complex conjugate of $a+ib$ is just $a-ib$. We denote the complex conjugate with a star: the complex conjugate of z is z^*. As another example, if $z = e^{i\theta}$, then $z^* = e^{-i\theta}$. The complex conjugate is often used to calculate the magnitude of a complex number. This is perhaps the easiest to see with a number expressed in polar form: if $z = Ae^{i\theta}$ (with A real), then $z^* = Ae^{-i\theta}$, so that $z^*z = A^2 = |z|^2$. So, in general, we have

$$|z|^2 = z^*z.$$

Finally, we introduce the notation for the real and imaginary parts of a complex number:

$$\text{If } z = a + ib, \text{ then Re } z = a \text{ and Im } z = b.$$

(Note that the i is not included in Im z.)

Self-test (answer below[7]): Show that, for any complex number z, Re $z = $ Re (z^*).

1.9   Complex exponential notation for oscillatory motion

Finally, we are ready to apply these ideas to the simple harmonic oscillator. We can very easily rewrite the solution using complex exponential notation:

$$x = A\cos(\omega_0 t + \varphi) = \text{Re}\left(Ae^{i(\omega_0 t + \varphi)}\right) = \text{Re}\left(e^{i\omega_0 t}Ae^{i\varphi}\right). \qquad (1.9.1)$$

Written this way, we can see that the complex plane vector that represents the system has length A and at $t = 0$ points in the direction given by the angle φ. This vector is then multiplied by $e^{i\omega_0 t}$, that is, it is rotated counterclockwise by the angle $\omega_0 t$, as shown in figure 1.9.1. Since this angle increases in time, the vector rotates around and around the origin. The "angular velocity" is the time derivative of this angle, that is, $\frac{d}{dt}(\omega_0 t + \varphi) = \omega_0$. The actual position of the oscillator is given by the real part of the vector, that is, the projection onto the horizontal axis. As the vector rotates in uniform circular motion, this projection changes sinusoidally. It is convenient to define

$$z \equiv Ae^{i(\omega_0 t + \varphi)} \qquad (1.9.2)$$

so that $x = \text{Re } z$.

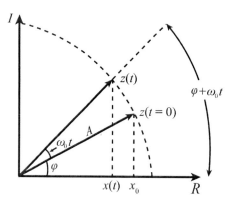

Figure 1.9.1 Complex plane representation of SHM.

---

7. Answer to self-test: Express z in Cartesian form: $z = a + ib$. The real part of z is just a. The complex conjugate is $z^* = a - ib$, and the real part of this is also a.

This method of portraying the motion brings out the physical significance of A and φ more clearly. From figure 1.9.1, we can see that

$$\varphi = \cos^{-1} \frac{x_0}{A}, \tag{1.9.3}$$

where x_0 is the initial position. It is also clearer, perhaps, that A is related to the total energy of the system. From the discussion surrounding figure 1.2.1, we know that the potential energy of a harmonic oscillator is given by

$$U(x) = \frac{1}{2}kx^2 + \text{const.}$$

Usually, it is convenient to choose the constant so that $U(x) = 0$ at the equilibrium position $x = 0$, so that

$$U(x) = \frac{1}{2}kx^2. \tag{1.9.4}$$

When $x = A$, the oscillator is at its maximum displacement, and so is momentarily at rest. Therefore, all the energy is in the form of potential energy, so that

$$E = \frac{1}{2}kA^2 \Leftrightarrow A = \sqrt{\frac{2E}{k}}, \tag{1.9.5}$$

where E is the total energy.

It is also easy to show the relationships between position, velocity, and acceleration using this complex plane picture. Take a moment now to convince yourself that the operations of taking time derivatives and taking the real part of a quantity "commute," that is, the order doesn't matter, that is,

$$\text{Re}\frac{dz}{dt} = \frac{d\,\text{Re}\,z}{dt}.$$

Therefore, we can write

$$\dot{x} = \text{Re}\,\dot{z} \quad \text{and} \quad \ddot{x} = \text{Re}\,\ddot{z}.$$

Plugging in for z from equation (1.9.2) gives

$$\dot{z} = i\omega_0 z \quad \text{and} \quad \ddot{z} = -\omega_0^2 z. \tag{1.9.6}$$

Using Euler's equation, we see that $e^{i\frac{\pi}{2}} = i$, so that multiplication by i rotates a complex plane vector counterclockwise by $\pi/2$. Equation (1.9.6) thus says that the complex plane vector representing the velocity, $\dot{z}$, is rotated by a constant angle $\pi/2$ "ahead" of the position vector z (and is scaled by the factor ω_0). Similarly, since $-1 = i \cdot i$, multiplication by -1 rotates a complex plane vector through $2 \cdot (\pi/2) = \pi$. Equation (1.9.6) thus says that $\ddot{z}$ is always an angle π ahead of z (and is scaled by ω_0^2). These relationships are shown in figure 1.9.2; bear in mind that the position, velocity, and acceleration have different units, so the relative lengths of the vectors in each picture are not meaningful. As shown in the upper left part of the figure, *for the important special case of zero initial velocity, $A = x_0$*. We will use this result again later.

This is a good time to point out that, although taking the real part does commute with taking the derivative, addition, and multiplication by a real number, taking the real part does *not* commute with multiplication by a complex number. For example,

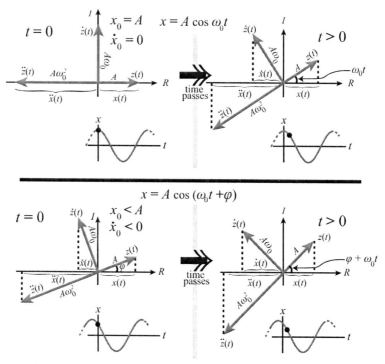

Figure 1.9.2 Phase relationships for SHM. Top left: initial velocity $= 0$ shown for $t = 0$. Top right: initial velocity $= 0$ shown for $t > 0$. Bottom: Similar pictures for initial velocity < 0.

if $z_1 = A_1 e^{i\varphi_1}$ and $z_2 = A_2 e^{i\varphi_2}$, then $\mathrm{Re}[z_1 z_2] = \mathrm{Re}\left[A_1 A_2 e^{i(\varphi_1 + \varphi_2)}\right] = A_1 A_2 \cos(\varphi_1 + \varphi_2)$, whereas $(\mathrm{Re}\, z_1)(\mathrm{Re}\, z_2) = A_1 A_2 \cos\varphi_1 \cos\varphi_2$. It is especially important to bear this in mind when calculating energies; for example, $KE = \frac{1}{2} m \dot{x}^2 = \frac{1}{2} m (\mathrm{Re}\,\dot{z})^2 \neq \frac{1}{2} m \mathrm{Re}(\dot{z}^2)$.

1.10   The complex representation for AC circuits

In section 1.5, we discussed one type of electrical oscillator. However, there are many other circuit examples in which the voltage and current vary sinusoidally in time. Such circuits are essential to the operation of virtually all analog (that is, nondigital) electronics, and the concepts involved in analyzing them are critical to the detailed understanding of *all* circuits. It is convenient to use a complex representation for the currents and voltages in AC circuits, and this approach is used essentially by all scientists who work with circuits and essentially by all electrical engineers. So, this section will provide a good chance to exercise the skills involving complex numbers that we have just reviewed.

The simplest possible AC circuit is shown in figure 1.10.1a. The signal generator on the left of the circuit is a device that produces a voltage difference $V_0 \cos \omega t$ between its two terminals. In this circuit, it is connected to a resistor. Since only voltage *differences* are physically important, we can define $V \equiv 0$ at any point in the circuit. For many actual circuits, the $V \equiv 0$ point is ground (literally the voltage of the dirt under your building). For the rest of this section, we'll use this convention; in the circuit shown, we

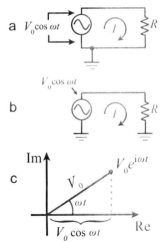

Figure 1.10.1 a: A signal generator (denoted by a sine wave inside a circle) connected to a resistor. b: Same circuit, drawn in the more conventional way using the ground symbol. c: Complex representation for an AC voltage.

have grounded the lower terminal of the signal generator. Again, this does not affect the operation of the circuit at all; it merely fixes the reference point with respect to which voltages are measured. The circuit can then be redrawn as shown in figure 1.10.1b; all points in the circuit with the ground symbol are connected together.

For a resistor, $V = IR \Leftrightarrow I = V/R$, so for this circuit $I = \dfrac{V_0}{R}\cos\omega t$.

Now, let's apply the complex representation. The voltage is

$$V = V_0 \cos\omega t = \mathrm{Re}\,\tilde{V}, \quad \text{where } \tilde{V} = V_0 e^{i\omega t}$$

is the complex version of the voltage, as shown in figure 1.10.1c.[8] Similarly, the current is

$$I = \frac{V_0}{R}\cos\omega t = \mathrm{Re}\,\tilde{I}, \quad \text{where } \tilde{I} = \frac{V_0}{R}e^{i\omega t}.$$

Comparing this with the definition of $\tilde{V}$, we can write the complex version of Ohm's Law,

$$\tilde{V} = \tilde{I}Z_R,$$

where $Z_R = R$ is the "impedance" of the resistor. The impedance is a generalized version of the resistance; we will see below that it can be complex, so that we could write it as $|Z|\,e^{i\varphi}$. Then, the general complex version of Ohm's law would be

$$\tilde{V} = \tilde{I}\,|Z|\,e^{i\varphi}.$$

Since multiplying a complex number by the factor $e^{i\varphi}$ rotates the complex plane vector by angle φ, we see that the complex phase φ of the impedance is the phase difference between the current and the voltage. For a resistor, Z is real, so that the current and the voltage are in phase, but we will see that, for inductors and capacitors, Z is imaginary,

---

8. We use the tilde ($\sim$) above a symbol to indicate "complex version of," so that $\tilde{V}$ is the complex version of V. Unfortunately, in many texts, $\tilde{V}$ is simply written as V, and one must remember that, to find the actual voltage, one must take the real part.

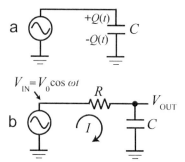

Figure 1.10.2 a: A signal generator connected to a capacitor. b: An RC low-pass filter.

so that the current and voltage are not in phase, meaning that the sinusoidally varying voltage across a capacitor or inductor reaches a peak at a different time than the sinusoidally varying current flowing through it. (See problems 1.19, 1.22, and 1.23 for more about these phase differences.) Note that the use of Z as a symbol for the impedance is meant to tip you off that it may be a complex quantity, so it is conventional not to include the tilde over the Z.

So far, this is not terribly exciting. Things get more interesting when we introduce a capacitor, as shown in figure 1.10.2a. We again use a signal generator to apply a voltage $V_0 \cos \omega t$ across the capacitor. This creates a time-dependent charge $Q(t)$ on the capacitor. To find the current, we use

$$Q = CV \xrightarrow{\text{d}/\text{d}t} \frac{dQ}{dt} = C\frac{dV}{dt}.$$

Since $I = \dfrac{dQ}{dt}$, we have

$$I = C\frac{dV}{dt} = C\frac{d}{dt}\left(V_0 \cos \omega t\right) = -CV_0\, \omega \sin \omega t.$$

Now, let's apply the complex representation. As before, we have

$$V = V_0 \cos \omega t = \text{Re } \tilde{V}, \text{ where } \tilde{V} = V_0 e^{i\omega t}.$$

The current is

$$I = -CV_0\omega \sin \omega t = \text{Re } \tilde{I}, \text{ where } \tilde{I} = iCV_0\, \omega e^{i\omega t}.$$

the complex version of Ohm's Law in this case is

$$\tilde{V} = \tilde{I}Z_\text{C},$$

where Z_C is the impedance of the capacitor.

Your turn (answer below[9]): What is Z_C in terms of C and ω?

---

9. $Z_\text{C} = \dfrac{\tilde{V}}{\tilde{I}} = \dfrac{V_0 e^{i\omega t}}{iCV_0\omega e^{i\omega t}} = \dfrac{1}{i\omega C}.$

To really see the power of this approach, we need to consider a more complicated circuit, such as that shown in figure 1.10.2b. No current is allowed to flow out of the circuit at the point labeled V_{OUT} ; instead, this is a place at which we will later calculate the voltage. Again, we apply a voltage $V_0 \cos \omega t$, this time to a series combination of a resistor and capacitor. What is the resulting current? It is related to the voltage across the resistor by $V_R = IR$, which is the real part of

$$\tilde{V}_R = \tilde{I} Z_R. \tag{1.10.1}$$

The total voltage across the series combination is the sum of the individual voltages:

$$V_{IN} = V_R + V_C, \text{ which is the real part of}$$

$$\tilde{V}_{IN} = \tilde{V}_R + \tilde{V}_C \Leftrightarrow \tilde{V}_R = \tilde{V}_{IN} - \tilde{V}_C \Rightarrow \tilde{V}_R = \tilde{V}_{IN} - \tilde{I} Z_C.$$

Substituting for $\tilde{V}_R$ from equation (1.10.1) gives

$$\tilde{I} Z_R = \tilde{V}_{IN} - \tilde{I} Z_C \Leftrightarrow \tilde{V}_{IN} = \tilde{I} \left(Z_R + Z_C \right).$$

So, the total impedance is $Z_{TOT} = Z_R + Z_C$, meaning that impedances in series add, just as one would expect from the behavior of resistors, i.e.,

$$Z_{series} = Z_1 + Z_2. \tag{1.10.2}$$

In problem 1.20, you can show that impedances in parallel combine as one would expect, that is,

$$\frac{1}{Z_{parallel}} = \frac{1}{Z_1} + \frac{1}{Z_2}. \tag{1.10.3}$$

Core example: The low-pass filter. The circuit in figure 1.10.2b is one of the most useful and common elements in analog circuitry. To understand why, let's calculate V_{OUT}:

$$\tilde{V}_{OUT} = \tilde{V}_C = \tilde{I} Z_C = \frac{\tilde{V}_{IN}}{Z_{TOT}} Z_C = \tilde{V}_{IN} \frac{Z_C}{Z_R + Z_C}.$$

You may recognize this as the equation for a voltage divider; the fraction of the total voltage $\tilde{V}_{IN}$ that appears across the capacitor equals the fraction of the total impedance that is due to the capacitor. Often, we are only interested in the amplitude of the output voltage expressed as a fraction of the amplitude of the input voltage. Recall that the amplitude of an oscillating quantity is the magnitude of the complex number that represents it, as shown in figure 1.10.1c. Therefore,

$$\frac{\text{amplitude of } V_{OUT}}{\text{amplitude of } V_{IN}} = \frac{\left| \tilde{V}_{OUT} \right|}{\left| \tilde{V}_{IN} \right|}.$$

You can show in problem 1.15 that, for any two complex numbers A and B, $\left| \frac{A}{B} \right| = \frac{|A|}{|B|}$. So,

$$\frac{\text{amplitude of } V_{OUT}}{\text{amplitude of } V_{IN}} = \left| \frac{V_{OUT}}{V_{IN}} \right| = \left| \frac{Z_C}{Z_R + Z_C} \right| = \frac{|Z_C|}{|Z_R + Z_C|}.$$

continued

Referring to section 1.8, we see that

$$|Z_C| = \left|\frac{1}{i\omega C}\right| = \frac{1}{\omega C} \text{ and } |Z_R + Z_C| = \left|R + \frac{1}{i\omega C}\right| = \sqrt{R^2 + \frac{1}{\omega^2 C^2}}.$$

Therefore,

$$\frac{\text{amplitude of } V_{\text{OUT}}}{\text{amplitude of } V_{\text{IN}}} = \frac{\dfrac{1}{\omega C}}{\sqrt{R^2 + \dfrac{1}{\omega^2 C^2}}} = \frac{1}{\sqrt{\omega^2 R^2 C^2 + 1}} = \frac{1}{\sqrt{1 + \left(\dfrac{\omega}{\omega_{\text{LO}}}\right)^2}},$$

where

$$\omega_{\text{LO}} \equiv \frac{1}{RC}.$$

The dependence of this ratio on the frequency of V_{IN} is shown in figure 1.10.3. You can see from these graphs, especially the log–log graph on the bottom, why this circuit is called a "low-pass filter." If the angular frequency of V_{IN} is well below ω_{LO}, then the amplitude at the output is the same as the amplitude at the input, whereas if the angular frequency of V_{IN} is well above ω_{LO}, then the output is dramatically smaller than the input.

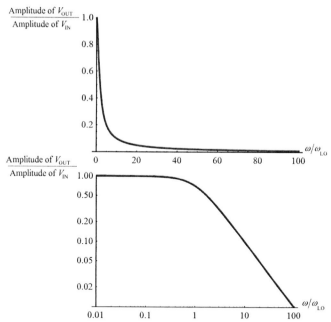

Figure 1.10.3 Amplitude ratio for the low-pass filter shown in figure 1.10.2b. Top: linear axes. Bottom: same data with log–log axes.

We can understand this behavior qualitatively, by remembering that the fraction of the input voltage that appears across the capacitor equals the fraction of the total impedance due to the capacitor. At low frequencies, the impedance of the capacitor is large, so the fraction is nearly equal to 1, whereas at high frequencies the impedance of the capacitor is low, so the fraction approaches zero.

Finally, let us consider an inductor. We use the signal generator to apply a voltage $V_0 \cos \omega t = \mathrm{Re}\left(V_0 e^{i\omega t}\right)$ across the inductor. The magnitude of the voltage across the inductor is

$$V_{\mathrm{L}} = L\frac{\mathrm{d}I}{\mathrm{d}t} \Leftrightarrow \frac{\mathrm{d}I}{\mathrm{d}t} = \frac{V_{\mathrm{L}}}{L}$$

$$\Rightarrow I = \frac{1}{L}\int V_{\mathrm{L}}\mathrm{d}t = \frac{1}{L}\int V_0 \cos \omega t\, \mathrm{d}t = \frac{V_0}{L\omega}\sin \omega t + \text{constant}.$$

If the voltage amplitude V_0 is zero, the current must be zero, so the constant must be zero. So,

$$I = \frac{V_0}{L\omega}\sin \omega t = \mathrm{Re}\,\tilde{I}, \quad \text{where } \tilde{I} = -\mathrm{i}\frac{V_0}{L\omega}e^{i\omega t},$$

since $\sin \omega t = \mathrm{Re}\left[-\mathrm{i}e^{i\omega t}\right]$. Thus, the impedance of the inductor is

$$Z_{\mathrm{L}} = \frac{\tilde{V}}{\tilde{I}} = \frac{V_0 e^{i\omega t}}{-\mathrm{i}\dfrac{V_0}{L\omega}e^{i\omega t}} = \mathrm{i}\omega L.$$

Summarizing:

$$\boxed{Z_{\mathrm{R}} = R \quad Z_{\mathrm{C}} = \frac{1}{\mathrm{i}\omega C} \quad Z_{\mathrm{L}} = \mathrm{i}\omega L.} \tag{1.10.4}$$

Concept test (answer below[10]): Does the circuit shown in figure 1.10.4 function as a low pass filter (for which $\dfrac{\text{amplitude of } V_{\mathrm{OUT}}}{\text{amplitude of } V_{\mathrm{IN}}} = 1$ for low frequencies and 0 for high frequencies), or instead does it function as a high-pass filter (for which $\dfrac{\text{amplitude of } V_{\mathrm{OUT}}}{\text{amplitude of } V_{\mathrm{IN}}} = 1$ for high frequencies and 0 for low frequencies)? You should be able to answer this question using qualitative reasoning, combined with equation (1.10.4).

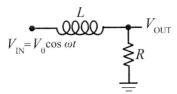

Figure 1.10.4 An RL filter.

---

10. The output voltage divided by the input voltage equals the impedance of the resistor divided by the total impedance. The impedance of the inductor is low for low frequency, meaning that most of the total impedance is in the resistor, so $V_{\mathrm{OUT}} \approx V_{\mathrm{IN}}$. On the other hand, at high frequencies, the impedance of the inductor is high, so the resistor represents a small fraction of the total impedance, and $V_{\mathrm{OUT}} \approx 0$. So, this is a low-pass filter.

1.11   Another important complex function: The quantum mechanical wavefunction

Our topics of study are waves and oscillations. However, one of the important reasons for mastering the concepts and techniques associated with these topics is that they apply directly to quantum mechanics. Therefore, although we will not study quantum mechanics *per se* in this book, we will point out some of the connections as we come to them.

As you may know, small particles such as electrons display many wave-like properties. The wave nature of such a particle is described by the "wavefunction" Ψ (the Greek capital letter psi). The wavefunction depends on both position and time, so we could write it as $\Psi(x, t)$, though usually we will simply write Ψ. One of the most remarkable things about quantum mechanics is that Ψ is *inherently complex!* All the information that can be known about the particle is contained in Ψ, and in a later course you will learn how to extract from Ψ quantities such as the momentum, angular momentum, and energy. One aspect of Ψ is relatively easy to understand: $|\Psi(x, t)|^2$ is called the "probability density," and is proportional to the probability of finding the particle near the position x. For example, for the probability density shown in figure 1.11.1, the particle is likely to be found near $x = -3$ or 1 nm (1 nm $= 10^{-9}$ m). This is called a "delocalized" wavefunction, since the particle might be found in two different places. You'll learn more about this in a later course.

Another example is that of an electron traveling at constant speed through a vacuum; this is called a "free electron." The wavefunction in this case is $\Psi(x, t) = \psi_0 e^{-i\omega t} e^{ikx}$, where ψ_0 (Greek lower case psi, with a naught subscript) is a constant, $k = \dfrac{\omega}{v_p}$ is called the "wavenumber," and v_p is the speed of the wave.[11] This wavefunction has an oscillatory dependence both on t and on x:

$$\Psi(x, t) = \psi_0 e^{-i\omega t} e^{ikx} = \psi_0 (\cos \omega t - i \sin \omega t) e^{ikx} = \psi_0 e^{-i\omega t} (\cos kx + i \sin kx).$$

Wavefunction for a free electron

This function is plotted in figure 1.11.2a and b.

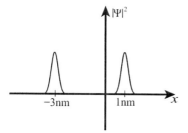

Figure 1.11.1 A delocalized wavefunction.

---

11. The wavenumber k is also equal to $2\pi/\lambda$, where λ is the "wavelength," that is, the repeat interval along the x-axis.

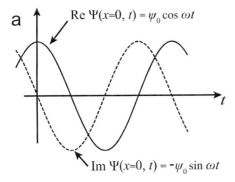

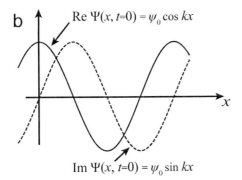

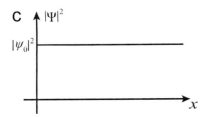

Figure 1.11.2 Real and imaginary parts of the wavefunction for an electron moving with a constant velocity in a vacuum (i.e., a "free electron"). a: dependence on time of the wavefunction at $x = 0$. b: dependence on position of the wavefunction at $t = 0$. c: The probability density is given by $|\Psi|^2$, showing that a free electron is completely delocalized.

Self-test (answer below[12]): Show for $\Psi(x, t) = \psi_0 e^{-i\omega t} e^{ikx}$ that the probability density $|\Psi|^2$ is equal to $|\psi_0|^2$.

The result of this self-test is remarkable—there is no space or time dependence in the probability density $|\Psi|^2$ for this case! This means that the particle is completely delocalized—it is just as likely to be found at $x = -100$ nm as at $x = +100$ km, or at any other point, as shown in figure 1.11.2c. This waveform is an idealized version, since no real particle could be so infinitely spread out. However, real particles can be highly delocalized, even over macroscopic distances.

We will encounter Ψ several more times in this text.

---

12. Answer to self-test: $|\Psi|^2 = \Psi^*\Psi = \left(\psi_0^* e^{i\omega t} e^{-ikx}\right)\left(\psi_0 e^{-i\omega t} e^{ikx}\right) = \psi_0^* \psi_0 = |\psi_0|^2$

1.12   Pure sinusoidal oscillations and uncertainty principles

In real life, oscillations usually only last for a finite period of time. For example, we might strike a piano key, hold it down for a short length of time (allowing the string inside the piano to vibrate), and then release the key (which immediately stops the vibration). If we strike the key at t_1 and release it at t_2, the resulting vibration of a particular point on the string might look as shown in figure 1.12.1.

It is important to realize that the waveform of figure 1.12.1 is *not* a pure sinusoidal oscillation, since it is not of the form (1.2.4): $x = A\cos(\omega t + \varphi)$. Equation (1.2.4) describes an oscillation which goes on infinitely in time, stretching back in time to $t \to -\infty$, and forward in time to $t \to +\infty$. This is not merely a semantic distinction. We will see in our study of Fourier Analysis (chapter 8) that a function of the form shown in figure 1.12.1 can be created by adding together a very large number of pure sinusoids, only one of which is at the angular frequency ω.

This means that, for a function such as that shown in figure 1.12.1, the angular frequency is not "well-defined," that is, the function cannot be characterized by a single angular frequency. (If it could, we could write it as $x = A\cos(\omega t + \varphi)$.) We don't have to wait for chapter 8 to see this – we can develop a qualitative argument now that shows it. Imagine that we try to determine the frequency of the waveform shown in figure 1.12.1 by counting the number of times it crosses zero. (This, in fact, is how frequencies are determined in most experiments, which usually rely on electronic "frequency counters.") There are two zero-crossings per period. Therefore, if we define

$$N \equiv \text{number of zero crossings and } \Delta t \equiv t_2 - t_1,$$

Then,

$$T = \frac{\Delta t}{(N/2)} \Rightarrow f = \frac{1}{T} = \frac{N}{2\Delta t}.$$

However, in any real signal, the beginning and the end are not defined with absolute crispness – there is always a question about exactly where we should begin counting the zero-crossings and where we should stop. We're only making a rough argument here, so let's say that there's an uncertainty of 1 in the value of N. In other words, we can't really be sure whether we should write $f = \dfrac{N}{2\Delta t}$ or $f = \dfrac{N+1}{2\Delta t}$. Thus, there is an uncertainty in f:

$$\Delta f = \frac{N+1}{2\Delta t} - \frac{N}{2\Delta t} = \frac{1}{2\Delta t}.$$

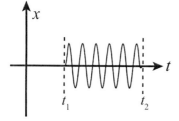

Figure 1.12.1 Motion of one point on a piano string.

We have used simple arguments about a simple way of determining the frequency. However, more sophisticated arguments give the same qualitative results: the shorter the time interval Δt, the greater the uncertainty in the frequency Δf. The more sophisticated arguments, which use a more careful definition of Δt and Δf, show that Δf is always at least as big as $\dfrac{1}{4\pi\,\Delta t}$. This is for an ideal circumstance, where the profile of the wave has an ideal shape. Of course, it is always possible to have a wave with a less ideal profile, or to be sloppy in doing the measurements, so that in general

$$\Delta f \geq \frac{1}{4\pi\,\Delta t} \Leftrightarrow$$

$$\boxed{\Delta f\,\Delta t \geq \frac{1}{4\pi}.} \tag{1.12.1}$$

Frequency–time uncertainty relationship
(valid for all types of waves and oscillations)

We can apply identical arguments to something (such as a wave on the surface of the ocean) that oscillates as a function of x instead of as a function of t. A pure sinusoidal oscillation as a function of t can be written

$$y(t) = A\cos(\omega t + \varphi) \Rightarrow$$

$$y(t) = A\cos\left(\frac{2\pi}{T}t + \varphi\right), \tag{1.12.2}$$

where T is the repeat interval in time (the period). (Here, we use $y(t)$ for the oscillating *function*, to avoid confusion with the *variable* x which is used for the position in the next equation. However, $y(t)$ can stand for exactly the same types of things as we've previously discussed, such as the position of an oscillating tree branch.) Similarly, we can write a pure sinusoidal oscillation as a function of x as

$$y(x) = A\cos\left(\frac{2\pi}{\lambda}x + \varphi\right), \tag{1.12.3}$$

where λ is the repeat interval in space (the "wavelength"). We see that equations (1.12.2) and (1.12.3) are isomorphic. (Recall that this means that they are exactly the same, except that different symbols appear.) In the isomorphism t becomes x and T becomes λ. We can use exactly the same arguments that lead us to equation (1.12.1), but apply them to oscillations as a function of x instead. Since $f = 1/T$, this gives

$$\boxed{\Delta\left(\frac{1}{\lambda}\right)\Delta x \geq \frac{1}{4\pi}.} \tag{1.12.4}$$

Uncertainty relationship between inverse wavelength and position
(valid for all types of waves and oscillations)

The uncertainty relations (1.12.1) and (1.12.4) apply to *any* type of oscillation or wave, whether it is a simple harmonic oscillator or the quantum mechanical wavefunction.

One of the fundamental relations of quantum mechanics relates the energy E of the particle to the frequency of oscillation of the wavefunction. (e.g., for a free electron, the wavefunction is $\Psi(x, t) = \psi_0 e^{-i\omega t} e^{ikx}$, and the frequency is $f = \dfrac{\omega}{2\pi}$.) This fundamental relation is

$$\boxed{\boxed{E = hf,}} \tag{1.12.5}$$

where $h = 6.6260690 \times 10^{-34}$ J s is called "Planck's constant." As you will learn in a later course, this relation comes from experimental results, and cannot be derived from arguments based on classical physics. As you read through this text, bear in mind that essentially everything we say about classical waves and oscillators applies to Ψ as well. In particular, essentially everything we say about the frequency of a classical oscillator applies to the frequency of Ψ as well, and that, by equation (1.12.5), this frequency is proportional to the energy of the particle.

Combining equation (1.12.5) with (1.12.1) gives

$$\Delta\left(\frac{E}{h}\right)\Delta t \geq \frac{1}{4\pi} \Rightarrow$$

$$\boxed{\Delta E \Delta t \geq \frac{h}{4\pi}.} \tag{1.12.6}$$

This is called the "energy–time uncertainty principle." It states that it is not possible to exactly determine the energy of a system if one can only observe it for a short time. If the energy of a system cannot be precisely determined on short time scales, then this "energy–time uncertainty principle" requires that we modify our understanding of the conservation of energy to allow for "quantum fluctuations." For example, it is possible for pairs of particles (one normal matter and one antimatter) to be spontaneously created out of the vacuum, which requires a tremendous amount of energy (on a particle scale). These particles can exist only for a fleeting moment, and then annihilate with each other, releasing the energy they had "borrowed" before anyone could notice it was missing! Perhaps surprisingly, the effects of these "virtual particles" can be observed experimentally, for example through the Casimir effect.[13]

The other fundamental relation of quantum mechanics is

$$\boxed{p = \frac{h}{\lambda},} \tag{1.12.7}$$

where p is the momentum of the particle. As with equation (1.12.5), this relation comes from experimental results, and cannot be derived from classical principles.

---

13. In the region between two metal plates, the density of virtual particles is lower than in the region outside the plates, resulting in a force that pushes the plates together. For more information, see *Precision Measurement of the Casimir Force from 0.1 to 0.9 μm*, by U. Mohideen and Anushree Roy, Phys. Rev. Lett. **81**, 4549 (1998). A summary is available at http://focus.aps.org/v2/st28.html

Combining equation (1.12.7) with (1.12.4) gives

$$\Delta\left(\frac{p}{h}\right)\Delta x \geq \frac{1}{4\pi} \Leftrightarrow$$

$$\boxed{\Delta p \Delta x \geq \frac{h}{4\pi}.}$$

(1.12.8)

This is the more famous "Heisenberg uncertainty principle," which states that it is not possible to simultaneously determine a particle's position and its momentum with absolute precision. Both of these important quantum mechanical uncertainty relations (1.12.6) and (1.12.8) are direct consequences of attributing a wave nature to particles and not the result of any other "quantum mechanical weirdness."

Despite all the above arguments, we are very often in the situation where the time interval Δt is much, much longer than the period T. In such a case, the frequency is fairly well defined (i.e., Δf is small), and we need not worry much about the concerns raised in this section.

Concept and skill inventory for chapter 1

After reading this chapter, you should fully understand the following terms:

Stable equilibrium (1.2)
Hooke's Law (1.2)
DEQ (1.2)
Simple Harmonic Motion (SHM) (1.2)
Harmonic approximation (1.2)
Amplitude, phase, frequency, angular frequency, period (1.2–1.3)
Adjustable constants (1.2)
Capacitor, inductor (1.5)
Kirchhoff's loop rule (1.5)
Isomorphism (1.5)
Taylor series (1.6)
Euler's equation (1.7)
Complex plane (1.8)
Magnitude of a complex number (1.8)
Complex conjugate (1.8)
Signal generator (1.10)
Ground for electrical circuits (1.10)
Complex version of Ohm's Law (1.10)
Impedance, as applied to electrical components (1.10)
Low-pass filter (1.10)
Log–log axes (1.10)
Quantum mechanical wavefunction (1.11)
Probability density (1.11)

Free electron (1.11)
Wavenumber (1.11)
Frequency–time uncertainty relation (1.12)
Inverse wavelength – position uncertainty relation (1.12)
Energy–time uncertainty relation (1.12)
Heisenberg (momentum–position) uncertainty relation (1.12)

You should know what happens when:

A system described by the harmonic approximation is displaced from equilibrium and
   then released (1.2–1.3)
A complex number is multiplied by $e^{i\alpha}$ (1.8)
An input voltage is applied to a low-pass filter (1.10)

You should understand the following connections:

Frequency, angular frequency, & period (1.2–1.3)
Mass on a spring & an LC oscillator (1.5)
Current in a circuit & charge on a capacitor; be able to tell whether $I = \dot{q}$ or instead
   $I = -\dot{q}$ (1.5)
Cosine representation & complex exponential representation for SHM (1.9)
x & z (1.9)
One-dimensional oscillation & rotation in the complex plane (1.9)
V & $\tilde{V}$ (1.10)

You should be familiar with the following additional concepts:

Dot notation for time derivatives (1.2)
Angular frequency for a mass/spring (1.3)
The frequency for SHM doesn't depend on amplitude (1.3)
Angular frequency for an LC oscillator (1.5)
First-order Taylor series approximation for sin (1.6)
Second-order Taylor series approximation for cos (1.6)
Taking the real part doesn't commute with multiplication by a complex number (1.9)
Impedance of resistors, capacitors, and inductors (1.10)

You should be able to:

Check a proposed solution to a DEQ to determine if it's correct and if there are
   restrictions on the parameters (1.2)
Find the amplitude and phase given the initial position and velocity (1.3)
Deal with the multi-valued aspect of the arctan function (1.3)
Use the isomorphism between the mass/spring and the LC oscillator to quickly adapt
   results from one system to the other (1.5)
Given a function, create the Taylor series for it (1.6)
Explain what it means to take the exponential of a complex number (1.7)

Go back and forth between Cartesian and polar representations for complex numbers (1.8)

Find the magnitude of a complex number (1.8)

Find the energy of an oscillator given either the spring constant and the amplitude *or* the mass and the maximum velocity (1.9)

Calculate $\left|\dfrac{V_{OUT}}{V_{IN}}\right|$ for any simple combination of resistors, inductors, and capacitors (1.10)

> **In addition to all of the above, you should be able to combine the concepts you've learned to address new situations.**

Problems

Note: Additional problems are available on the website for this text.

Instructor: Ratings of problem difficulty, full solutions, and additional support materials are available on the website.

1.1 What is the difference between a stable and unstable equilibrium? Give examples of each type of equilibrium from everyday life.

1.2 **Plasma oscillations.** To turn a gas into a "plasma," one or more electrons must be completely separated from each nucleus. This can be accomplished by applying a very strong electric field (as in a spark or a lightning bolt), by the absorption of ultraviolet light (as happens in the "ionosphere," one of the upper layers of our atmosphere), or by heating to a very high temperature (as occurs in the sun). We will model a plasma as a "gas" of electrons with number density n. (In other words, there are n electrons per cubic meter.) Each electron has charge $-e$ and mass m_e. Occupying the same volume is a gas of positively charged ions, each with charge $+e$. Because the ions were created by removing the electrons from neutral atoms, the number density of ions is also n. (In other words, there are n ions per cubic meter.)

Consider a cube of plasma, with sidelength ℓ. Somehow (e.g., by applying an electric field), the electrons are all displaced a small distance x to the right. This creates a layer of negative charge on the right, and leaves behind a layer of positive charge on the left, as shown in figure 1.P.1. For simplicity, treat each of these layers as an infinite thin sheet of charge. The layers create an electric field throughout the plasma, pointing to the right. This field exerts a force to the left on the electrons filling the central region of the figure, and to the right on the ions. However, we'll focus on the electrons because they are so much lighter, and assume that the ions remain stationary.

When we turn off the original force that caused the displacement x of the cube of electrons, the electric field from the surface charge layers causes the electrons to spring back toward the left. However, because of their finite mass, they overshoot the equilibrium position (corresponding to

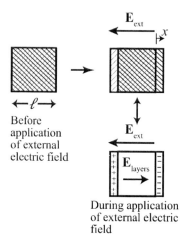

Before
application
of external
electric field

During application
of external electric
field

Figure 1.P.1 Model for the ionosphere. The figure shows a cross-sectional view of a cube of sidelength ℓ. The electrons are represented by the region that is shaded with down-slanting lines, while the positive charges are represented by the region shaded with up-slanting lines. E_{layers} is the field due to the thin layers of charge on the left and right surfaces.

displacement $x = 0$), move to the left of the ions, then are pulled back to the right, and so on, in an oscillatory motion. In this problem, you'll calculate the frequency of this oscillation.

(a) Explain why the field produced by the combination of the two charge layers in the region between them is $E = \dfrac{nex}{\varepsilon_0}$. (You may need to refer back to your intro. E&M textbook. Remember that we're treating the layers of charge as infinite sheets.)

(b) Explain why this leads to a restoring force on the electrons $F = -\dfrac{n^2 e^2 \ell^3}{\varepsilon_0} x$. (Remember that $x \ll \ell$, so that the total charge of electrons that experiences the electric force is not significantly changed by the small number in the displaced region x.)

(c) Explain why this means that the oscillation frequency is $\omega = \sqrt{\dfrac{ne^2}{m_e \varepsilon_0}}$. This is called the plasma frequency. Only radio waves with frequencies significantly higher than this can propagate through the plasma. Lower frequency waves are instead reflected. Thus, AM radio waves (which are low frequency) can bounce off the ionosphere, leading to very long transmission range under good conditions (at night), whereas FM radio waves (higher in frequency) don't bounce off the ionosphere, and so the transmission range is much more limited.

1.3 Derive equations (1.3.4a) and (1.3.4b) from the equations immediately preceding them.

1.4 Consider the potential energy described in problem 1.14. For low amplitudes, the motion of the object is well described by simple harmonic motion, so that the period is independent of amplitude. However, once the amplitude gets high enough this is no longer true. As the amplitude increases, does the period increase or decrease? Explain your reasoning thoroughly, and assume that the amplitude is always less than π/β.

1.5 A particle of mass m moves in the potential energy $U = \alpha x^2 + \beta x^4$, where α and β are both positive. **(a)** What is the angular frequency of oscillation for small amplitudes? **(b)** If the amplitude is increased far enough, one finds that the angular frequency starts to depend on amplitude. Does the angular frequency increase or decrease as the amplitude is increased? Explain your reasoning.

1.6 We can rewrite equation (1.3.4a) to get $\omega_0 = \dfrac{v_0}{\sqrt{A^2 - x_0^2}}$. This makes it seem that ω_0 depends on the amplitude and initial conditions. Explain this seeming contradiction with equation (1.3.3).

1.7 Before doing this problem, read footnote 4 in section 1.3 about linear homogeneous DEQs. Show that, as claimed in that footnote, $A_1 \cos \omega_0 t + A_2 \sin \omega_0 t = A \cos(\omega_0 t + \varphi)$, with $A = \sqrt{A_1^2 + A_2^2}$ and $\varphi = \tan^{-1}\left(-A_2/A_1\right)$.

1.8 A mass sits on a platform which oscillates vertically in simple harmonic motion at a frequency of 5 Hz. Show that the mass loses contact with the platform when the amplitude exceeds 1 cm. (Assume $g = 9.8$ m/s^2.) *Hints: The mass loses contact with the platform if the downward acceleration of the platform exceeds g. To keep the math simple, choose a zero phase factor, i.e., choose $x = A \cos \omega_0 t$.*

1.9 **Home experiment (work with only one other student on this, and include his/her name with your problem):** The light emitted by a nonflatscreen TV makes a good stroboscope. A given point on the screen is actually dark most of the time; it is lit a small fraction of the time at a regular repetition rate, which we'll call f_{TV}. The object of this experiment is to measure f_{TV}; I'll tell you that it's either 30 or 60 Hz. As a very crude measurement, wave your finger in steady oscillation in front of the screen at a rate of about 4 Hz, for example. Your finger will block the light from the screen wherever it happens to be when the screen flashes on. Roughly measure (as best you and your partner can) the amplitude of your finger's oscillation. Then measure the separation between successive finger shadows at the point of maximum velocity. Assume the motion is sinusoidal. Calculate the maximum velocity of the finger, given the amplitude and finger frequency. Put all of this together to find f_{TV}. (Recall that you're told that it's either 30 or 60 Hz, so your measurements need only be precise enough to get a rough answer which allows you to determine which of these two possibilities is correct.)

1.10 Read the aside about the arctan function in section 1.3. Explain why a more complete version of equation (1.3.4b) would be $\varphi = \tan^{-1}\left(\dfrac{-v_0}{\omega x_0}\right) + \begin{cases} 0 \text{ if } x_0 > 0 \\ \pi \text{ if } x_0 < 0 \end{cases}$.

1.11 Before doing this problem, read footnote 4 in section 1.3 about linear homogeneous DEQs. A particle far from the Earth feels the pull of gravity $F = -\dfrac{GMm}{x^2}$, where G is the gravitational constant, M the mass of the Earth,

m the mass of the particle, and x is the distance from the center of the Earth to the particle. Assume that the particle moves along a straight line that passes through the center of the Earth, that gravity is the only force acting on it, and that for the time period we are considering the particle doesn't touch the surface of the Earth (i.e., that it's out in space).

(a) Write a DEQ of motion for the particle. (As an example of what I mean by a DEQ of motion, for the simple harmonic oscillator it would be $\ddot{x} + \omega_0^2 x = 0$.)

(b) If $x_1(t)$ is one solution of the DEQ and $x_2(t)$ is another, is the combination $x_1 + x_2$ necessarily a solution? Why or why not?

1.12 Analogy between a capacitor and a spring. (a) How is adding charge to a capacitor like compressing a spring? (b) Why is it that the inverse of C isomorphic to k, rather than just C itself?

1.13 In research-level theoretical physics, it is almost never possible to get an exact solution because of the complexity of the problems being considered. Therefore, it is essential to make appropriate approximations, so as to get physical insight. The Taylor series is central to many of these approximations. You have already seen two of the three most common applications of the Taylor series: $\sin \theta \cong \theta$ (for small θ) and $\cos \theta \cong 1 - \dfrac{\theta^2}{2}$ (for small θ). In this problem, you will demonstrate the third of the three most common applications. Show that $(1 + x)^n \cong 1 + nx$ for $x \ll 1$. (Note that this works whether n is positive or negative, integer or fractional.)

1.14 The potential energy for a particular object is $U(x) = -L \cos \beta x$, where L and β are both > 0. (This potential energy function is important in the study of superconductivity.)

(a) Make a sketch of this potential energy from $x = -\dfrac{2\pi}{\beta}$ to $x = +\dfrac{2\pi}{\beta}$. Indicate the scale on the vertical axis.

(b) The object has mass m and total energy $E_{\text{TOT}} = -L + G$, where $0 < G \ll L$. (The symbol "$\ll$" means "much less than.") Add a dashed line to your sketch indicating this total energy.

(c) At $t = 0$, the object is at $x = 0$. Show that its motion can be approximated by a simple harmonic oscillation, and find the approximate frequency of oscillation. *Hint: recall that the Taylor series expansion for $\cos \theta$ is $\cos \theta = 1 - \dfrac{\theta^2}{2!} + \dfrac{\theta^4}{4!} - \cdots$, so that for $\theta \ll 1$, $\cos \theta \cong 1 - \dfrac{\theta^2}{2!}$.*

1.15 Let C_1 and C_2 be complex numbers. Show that $\left| \dfrac{C_1}{C_2} \right| = \dfrac{|C_1|}{|C_2|}$. *Reminder: $|C_1|$ means "magnitude of C_1"*

1.16 For each of the following, express the quantity shown *either* in the form $a + ib$ (i.e., Cartesian representation) *or* in the form $Ae^{i\alpha}$ (i.e., polar representation), *whichever you find easier for each part of the problem.*

In case you might be confused by the way I've written things: "i 4.5" means "i times 4.5."

- **(a)** $(3.2 + i\ 6.7) + (5.6 - i\ 4.5)$
- **(b)** $6.1\ e^{i\ 1.2} + 1.2\ e^{i\ 1.7}$
- **(c)** $(3.2 + i\ 6.7)(5.6 - i\ 4.5)$
- **(d)** $(6.1\ e^{i\ 1.2})(1.2\ e^{i\ 1.7})$
- **(e)** What point is this problem trying to get across?

1.17 Let $z_1 = 8e^{i\pi/6}$ and $z_2 = \sqrt{2}e^{i3\pi/4}$.

- **(a)** Represent z_1 and z_2 in the complex plane.
- **(b)** Find the real and imaginary parts of z_1 and z_2.
  Express each of the following in the form $Ae^{i\varphi}$, where A and φ are real:
- **(c)** $z_1 + z_2$
- **(d)** $z_1^3 z_2^2$

1.18 A mass $m = 10$ kg is oscillating on a spring with $k = 10$ N/m with little damping. The displacement of the mass can be described by $x(t) = \text{Re}\left\{Ce^{i\omega t}\right\}$, where $C = (1 - i)$ cm.

- **(a)** What is the value of ω?
- **(b)** What is the amplitude of the motion?
- **(c)** The solution can also be expressed in the form $x(t) = A\cos(\omega t + \varphi)$. What is the value of φ ?
- **(d)** Describe the initial conditions of the motion, that is, specify the position and velocity at $t = 0$. As for all numbers in physics (except dimensionless quantities) be *sure* to include units!
- **(e)** Sketch two graphs, one of position *versus* time and the other of velocity *versus* time. Be sure to label both axes of each graph quantitatively.
- **(f)** What is the energy of the oscillator?

1.19 For an RC low-pass filter (figure 1.10.2b), show that $\left|\dfrac{V_{OUT}}{V_{IN}}\right|$ drops by a factor of 10 for each factor of 10 increase in ω, for $\omega \gg \omega_{LO}$.

1.20 Using methods similar to those leading to equation (1.10.2), show that the complex impedances of two circuit elements in parallel combine in the way specified by equation (1.10.3).

1.21 A voltage $V(t) = V_0 \cos \omega t$ is applied across a capacitor with capacitance C. **(a)** Without using a symbolic algebra program or graphing calculator, make a sketch with curves for both $V(t)$ and $I(t)$, showing two full periods of oscillation. Label your sketch quantitatively. **(b)** For a capacitor, does the current "lead" the voltage (meaning that the current reaches a peak before the voltage does), or does the voltage lead the current? **(c)** Using simple ideas about charging of the capacitor and the connection between voltage and charge, explain why your answer to part (b) makes sense.

1.22 A voltage $V(t) = V_0 \cos \omega t$ is applied across an inductor with inductance L. **(a)** Without using a symbolic algebra program or graphing calculator, make a sketch with curves for both $V(t)$ and $I(t)$, showing two full periods of oscillation. Label your sketch quantitatively. **(b)** For an inductor, does the current "lead" the voltage (meaning that the current reaches a peak before the voltage does), or does the voltage lead the current? **(c)** Using elements from the isomorphism between an LC oscillator and a mass/spring system, explain why your answer to part (b) makes sense.

1.23 A voltage $V(t) = V_0 \cos \omega t$ is applied to the input of an RC low-pass filter (figure 1.10.2b). At the output, one observes a voltage $V_{\text{OUT}} \cos(\omega t + \varphi)$. **(a)** Find φ as a function of ω, R, and C. **(b)** Using whatever software you like, make two graphs of $-\varphi$ as a function of $\omega/\omega_{\text{LO}}$, ranging from $\omega/\omega_{\text{LO}} = 0.01$ to 100, one plot with linear axes and one with logarithmic axes.

1.24 **The RC high-pass filter.** For the circuit shown in figure 1.P.2, a sinusoidal input voltage is applied relative to ground, and the resulting output voltage is measured relative to ground. If a sinusoidal voltage $V_{\text{in}} = V_i \cos \omega t$ is applied to the input, one observes a sinusoidal voltage $V_{\text{out}} = V_o \cos(\omega t + \varphi)$ at the output (relative to ground). **(a)** Show that $\dfrac{\text{amplitude of output voltage}}{\text{amplitude of input voltage}} =$

$$\frac{V_o}{V_i} = \frac{1}{\sqrt{1 + \left(\frac{\omega_{\text{HI}}}{\omega}\right)^2}}, \text{ where } \omega_{\text{HI}} \equiv \frac{1}{RC}. \text{ (b) Show that the phase shift}$$

of the output relative to the input is $\varphi = \tan^{-1}\left(\dfrac{\omega_{\text{HI}}}{\omega}\right)$. **(c)** Given that φ is positive, does the output "lag" the input (i.e., do the peaks in the output voltage occur after the peaks in the input), or does the output "lead" the input (i.e., do the peaks in the output occur before the peaks in the input)?

1.25 **More on the RC high-pass filter.** Read all of problem 1.24 before beginning this problem. **(a)** Using a suitable computer program, make two plots of $\dfrac{\text{amplitude of output voltage}}{\text{amplitude of input voltage}} = \dfrac{V_o}{V_i}$ *versus* $\dfrac{\omega}{\omega_{\text{HI}}}$. In your first plot, use linear scales for both axes. In your second plot, use logarithmic scales for both axes. For both plots, let $\dfrac{\omega}{\omega_{\text{HI}}}$ vary from 0.01 to 100. **(b)** Make two plots of φ versus $\dfrac{\omega}{\omega_{\text{HI}}}$. In your first plot, use linear scales for both axes. In your second plot, use a linear scale for the φ axis and a logarithmic scale for the $\dfrac{\omega}{\omega_{\text{HI}}}$ axis.

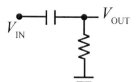

Figure 1.P.2 The RC high-pass filter.

For both plots, let $\dfrac{\omega}{\omega_{HI}}$ vary from 0.01 to 100. **(c)** For $\omega \ll \omega_{HI}$, show that $\dfrac{\text{amplitude of output voltage}}{\text{amplitude of input voltage}} = \dfrac{V_o}{V_i}$ drops by a factor of 10 for each factor of 10 decrease in ω.

1.26 **More on the RC low-pass filter. (a)** Consider the RC low-pass filter shown in figure 1.10.2b. If a voltage (relative to ground) $V_{in} = V_i \cos \omega t$ is applied to the input, one observes a sinusoidal voltage $V_{out} = V_o \cos(\omega t + \varphi)$ at the output (relative to ground). Show that the phase shift of the output voltage relative to the input is $\varphi = \tan^{-1}\left(-\dfrac{\omega}{\omega_{LO}}\right)$, where $\omega_{LO} \equiv \dfrac{1}{RC}$. **(b)** Given that φ is negative, does the output "lag" the input (i.e., do the peaks in the output voltage occur after the peaks in the input), or does the output "lead" the input (i.e., do the peaks in the output occur before the peaks in the input)?

1.27 **The RL low-pass filter.** One can use resistors and capacitors to build either low-pass filters (see section 1.10) or high-pass filters (see problems 1.24 and 1.25). It is also possible instead to use resistors and inductors for these tasks; we'll explore the low-pass version in this problem. In practical circuits, RC filters are much more common than RL filters, partly because it is easier to build an essentially ideal capacitor than to build an ideal inductor. (There is always resistance in the wires used to wind an actual inductor, and there is always capacitance between the windings.) Also, inductors are generally bulkier and more expensive than the capacitors that could be used to build comparable filters. However, one does see RL filters in some high frequency applications. Consider the circuit shown in figure 1.10.4. If a voltage (relative to ground) $V_{in} = V_i \cos \omega t$ is applied to the input, one observes a sinusoidal voltage $V_{out} = V_o \cos(\omega t + \varphi)$ at the output (relative to ground). **(a)** Show that $\dfrac{\text{amplitude of output voltage}}{\text{amplitude of input voltage}} = \dfrac{V_o}{V_i} = \dfrac{1}{\sqrt{1 + \left(\dfrac{\omega}{\omega_{LO}}\right)^2}}$, where $\omega_{LO} \equiv \dfrac{R}{L}$. **(b)** Show that the phase shift of the output relative to the input is $\varphi = \tan^{-1}\left(-\dfrac{\omega}{\omega_{LO}}\right)$. (Note that, except for the definition of ω_{LO}, these expressions are the same as for the RC low-pass filter.)

1.28 An electron in an atom can be excited from its original "ground state" to a well-defined and reproducible higher energy metastable "excited state." The electron can then "fall" back down to the ground state, emitting a photon in the process which has energy equal to the difference between the ground state and the excited state. (As it turns out, this energy is characteristic of the type of atom, and so analysis of such photons can be used for determining the elements which are present in a sample.) Many of these excited states have only a very short lifetime. For such a state, if the experiment is performed carefully, one can determine that the photons emitted from a large sample of material do not all have exactly the same energy, but instead there is a small spread. If a particular excited state has a lifetime of 10^{-7} s, about how big a spread in photon energy would you expect? *Hint: Use the energy-time uncertainty relation; this is meant to be a very easy problem.*

2 Examples of Simple Harmonic Motion

And I saw the cantilever jutting through the mist, resplendent in the light of dawn, oscillating jauntily like a promise of joy.

–Marian McKenzie

2.1 Requirements for harmonic oscillation

In this chapter, we will explore several examples of the remarkable variety of systems that show the harmonic oscillations described in chapter 1. There are two basic requirements for a system to oscillate: (1) If the system is disturbed from equilibrium, there must be something (such as a force) that tends to bring it back toward equilibrium. For the oscillations to be of the sinusoidal form described in chapter 1, this restoring drive must be proportional to the displacement from equilibrium, for example, the spring force F is proportional to x: $F = -kx$. (2) As the system moves toward equilibrium, there must be something (such as inertia) which tends to make the system overshoot the equilibrium point.

Saying the same thing mathematically, if the system is described by a differential equation of the form

$$F = -kx \Rightarrow \underbrace{m\frac{d^2x}{dt^2}}_{\substack{\text{inertial term:}\\\text{satisfies}\\\text{condition 2}}} = \underbrace{-kx}_{\substack{\text{restoring term:}\\\text{satisfies}\\\text{condition 1}}}, \qquad (2.1.1)$$

then it will exhibit harmonic oscillations.

Sometimes, it is easier to consider the energy of a system. If the system can be described by an equation of the form

$$\underbrace{\frac{1}{2}m\left(\frac{dx}{dt}\right)^2}_{\substack{\text{kinetic}\\\text{energy}}} + \underbrace{\frac{1}{2}kx^2}_{\substack{\text{potential}\\\text{energy}}} = \text{constant}, \qquad (2.1.2)$$

then it will exhibit harmonic oscillation.

In other words, if we can show that a system obeys an equation either of the form (2.1.1) or of the form (2.1.2), then we can immediately conclude that

$$x = A\cos(\omega_0 t + \varphi), \text{ where } \omega_0 = \sqrt{\frac{k}{m}}. \qquad (2.1.3)$$

2.2 Pendulums

Professor Roger Newton, author of the book *Galileo's Pendulum*, recounts this wonderful legend about Galileo, the world's first experimental physicist, and a revelation that occurred during a church service in 1581:

> He was seventeen and bored listening to the Mass being celebrated in the cathedral of Pisa. Looking for some object to arrest his attention, the young medical student began to focus on a chandelier high above his head, hanging from a long, thin chain, swinging gently to and fro in the spring breeze. How long does it take for the oscillations to repeat themselves, he wondered, timing them with his pulse. To his astonishment, he found that the lamp took as many pulse beats to complete a swing when hardly moving at all as when the wind made it sway more widely.[1]

This description of one of the first quantitative observations of experimental physics shows the historic importance of pendulums in physics. We will see in chapter 4 that pendulums provide an excellent illustration of chaos theory. Pendulums are common in everyday life, from a baby's swing to a grandfather clock, from a fair ride to a wrecking ball.

We recognize Galileo's observation that the period is independent of the amplitude, as a characteristic of simple harmonic motion. In the case of the pendulum, although there is an obvious mass, there is no obvious spring. Yet, since it does have a position of stable equilibrium, we should be able to model the potential energy near this position as a parabola, **and so we should be able to find an "effective spring constant" that arises from the combination of gravity and the tension in the string.**

Let's consider an arbitrary rigid object of mass m that can rotate in the x–y plane about a pivot P, as shown in figure 2.2.1. We'll show that, for small amplitudes of swing, the energy takes the form (2.1.2). In the figure, we displace the pendulum by an angle θ from equilibrium. The potential energy is determined by the position of the center of mass (marked CM in the figure): $U = mgy$, where y is the height of the CM above its equilibrium position.

Your turn: Show that, if θ is small, then

$$U \cong \tfrac{1}{2}\, mg\, \ell_{CM}\, \theta^2. \qquad (2.2.1)$$

---

1. Roger G. Newton, *Galileo's Pendulum: From the Rhythm of Time to the Making of Matter,* Harvard University Press, Cambridge, MA, 2004, p. 1.

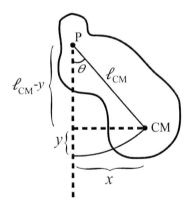

Figure 2.2.1 A pendulum of arbitrary form.

Hint: Use the Taylor expansion for $\cos\theta$, which you derived in section 1.6:

$$\cos\theta = 1 - \frac{\theta^2}{2!} + \frac{\theta^4}{4!} - \cdots \qquad (1.6.5)$$

(If θ is small, this means that $\cos\theta \cong 1 - \frac{\theta^2}{2!}$.)

The motion is a pure rotation about P, so the kinetic energy is $K = \frac{1}{2}I\omega^2$, where I is the moment of inertia for rotations about P and $\omega = \dot\theta$ is the angular velocity. Combining this with your result (2.2.1) gives the total energy:

$$E = U + K = \tfrac{1}{2}mg\,\ell_{CM}\,\theta^2 + \tfrac{1}{2}I\dot\theta^2 . \qquad (2.2.2)$$

Since the total energy is constant, this has exactly the same form as

$$(2.1.2): \tfrac{1}{2}kx^2 + \tfrac{1}{2}m\dot x^2 = \text{constant},$$

so that we immediately know that the pendulum displays simple harmonic motion, that is, that

$$\boxed{\theta = \theta_0 \cos\left(\omega_0 t + \varphi\right), \text{ where } \omega_0 = \sqrt{\frac{mg\,\ell_{CM}}{I}}.} \qquad (2.2.3)$$

Motion of a pendulum (mass need not be concentrated at a point).

I = moment of inertia about pivot, ℓ_{CM} = length from pivot to CM

Core example: the simple pendulum. The simplest example of a pendulum is a compact mass (the "pendulum bob") at the end of a thin rod of length ℓ; we assume the mass of the rod is negligible compared to that of the bob. In this case, the moment of inertia about the pivot point is $I = m\ell^2$. Plugging this into equation (2.2.3) gives

$$\boxed{\omega_{\text{simple pendulum}} = \sqrt{\frac{g}{\ell}}.} \qquad (2.2.4)$$

For more complicated objects, one often uses the parallel axis theorem, which you may have seen proved in an introductory physics book:

$$I = I_{\text{CM}} + mh^2,$$ (2.2.5)

The parallel axis theorem

where I_{CM} is the moment of inertia for rotations about the center of mass and h is the distance from the pivot point P to the center of mass. By breaking a complicated object up into smaller symmetrical objects and applying the parallel axis theorem, one can compute I of the complicated object.

Although the harmonic motion of the pendulum is most easily seen by considering the time dependence of θ (as we have done earlier), we can also show that there is a harmonic variation in the horizontal position x. For a simple pendulum $I = m\ell^2$, so that equation (2.2.2) becomes

$$E = \tfrac{1}{2}mg\ell\theta^2 + \tfrac{1}{2}m\ell^2\dot{\theta}^2 = \tfrac{1}{2}\frac{mg}{\ell}\ell^2\theta^2 + \tfrac{1}{2}m\ell^2\dot{\theta}^2 = \tfrac{1}{2}\frac{mg}{\ell}(\ell\theta)^2 + \tfrac{1}{2}m\left[\frac{\mathrm{d}}{\mathrm{d}t}(\ell\theta)\right]^2.$$

As we can see from the figure, in the limit of small displacements, the arc length $\ell\theta \cong x$, so that

$$E \cong \tfrac{1}{2}\left(\frac{mg}{\ell}\right)x^2 + \tfrac{1}{2}m\dot{x}^2.$$

Since this has the same form as the energy of a mass/spring system, $E = \tfrac{1}{2}kx^2 + \tfrac{1}{2}m\dot{x}^2$, we see that the effective spring constant for the pendulum is $k_{\text{pendulum}} = \dfrac{mg}{\ell}$. This means that the net restoring force, which is created by a combination of gravity and the string tension, is

$$\boxed{F_{\text{pendulum}} = -k_{\text{pendulum}}\,x = -\frac{mg}{\ell}x.}$$ (2.2.6)

"Pendulum force" resulting from the combination
of gravity and tension for a simple pendulum.

We will use this result again in chapter 5.

2.3   Elastic deformations and Young's modulus

All materials are at least a little stretchy, although the amount an object can be stretched before breaking is often too small to see with the naked eye. This stretchiness, and the vibrations that occur when an object is stretched or twisted and then released, determine the engineering limits of building materials, the performance limits of automotive components, and the behavior of a new class of devices known as Micro ElectroMechanical Systems (MEMS).[2] In the remainder of this chapter, we'll explore

---

2. These devices, which combine mechanical motion with electrical actuation or sensing, are fabricated using techniques of photolithography, electron beam lithography, and various types of

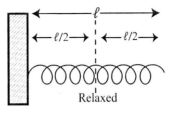

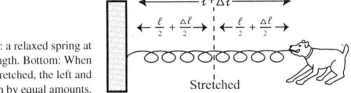

Figure 2.3.1 Top: a relaxed spring at its equilibrium length. Bottom: When the spring is stretched, the left and right halves stretch by equal amounts.

various types of elastic (i.e., reversible) deformations, and their importance in science and everyday life.

The simplest way to deform an object is to stretch or compress it. Consider a long object of uniform cross-section, such as a beam, which is anchored at the left end. If a force is applied to the right end, one always observes that the beam stretches by an amount proportional to the force. This comes as no surprise – before the force is applied, the beam is in equilibrium, and by the arguments in chapter 1 any displacement from equilibrium is countered by a force $F_{\text{spring}} = -kx$. Therefore, to produce a displacement x, we must apply $F_{\text{applied}} = -F_{\text{spring}} = kx$.

The spring constant k depends on the material from which the beam is made; diamond is stiffer than rubber. However, k also depends on the cross-sectional area and length of the beam. We wish to divide out these geometric dependencies to get a parameter that describes the springiness or stiffness of the material itself. First, we consider the dependence on length. How does k change if the beam is cut to half its length? Since we're modeling the beam as a spring, this is equivalent to asking what happens to the spring constant of a spring when it is cut in half.

Imagine a spring of equilibrium length ℓ which is attached to a wall on the left side. A dog pulls on the right end, stretching it by an amount $\Delta\ell$, as shown in figure 2.3.1. The force exerted by the spring on the dog is

$$F_{\text{by spring}} = -k\,\Delta\ell$$

$$\Leftrightarrow k = -\frac{F_{\text{by spring}}}{\Delta\ell} = -\frac{F_{\text{by spring}}}{\text{extension}}. \qquad (2.3.1)$$

anisotropic (meaning directionally dependent) etching. The sizes of the devices range from the diameter of a human hair down to the molecular regime, allowing extremely fast response times and, for devices designed to detect trace chemicals, extraordinary sensitivity. We will discuss MEMS devices in sections 2.6 and 3.4. You can learn more about MEMS in problem 2.14, and in the website for this text, under the entry for this section.

The dog must be exerting an equal and opposite force $k\,\Delta\ell$ on the right end of the spring. Since the right half of the spring doesn't accelerate, we know there must be another force to balance this; this force is exerted by the left half on the right half

$$F_{\substack{\text{by left}\\\text{half}}} = -k\,\Delta\ell.$$

The extension of the left half is $\dfrac{\Delta\ell}{2}$. Therefore, by analogy with equation (2.3.1), the spring constant of the left half is

$$k_{\substack{\text{left}\\\text{half}}} = -\frac{F_{\substack{\text{by left}\\\text{half}}}}{\text{extension}} = -\frac{(-k\,\Delta\ell)}{(\Delta\ell/2)} = 2k.$$

Therefore, when a spring is cut in half, the spring constant gets doubled. (Another way to see this: for the same extension, the coils of a short spring are distorted more than the coils of a long spring, therefore the shorter spring exerts a bigger force.)

Of course, this also means that, if the length of the spring (or in our case the length of the beam) is doubled, the spring constant is halved. Thus, $k \propto \dfrac{1}{\ell}$. (The symbol $\propto$ means "proportional to.")

> **Your turn (answer[3] at bottom of page)**: Explain why k is proportional to the cross-sectional area A of the beam.

Putting these results together, we can write $k \propto \dfrac{A}{\ell}$ or

$$\boxed{k = E\frac{A}{\ell},} \tag{2.3.2}$$

where the proportionality constant E is called "Young's Modulus,"[4] and ℓ is the *equilibrium length* of the beam.

We can also write

$$F_{\text{spring}} = -kx = -\frac{EA}{\ell}x \Leftrightarrow \frac{F_{\text{spring}}}{A} = -E\frac{x}{\ell}.$$

The force applied *to* the beam, F_{applied}, is equal and opposite to the force F_{spring} applied *by* the beam, so that

$$\frac{F_{\text{applied}}}{A} = E\frac{x}{\ell}. \tag{2.3.3}$$

---

3. Answer: We can divide the beam into N smaller beams running in parallel along the length. The force from a single one of these would be $F_{\text{small}} = -k_{\text{small}}\,x$, where x is the amount by which the beam is stretched. The force from the entire beam is the total of the forces from the small beams: $F_{\text{TOT}} = NF_{\text{small}} = -\left(Nk_{\text{small}}\right)x$, so that the total spring constant is $k = Nk_{\text{small}}$. If the cross-sectional area A is doubled, then N doubles, so $k \propto A$.

4. This is named after Thomas Young, who is most famous for his 1801 two-slit experiment, which established the wave nature of light. Young was also a physician, and figured out how the eye focuses on objects at different distances. He was familiar with twelve languages by the age of fourteen, and later was involved in the translation of the Rosetta stone.

The quantity

$$\sigma \equiv \frac{F}{A} \tag{2.3.4}$$

is called the stress. It is usually best to think of the F in this relation as being $F_{applied}$. The stress has units of pressure; in SI units, 1 Pa ("Pascal") $= 1 \text{ N/m}^2$. The quantity

$$\varepsilon \equiv \frac{x}{\ell} \tag{2.3.5}$$

(the amount by which the beam stretches divided by its equilibrium length) is called the strain. Combining equations (2.3.3)–(2.3.5),

$$\sigma = E\varepsilon. \tag{2.3.6}$$

In idiomatic English, the terms "stress" and "strain" are used in very similar ways, for example, "The stress of this job is killing me." or, "I can't stand the strain of this responsibility." We can see above that, as used in physics and engineering, these are quite different quantities, with different units. The stress is usually best thought of as something applied to the sample, while the strain is usually best thought of as the response of the sample to the stress. It might help you to remember that "stress" sounds like the first part of "pressure."

Figure 2.3.2 shows a graph of strain versus stress for a typical metal beam. Rearranging equation (2.3.6) gives $\varepsilon = \sigma/E$, so that the slope of this graph is $1/E$. For low stress, the graph is linear, as predicted by equation (2.3.6). However, once the stress becomes large enough, the beam no longer follows this relation. As shown, when the strain reaches the "elastic limit" (typically about 0.2% for metals), the spring constant decreases, then the beam "yields" and stretches a great deal with no additional increase in the applied stress. Then (again for metal beams), the microscopic structure changes in a process called "strain hardening," and finally the beam breaks. Table 2.3.1 lists typical values of three important material parameters.

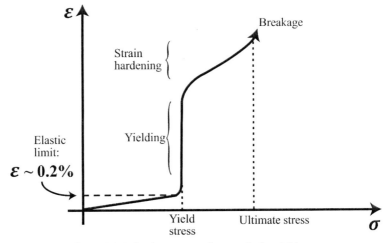

Figure 2.3.2 Strain *vs.* stress for a typical metal beam.

Table 2.3.1. Typical values of material parameters:

Material	Young's Modulus E $(10^9$ N/m$^2)$	Yield stress $(10^6$ N/m$^2)$	Shear Modulus G $(10^9$ N/m$^2)$
Aluminum	75	300	28
Brass	100	250	40
Steel	200	400	75
Concrete (compression only)	20	40[a]	
Rock (compression only)	50	100[a]	
Plastic	2	20	0.1
Rubber	0.002	3	0.0005
Wood	10	40	
Carbon nanotube	1000	60000	

[a]These materials do not exhibit yielding. The number quoted is the ultimate stress. i.e. the stress at breakage.

Self-test (answer below[5]): From table 2.3.1, we see that the values of E are a few hundred times larger than the values of the yield stress. How are E and the yield stress related?

Example: Scientists often need to isolate sensitive scientific equipment, such as atomic force microscopes (AFMs), from building vibrations. An economical and effective way to do this is to place the equipment on a platform that is suspended on soft springs. As we'll see in chapter 4, the effectiveness of this vibration isolation is best when the vibration frequency of the system, $\omega_0 = \sqrt{\frac{k}{m}}$ is as low as possible. A scientist wants to suspend an AFM of mass 5.4 kg (including the mass of the platform) using a rubber bungee cord of equilibrium length 1.2 m. If she wants $\omega_0 = 10.0$ rad/s, what is the required diameter for the rubber cord? (Assume the mass of the cord is negligible compared to that of the AFM.)

Solution: $\omega_0 = \sqrt{\frac{k}{m}} \Rightarrow k = \omega_0^2 m$. Plugging in the numbers gives $k = 540$ N/m. From equation (2.3.2), we have $k = E\frac{A}{\ell} \Leftrightarrow A = \frac{k\ell}{E}$. From table 2.3.1, the Young's Modulus for rubber is $E = 0.002 \times 10^9$ N/m^2. Plugging in $\ell = 1.2$ m gives $A = 3.2$ cm^2. Since $A = \pi (d/2)^2$, where d is the diameter, this gives $d = 2.0$ cm.

The cord stretches a distance given by $|F| = kx = mg \Leftrightarrow x = \frac{mg}{k} = 9.8$ cm when the load is applied to it. The stress under load is approximately $\sigma = \frac{mg}{A} = 1.6 \times 10^5$ N/m^2, well below the yield stress.[6]

---

5. Answer to self-test: The yield stress is approximately equal to the Young's modulus *multiplied by* the elastic limit, which is about 0.2%.

6. As the cord stretches, its cross-sectional area decreases somewhat, so this value for the stress is less than the true value. We can get a different estimate by assuming that the volume of the cord remains constant, so that $A\ell = A_{\text{stress}} (\ell + x) \Leftrightarrow A_{\text{stress}} = \frac{A\ell}{\ell+x}$, where A_{stress} is the cross-sectional area with the load applied. Plugging in the numbers gives $A_{\text{stress}} = 3.0$ cm^2, so that our new estimate of the stress under load is $\sigma = \frac{mg}{A_{\text{stress}}} = 1.8 \times 10^5$ N/m^2. This is greater than the actual stress, because the volume actually increases when a beam (or in this case a bungee cord)

2.4   Shear

If we apply a force perpendicular to the face of a solid object, then it is stretched or compressed, and the force over the area is the stress, as discussed in section 2.3. The force can be spread uniformly over the surface, resulting in a uniform pressure or stress $\sigma = F/A$. A force that is instead applied parallel to the face is called a shear force. If this force is applied uniformly over the top surface of the cube shown in figure 2.4.1a, and this face has area A, then the shear stress is

$$\sigma_{\text{shear}} \equiv \frac{F_{\text{applied}}}{A}. \tag{2.4.1}$$

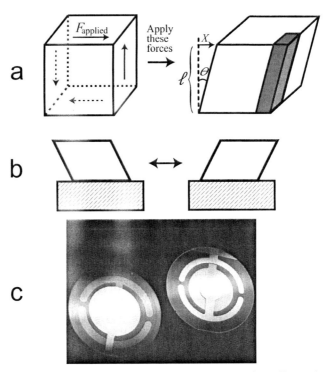

Figure 2.4.1 a: A force applied parallel to a face is called a shear force. For static equilibrium, forces must be applied to three other faces of the cube, as shown. This results in shear strain, as shown on the right. By imagining that we divide the cube into small parallel beams, such as that shown shaded on the right, we can see that the force with which the cube resists the shear deformation is proportional to the area of the top surface. b: Shear oscillations (side view). c: Front and back surfaces of a crystal from a quartz crystal microbalance, with gold electrodes applied. Image from Wikimedia Commons.

is put under stress. For most materials, the volume increases by a fraction in the range $\varepsilon/3$ to $\varepsilon/2$. So, the correct value for σ is between this overestimate of 1.8×10^5 N/m^2 and the underestimate of 1.6×10^5 N/m^2 given in the main text above. For metal beams, the changes in volume and in cross-sectional area are much smaller, and can often be ignored.

For static equilibrium to be achieved, this force to the right on the top surface must be balanced by an equal force to the left on the bottom surface. (Again, we assume this force is applied uniformly over the surface.) However, this pair of forces would tend to rotate the cube clockwise. So, to provide zero net torque about the center of the cube, we need forces on the left and right sides, as shown in figure 2.4.1a, and again we assume these are spread uniformly over the surface. This combination of forces produces a condition of pure shear, resulting in the deformation shown in the right side of figure 2.4.1a. If we imagine slicing the cube into thin horizontal slices, each slice is moved a little to the right of the one below it; this is called "shear strain."

By now, it will not surprise you that the solid resists an applied shear force with a force $F_{\substack{spring \\ shear}} = -k_{shear}x$. We can see from the figure that the displacement x is proportional to the length ℓ. Since $k_{shear} = -\dfrac{F_{\substack{spring \\ shear}}}{x} = \dfrac{F_{applied}}{x}$, we see that $k_{shear} \propto \dfrac{1}{\ell}$. By the same arguments as in section 2.3, we could imagine dividing the solid into smaller parallel beams, each of which resists the shear force, so that $k_{shear} \propto A$. Combining these results, we get $k_{shear} \propto \dfrac{A}{\ell}$, or

$$k_{shear} = G\frac{A}{\ell}, \tag{2.4.2}$$

where the proportionality constant G is called the Shear Modulus.[7]

As in section 2.3, we can also write

$$k_{shear}x = F_{applied} \Rightarrow G\frac{A}{\ell}x = F_{applied} \Leftrightarrow \frac{F_{applied}}{A} = G\frac{x}{\ell}.$$

We define the shear strain to be

$$\gamma \equiv \frac{x}{\ell}. \tag{2.4.3}$$

Therefore,

$$\sigma_{shear} = G\gamma. \tag{2.4.4}$$

In addition to the shear modulus, we can also define the shear yield stress and the shear ultimate stress, which play roles analogous to those discussed in section 2.3. The values of all three numbers are typically about half those of the corresponding numbers for stretching/compression. (Some values are listed in table 2.3.1.)

Shear is particularly important in the design of earthquake-resistant buildings. In many earthquakes, there is a strong lateral oscillation of the ground, resulting in large shear stress applied to the building. Buildings which are not specifically designed to withstand this type of stress are severely damaged by it.

---

7. This is named after William Shear (1701–1785), who also invented scissors. He received a doctorate from Oxford at the age of 14, and ... OK, just kidding.

Since we can define a spring constant for shear, we see that an object can undergo shear oscillations, as shown in figure 2.4.1b. You can easily produce such oscillations by placing a cube of jello on a plate and moving the plate sideways (simulating an earthquake).

Connection to current research: Quartz Crystal Microbalances

A piezoelectric material responds to an applied voltage by changing its shape. The most commercially important examples are quartz crystals (which exhibit a small but very reproducible shape change) and lead zirconium titanate ceramics (which produce a larger but less reproducible change). In a quartz crystal microbalance (QCM), a quartz crystal in the shape of a thin disk is prepared in such a way that it can be excited into shear oscillations by the application of appropriate voltages to gold electrodes on the crystal surface, as shown in figure 2.4.1c.

Because the mass is distributed through the thickness of the crystal, the quantitative analysis of these oscillations is beyond the level of this book. However, the frequency of oscillation is given by $\omega_0 = \sqrt{\dfrac{k_{eff}}{m_{eff}}}$, where k_{eff} is an effective spring constant (proportional to the shear modulus), and m_{eff} is an effective mass.

The most common use for QCMs is for monitoring vacuum depositions of thin films. For example, one might wish to coat a silicon wafer with a layer of copper, to form connecting wires between micro-fabricated circuits. One can apply such a coating in several ways, including "thermal evaporation." In this technique, the wafer is mounted face down in a vacuum chamber, and almost all the air is pumped out. Below the wafer, a pellet of copper is heated until it melts. Once the copper is liquid, it starts to evaporate. The thermal energy of the copper atoms that evaporate off the liquid is quite high so that they travel in straight lines until they strike the wafer. The thickness of the copper film thus applied to the wafer is measured using a QCM, which is mounted just to the side of the wafer. The copper atoms coat the surface of the quartz crystal (as well as the silicon wafer), increasing its mass and thus changing ω_0. This change can be measured so accurately that it is routine to measure thicknesses down to one atomic layer!

QCMs can also be used in aqueous solutions. Typically, the surface of the QCM sensor crystal is first coated with a receptor chemical, such as the antibody for a particular virus. When the virus is introduced into the solution, it binds to the antibody resulting in an increase in the effective mass of the oscillating crystal. This technique can be used to make sensors, to study the progress of chemical reactions, to sort through possible anti-cancer drugs, and for many other applications.

2.5   Torsion and torsional oscillators

As a child, you probably played with a yo-yo, and had trouble with it when the string got twisted up. You may have held the top of the string, with the yo-yo resting at the bottom of the string, and watched it slowly untwist, picking up rotational velocity as

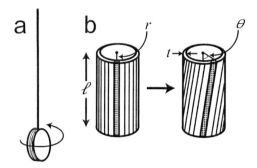

Figure 2.5.1 a: A yo-yo twisting on the end of its string. b: Twisting a thin-walled tube.

it went, as shown in figure 2.5.1a. You may have noticed that the yo-yo didn't stop rotating when the string was untwisted, but instead kept rotating, twisting the string in the other direction. Eventually, the yo-yo came to rest briefly before starting to rotate in the opposite direction because of this new twist. Your yo-yo was acting as a torsional oscillator. As we'll explore, more sophisticated versions have been very important in the history of physics and continue to be important in current research on superfluids, supersolids, and other topics.

We begin by considering a thin-walled tube, as shown in figure 2.5.1b. Think of it as a collection of thin strips, one of which is highlighted. We wish to calculate the potential energy stored in this system when we twist it, and show that this has the form of the energy stored in a spring, $U = \frac{1}{2}kx^2$. From there, we will be able to find the angular frequency of oscillation.

If we twist the top of the tube by a small angle θ, the strip experiences a shear strain given by $\dfrac{\text{lateral displacement}}{\text{length}} = \dfrac{x}{\ell} = \dfrac{r\theta}{\ell}$. Therefore, it produces a force

$$F_{\substack{\text{spring} \\ \text{shear}}} = -k_{\text{shear}}x = -k_{\text{shear}}r\theta.$$

Using equation (2.4.2): $k_{\text{shear}} = GA/\ell$ we then have

$$F_{\text{spring}} = -G\frac{A_{\text{strip}}}{\ell}r\theta,$$

where A_{strip} is the cross-sectional area of the strip. The resulting torque is

$$\tau_{\substack{\text{spring} \\ \text{strip}}} = rF_{\substack{\text{spring} \\ \text{shear}}} = -G\frac{A_{\text{strip}}}{\ell}r^2\theta.$$

To find the total torque from all the strips, we simply replace A_{strip} by the total cross-sectional area of the tube, $A = 2\pi rt$ (valid for $t \ll r$), where t is the wall thickness of the tube:

$$\tau_{\substack{\text{spring} \\ \text{tube}}} = -2\pi G\frac{r^3 t}{\ell}\theta. \tag{2.5.1}$$

To create the twist, we must apply a torque of equal magnitude and opposite sign:

$$\tau_{applied} = 2\pi G \frac{r^3 t}{\ell} \theta.$$

The work that we do in creating the twist is stored as potential energy. Recall from your earlier study of mechanics that the work done in a rotation is $W = \int \tau \, d\theta$, which is analogous to the work done in a translation, $W = \int F \, dx$. Therefore,

$$U = W_{\text{on system}} = \int_0^\theta \tau_{applied} \, d\theta = 2\pi G \frac{r^3 t}{\ell} \int_0^\theta \theta \, d\theta = \tfrac{1}{2} \left(2\pi G \frac{r^3 t}{\ell} \right) \theta^2, \qquad (2.5.2)$$

which, as hoped, has the same form as the potential energy of a conventional spring.

To complete our argument and find the angular frequency of oscillation, we must consider the rotational kinetic energy. In most cases of interest, the object being twisted (the string of the yo-yo, or the tube) is used to support a much more massive object, such as the body of the yo-yo, which is called the rotor. For simplicity, we assume that the rotational kinetic energy of the rotor is much larger than that of the twisted element that supports it, so that

$$K = \tfrac{1}{2} I_{\text{rotor}} \omega^2 = \tfrac{1}{2} I_{\text{rotor}} \left(\frac{d\theta}{dt} \right)^2 .$$

Therefore, the total energy is

$$E = K + U = \tfrac{1}{2} I_{\text{rotor}} \left(\frac{d\theta}{dt} \right)^2 + \tfrac{1}{2} \left(2\pi G \frac{r^3 t}{\ell} \right) \theta^2. \qquad (2.5.3)$$

This is isomorphic to (i.e., is identical to except for a change of symbols) equation (2.1.2):

$$E = \tfrac{1}{2} m \left(\frac{dx}{dt} \right)^2 + \tfrac{1}{2} k x^2,$$

so that we can rewrite $\omega_0 = \sqrt{\frac{k}{m}}$ for this case as

$$\omega_{\substack{\text{tube} \\ \text{torsion}}} = \sqrt{\frac{2\pi G r^3 t}{I_{\text{rotor}} \ell}}. \qquad (2.5.4)$$

In problem 2.12, you can show that, if the rotor is supported by a solid cylinder (e.g., a wire or a string) instead of a tube, the torque and angular frequency of oscillation are instead given by

$$\tau_{\substack{\text{spring} \\ \text{wire}}} = -\pi G \frac{r^4}{2\ell} \theta \qquad \omega_{\substack{\text{wire} \\ \text{torsion}}} = \sqrt{\frac{\pi G r^4}{2 I_{\text{rotor}} \ell}}. \qquad (2.5.5)$$

Perhaps the most famous use of a torsional oscillator occurred in 1798, when Henry Cavendish[8] used one to measure the gravitational attraction between two pairs of lead

---

8. Henry Cavendish made important contributions to chemistry as well as physics. He was the first to isolate hydrogen gas, and showed that water is made from hydrogen and oxygen. He performed

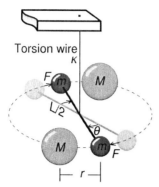

Figure 2.5.2 Schematic diagram of the Cavendish experiment. (Figure by Chris Burks.)

spheres, as shown in figure 2.5.2. The larger sphere in each pair was 1 ft (30.5 cm) in diameter, and the smaller was 2 in. (51 mm). To isolate the experiment from air currents, the instrument was housed inside a wooden box inside a specially built brick shed. The operation and observations could all be performed by Cavendish from outside the shed, using ropes, pulleys, and telescopes. To measure the gravitational interaction, he used the apparatus in "torsion balance" mode: he brought the large spheres close to the small ones, and measured how much the gravitational attraction caused the wire to twist; this was a static, not oscillating, experiment. However, to determine the torsional spring constant of the wire (the quantity $\dfrac{\pi G r^4}{2\ell}$ in equation (2.5.5)), he moved the large balls far away, and set the rotor (composed of the two small balls and the horizontal crossbar that connects them) into oscillation. By measuring the resulting oscillation frequency, he could then deduce the spring constant.

Connection to current research: Torsional oscillators are used to investigate the phase transitions of liquid helium to a superfluid and to a supersolid. See problem 2.13 to learn more.

2.6 Bending and Cantilevers

When you jump on the end of a diving board, setting up a beautiful dive or an enthusiastic cannonball, you apply a force straight down. Although this is parallel

---

many important experiments on electricity, including a precise demonstration of the inverse square law for the force between two charges, but never published most of his work. He was very shy, especially around women; he communicated with his women servants via notes. His main social outlet was attendance at scientific meetings. However, even once his scientific reputation was well-established, he could sometimes be observed standing outside the door of a meeting, trying to work up enough courage to go in. He was fabulously wealthy; upon his death his estate amounted to nearly a million pounds sterling, a tremendous sum. (At that time, a gentleman could live comfortably on 500 pounds a year.)

to the (vertical) end face of the board, it is neither spread uniformly over this face nor are there forces applied to the sides of the board to create the pure shear condition shown in figure 2.4.1a. Therefore, the resulting deformation of the board is not pure shear, although shear is involved along with other types of distortion. We consider the situation in figure 2.6.1a, in which the left end of the board is anchored to a wall; this configuration is called a cantilever.

The board, of course, tends to spring back to its original position. We want to find the spring constant k, defined by

$$F_{\text{applied}} = -F_{\text{spring}} = kd, \qquad (2.6.1)$$

where d is the displacement of the end of the board from equilibrium.

The full analysis is beyond the level of this text, however we can understand the way that the spring constant depends on the dimensions of the board with a simplified analysis. When the board is bent, the top half is stretched while the bottom half is compressed; these distortions produce part of the restoring spring force. The board is also subject to shear, which affects the displacement and the restoring force. In our analysis later, we will neglect the effects of shear, so that the final proportionality constant will not be correct, but the functional dependence on the dimensions of the board will be.

In static equilibrium, a small part of the top half of the board (the shaded section in the figure) must feel equal forces from the left and from the right. This is true for any similar section along the length of the board. Thus, the amount of stretching (i.e., the "tensile strain") in the top half must be uniform along the length. A constant stretch along the length is only possible if the board is in the shape of a circular arc. (The actual shape is more complicated, because of the effects of shear.) The radius of the arc (called the "radius of curvature") for a line along the center of the board is defined as r. The board has a length L, a thickness t, and a width w (the dimension perpendicular to the page in figure 2.6.1). Figure 2.6.1b shows an enlarged view of a section of the board. Along the centerline, the length of this section is $r\theta$, while a line a distance z above the center is stretched to a length $(r+z)\theta$. Thus, the tensile strain at the position z is

$$\varepsilon = \frac{\text{amount of stretch}}{\text{equilibrium length}} = \frac{[(r+z)\theta - r\theta]}{r\theta} = \frac{z}{r}.$$

Using equation (2.3.6), $\sigma = E\varepsilon$ we see that the stress at the position z is

$$\sigma = \frac{Ez}{r}.$$

This stress is the result of forces applied to the left end of the board by the wall, as shown in figure 2.6.1a. The force applied to an infinitesimally thick horizontal slice of the board (figure 2.6.1c) is the stress times the cross-sectional area of the slice:

$$dF = \sigma_{\text{av}} w \, dz = \frac{Ez}{r} w \, dz.$$

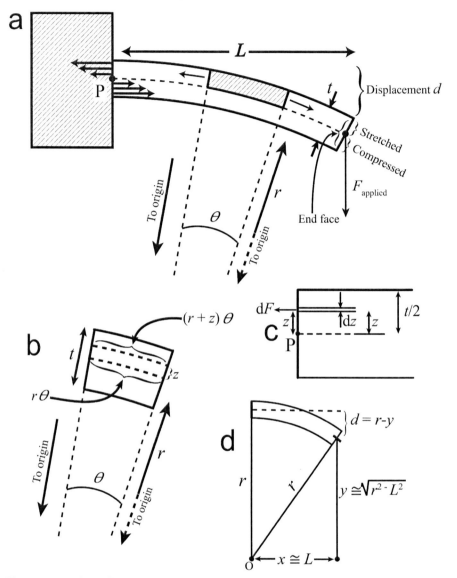

Figure 2.6.1 a: A cantilever is bent. The top half is stretched, while the bottom half is compressed. The highlighted segment must feel equal forces from the left and right. b: Enlarged view of a section of the cantilever. c: Enlarged view of the end of the cantilever that is attached to the wall. d: Relation between displacement and radius of curvature.

This creates a torque $z\,dF$ around the point P at the center of the board:

$$d\tau = z\,dF = \frac{Ez^2}{r}w\,dz.$$

The total torque applied to the top half of the board is then

$$\tau_{\text{top}} = \int_{\text{top}} d\tau = \int_0^{t/2} \frac{Ez^2}{r}w\,dz = \frac{Et^3 w}{24r}.$$

A torque of the same magnitude arises from the forces applied to the bottom half, so that the total torque around point P arising from the forces exerted by the wall is

$$\tau_{wall} = 2\tau_{top} = \frac{Et^3 w}{12r}.$$

For static equilibrium, this must be equal in magnitude to the torque produced by the force applied at the end of the board, $F_{applied} = kd$. For small deflections, this torque is

$$\tau_{applied} = F_{applied} L = kd\,L.$$

Equating the magnitudes of the torques gives

$$kd\,L = \frac{Et^3 w}{12r} \quad \Leftrightarrow$$

$$k = \frac{Et^3 w}{12rd\,L}. \tag{2.6.2}$$

Now, we must find the radius of curvature r in terms of the displacement d. Recall that we have approximated the shape of the bent board as a circle. We define the origin to be at the center of the circle. The end of the board is at the position (x, y), where for small displacement $x \cong L$. As we can see from figure 2.6.1d, the displacement is

$$d = r - y \cong r - \sqrt{r^2 - L^2} = r - r\sqrt{1 - \frac{L^2}{r^2}}.$$

For small displacements, the curvature is slight, that is, $r \gg L$. So, we can approximate the square root using a Taylor series approximation: $\sqrt{1+x} \cong 1 + x/2$ for $x \ll 1$. Therefore,

$$d \cong r - r\left(1 - \frac{L^2}{2r^2}\right) = \frac{L^2}{2r}. \tag{2.6.3}$$

Substituting this into equation (2.6.2) gives

$$k = \frac{Et^3 w}{12r\dfrac{L^2}{2r}L} = \frac{Et^3 w}{6L^3}. \tag{2.6.4}$$

The full analysis, including effects of shear, gives the exact result

$$\boxed{k_{cantilever} = \frac{Et^3 w}{4L^3}.} \tag{2.6.5}$$

This is proportional to $1/L^3$; a factor of $1/L^2$ comes from the fact that the board is curved, so that the displacement is geometrically proportional to L^2, equation (2.6.3). The third factor of L comes from the moment arm for the torque exerted by $F_{applied}$. The spring constant is also proportional to $t^3 w = t^2\,(tw)$. The first two factors of t come from the fact that the strain is proportional to the thickness, and from the moment arm for the force from the wall which provides the stress leading to this strain. The factor (tw) is the cross-sectional area of the board—a board with larger area of course has a larger spring constant.

The angular frequency of vibration is

$$\omega_0 = \sqrt{\frac{k_{\text{cantilever}}}{m_{\text{effective}}}}. \tag{2.6.6}$$

If a compact mass m is attached to the end of the cantilever, and m is much greater than the mass of the cantilever itself, then the effective mass $m_{\text{effective}}$ is simply m. (This is the same approximation we use for a conventional mass on a spring when we neglect the mass of the spring itself.) However, in many applications there is no such extra mass, and the effective mass is due to the cantilever itself. In this case, one can show[9] that the effective mass is

$$m_{\text{effective}} = 0.243\,m, \tag{2.6.7}$$

where m is the mass of the cantilever.

Connection to current research: Cantilevers are widely used as sensors in Nano ElectroMechanical Systems (NEMS). One approach is to coat the cantilever with a "receptor" molecule, which can bind the target molecule that is to be sensed. When this binding occurs, the effective mass of the cantilever increases, so the frequency of vibration goes down. In an extreme example of this approach, Professor Alex Zettl and his research group have used a carbon nanotube cantilever to detect tiny amounts of gold which are deposited onto the cantilever. Because the mass of the cantilever itself is so small, this system has a detection threshold close to a *single gold atom*! As you can see from figure 2.6.2, the addition of 51 gold atoms produces a change in frequency that is well above the noise level.

For further information about the topics in this chapter, see *Mechanics of Materials, 4th Ed.*, by James M. Gere and Stephen P. Timoshenko, PWS Publishing, Boston, 1997.

Concept and skill inventory for chapter 2

After reading this chapter, you should fully understand the following terms:

Pendulum force (2.2)
Parallel axis theorem (2.2)
Young's Modulus (2.3)
Stress, strain (2.3)
Yield stress, elastic limit, ultimate stress (2.3)
Shear stress, shear strain (2.4)
Shear modulus (2.4)

---

9. *Dynamics of Vibrations*, by Enrico Volterra and E. C. Zachmanoglou, Charles E. Merrill Books, Columbus, OH, 1965, p. 319.

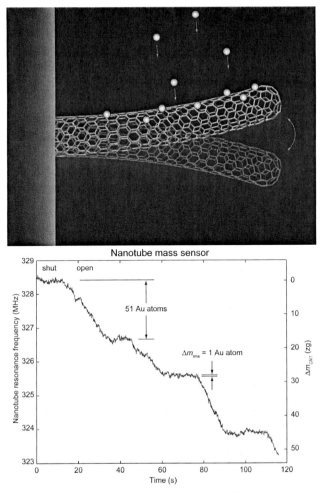

Figure 2.6.2 Deposition of gold atoms onto a cantilever made from a carbon nanotube (shown schematically at the top) increases the mass of the cantilever, thus decreasing the vibration frequency (left vertical scale of the graph). During the shaded time periods (e.g. 60 s to 76 s), no atoms are deposited, so the frequency is constant. During the unshaded periods, atoms are added, and the frequency decreases. Images *Courtesy Zettl Research Group, Lawrence Berkeley National Laboratory and University of California at Berkeley.*

Rotor (2.5)
Cantilever (2.6)

You should know what happens when:

A spring is cut in half (2.3)

You should be able to:

Recognize by considerations of either force or energy when a system will oscillate, and be able to identify the effective mass and spring constant (2.1).

Calculate the oscillation frequency for any pendulum made up from symmetrical objects (2.2)

Calculate the oscillation frequency for extension/compression (2.3), shear (2.4), torsional (2.5), and cantilever (2.6) oscillators

In addition to all of the above, you should be able to combine the concepts you've learned to address new situations.

Problems

Note: Additional problems are available on the website for this text.

Instructors: Difficulty ratings for the problems, full solutions, and important additional support materials are available on the website.

2.1 True or False? If true, explain. If false, give a corrected version.

If the net force on an object is zero at some position, and the object is moved a short distance away and then released, it will then oscillate in harmonic motion.

2.2 The Sears Tower in Chicago was the tallest building in the world for 22 years and still holds the record for the highest antennas on top of a building. The building itself is 442 m high. The building sways considerably in the famous winds of Chicago; on a typical day, the top floors sway laterally by up to 15 cm, causing the toilets to slosh and occasionally giving people motion sickness. The total mass of the tower is 2.02×10^8 kg. The average cross-sectional area is equivalent to a square 63 m on a side. If the tower is hit by a sudden gust of wind (which then suddenly stops), the tower is observed to sway back and forth with a period of 8 s. Model the building as a cantilever with square cross-section (63 m on a side) and length of 442 m. **(a)** If we pretend the building is made from a uniform slab of material, what is the Young's modulus of this material? **(b)** You should have found a rather low value, which is not surprising given that the volume of the Sears Tower is mostly air. To get a reasonable comparison, multiply your result by the ratio of the density of structural steel (7,850 kg/m^3) to the average density of the Sears Tower. You should still get a Young's modulus which is considerably less than that of steel, but this is reasonable since much of the weight of the tower does not contribute to its rigidity.

2.3 Radioactive materials emit three kinds of particles. When radioactivity was first discovered, the identity of these particles wasn't known, so they were named α, β, and γ particles. We now know that α-particles are helium nuclei, that is, two neutrons and two protons combined into a nucleus, β-particles are electrons, and γ-particles are high-energy photons. In this problem, treat each of these as a point particle. An α-particle is *fixed* at the origin. An electron is *fixed* at $x_0 = 2.00$ nm. **(a)** A negative fluorine ion is written F$^-$. This is just a fluorine atom with an extra electron, and has essentially the same mass as a fluorine atom. If the F$^-$ is placed at a position x_{eq}, which is to the right of x_0, the total force on it is zero. What is the value of x_{eq}? Treat all three particles as point charges. **(b)** Make a qualitative argument for why this is a position of stable equilibrium for the F$^-$. **(c)** If the F$^-$ is moved *slightly* from this position and then released, it oscillates. What is the frequency of oscillation? (Hint: it is of order 10^{10} Hz. Begin by finding the effective spring constant k_{spring}. You are expected to find the mass of an F atom by using the web or another reference source.)

2.4 You must do exercise 2.3 before doing this exercise. We allow the F$^-$ to move relative to its equilibrium position x_{eq}. **(a)** What is the potential energy function $U(x)$ for the F$^-$, assuming it is to the right of $x = 2.00$ nm? (Set $U \equiv 0$ at $x = x_{eq}$.) **(b)** Using a suitable computer program, graph the potential energy for the F$^-$ over the x-range from $x_{eq} - 1$ nm to $x_{eq} + 1$ nm, and superimpose on this the graph of the harmonic (i.e., parabolic) approximation to this potential energy using your effective spring constant from exercise 2.3.

2.5 The Lennard–Jones Potential. Two neutral atoms have an attractive interaction due to a dynamic rearrangement of the electron clouds, called the Van der Waals attraction. It is frequently modeled using the Lennard–Jones potential (proposed in 1924 by Sir John Lennard–Jones):

$$U(r) = 4\varepsilon\left[\left(\frac{\sigma}{r}\right)^{12} - \left(\frac{\sigma}{r}\right)^{6}\right],$$

where r is the distance between the two atoms and ε and σ are constants. This function is shown in figure 2.P.1. The positive term represents the repulsion experienced at very short distances, which increases extremely rapidly as the atoms are brought close together, while the negative term represents the attraction, which is more important at longer distances. **(a)** For what value of r is the potential energy a minimum? (Express your answer in terms of σ.) **(b)** For the interaction of two argon atoms, $\varepsilon = 1.654 \times 10^{-21}$ J and $\sigma = 3.405$ Å, where 1 Å (pronounced "Angstrom") is 10^{-10} m, and is named after Anders Ångström, one of the founders of spectroscopy. Assume these numerical values are exact. What is the value of U at the minimum in this case? (This is called the "bond energy" since it is the difference in potential energy between the bonded atoms and the situation with the atoms very far apart.) Quote your answer to six significant digits, and express your answer in electronvolts, where 1 eV $= 1.60217653 \times 10^{-19}$ J is the amount of energy an electron acquires as it moves through a voltage difference of 1 V. A typical

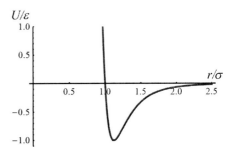

Figure 2.P.1 The Lennard-Jones potential.

covalent bond has a bond energy of 2–4 eV, so we can see that Van der Waals bonds are much weaker. **(c)** What is the value of U for an r which is 0.1 Å larger than the minimum? (Again, express your answer in electronvolts, and quote to six significant digits.) **(d)** Using your numbers from parts (b) and (c), estimate the effective spring constant k for this system? **(e)** If the system is displaced from equilibrium, by pulling the atoms further apart than the equilibrium distance you found in part (b) and then releasing them, both atoms oscillate relative to the center of mass. This means that the center of the "spring" that connects them doesn't move. So, you can picture this as each atom oscillating on a spring half as long connected to a brick wall. Recalling that a spring half as long has twice the spring constant, what is the frequency of oscillation? (You are expected to look up the mass of an Argon atom on the web or other reference source.)

2.6 What is wrong with the following reasoning: For a pendulum made from a thin rod, we can consider all the mass to be concentrated at the center of mass. Therefore, the effective length is $\ell/2$, so the angular frequency is

$$\omega_0 = \sqrt{\frac{g}{\ell/2}} = \sqrt{\frac{2g}{\ell}}.$$

2.7 **(a)** If one wishes to make a pendulum with the longest possible period, how should the mass be distributed along the length of the pendulum? Explain your answer. **(b)** Your friend thinks it would be keen to make a pendulum with a period of 1 h. Is this a practical idea? (Explain your reasoning.)

2.8 A pendulum is made from a light string that is 0.750-m long, with a small bob of mass 0.500 kg at the end. The pendulum swings with an x-amplitude of 5 cm. What is the energy present in this oscillation?

2.9 The tallest skyscraper (defined as having the highest occupied floor) in the world is Taipei 101, in Taipei, Taiwan. The highest occupied floor is 439.2 m above ground. A pendulum is suspended from this height (using a cantilever to get the support point out laterally away from the building). The pendulum is made from a steel rod that is 1 cm in diameter, with a 30-cm radius steel sphere attached at the bottom. The bottom of the sphere is 10 cm above the ground. Assuming wind forces can be ignored, what is the period of this pendulum?

2.10 Assuming that wind forces can be neglected, what is the highest that one can make a solid concrete tower of uniform cross-section (Take the density of concrete to be 2,400 kg/m^3.)?

2.11 You are part of the design team for a manufacturer of air conditioning units. A particular large unit is to be mounted on four rubber feet, one at each corner. You are charged to decide on the size of these feet. The mass of the unit is 1,000 kg. Its weight is equally distributed on the four feet. **(a)** What is the minimum area for each of the four feet? (Assume the yield stress for rubber in compression is the same as that in extension.) To play it safe, you decide on an area that is three times the minimum. **(b)** Your boss tells you, "We have to be careful to avoid resonance with the frequency of the motors inside the unit." (You'll learn more about resonance in chapter 4.) She continues, "So, make sure the frequency of vertical oscillations is less than 5 Hz." Does this impose a minimum or a maximum limit on the thickness of the feet? (Assume the mass of the rubber is negligible compared to that of the air conditioning unit.) Explain. (c) Based on what your boss has told you, choose a thickness for the feet. (d) If the air conditioning unit moves laterally in sinusoidal motion, this imposes forces on the rubber that we will approximate as pure shear. What would the frequency of such oscillations be?

2.12 For a torsional oscillator, show that, if the rotor is supported by a solid cylinder (e.g., a wire or a string), the torque and angular frequency of oscillation are given by

$$\tau_{\substack{\text{spring} \\ \text{wire}}} = -\pi\,G\frac{r^4}{2\ell}\theta \qquad \omega_{\substack{\text{wire} \\ \text{torsion}}} = \sqrt{\frac{\pi\,Gr^4}{2I_{\text{rotor}}\ell}}.$$

(The rotor is assumed to have a much greater moment of inertia than the wire.)

2.13 **Supersolid helium** At very low temperatures, helium exhibits a variety of astounding behaviors. At atmospheric pressure, helium gas liquefies when the temperature is reduced below 4.2 K. When the temperature is reduced below about 2 K, the liquid becomes a "superfluid" meaning that it has zero viscosity, so that objects can move through it with zero drag. This effect was demonstrated experimentally using torsional oscillators. Unlike all other materials, helium remains liquid all the way down to zero Kelvin, unless it is placed under a pressure of at least 25 atmospheres. Above 25 atmospheres, it does become solid. In 2004, E. Kim and M. H. W. Chan (Science **305**, 1941–1944) used a torsional oscillator to demonstrate that, below about 0.23 K the solid becomes a "supersolid," which can flow without resistance in much the same way that the superfluid can. (It is difficult to conceive of how a solid could behave this way, but the experimental evidence is fairly clear.) Figure 2.P.2 shows a schematic diagram of the torsional oscillator they used. The rotor is suspended on a hollow tube made of beryllium–copper (BeCu), a very springy alloy. The tube has an inner diameter of 0.400 mm and an outer diameter of 2.20 mm. The rotor itself consists of a solid cylinder of BeCu 12.8 mm in diameter and 5.00 mm high. Around this is a hollow annular region that is filled with the helium, and around this is a hollow aluminum cylinder that has an inner diameter of 15.0 mm, an outer diameter of 17.0 mm, and a height that is also 5.00 mm. A disk of BeCu that is 1-mm thick and has a diameter of 17.0 mm is attached to the top, and a second such disc is attached to the bottom, sealing off the annular region between the BeCu cylinder and the aluminum cylinder. The annular region is filled with solid helium. Take the density of BeCu to be 8,350 kg/m³, that of aluminum to be 2,700 kg/m³, and that of solid helium to be 172 kg/m³. Take the shear modulus of BeCu to be 4.83×10^{10} N/m². Assume all the preceding values are exact. Assume that the moment of inertia for the rotor assembly (including top and bottom caps, both cylinders, and the annular region filled with helium) is much larger than that of the BeCu tube from which it is suspended. (a) When the helium is in the normal solid state, what is the oscillation frequency f for

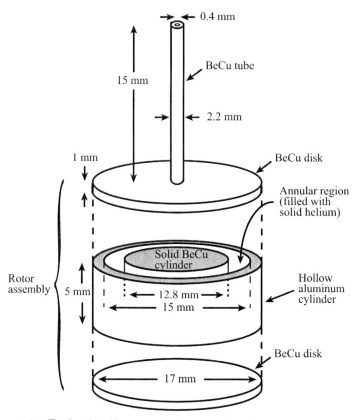

Figure 2.P.2 Torsional oscillator used to detect the transition to supersolid helium.

this system, calculated to five significant figures? (Recall that the moment of inertia for a disk or solid cylinder is $\frac{1}{2}MR^2$). **(b)** When the helium makes the transition to the supersolid state, it no longer rotates along with the rest of the rotor, so it doesn't contribute to the moment of inertia. What is the oscillation frequency under this condition, calculated to five significant figures? (It was by detecting this change in oscillation frequency that Kim and Chan demonstrated the transition to supersolid helium.)

2.14 Micro ElectroMechanical Systems (MEMS). You are designing a sensor to detect biological pathogens in drinking water. Your sensor is based on a cantilever which is micromachined from silicon, with a length of 300 μm, a width of 100 μm, and a thickness of 1 μm. (Assume these dimensions are exact.) Because silicon is a crystal, the relevant value of Young's modulus depends on the direction relative to the crystal axes; assume it is exactly 165 GPa for this problem. Assume the density of silicon is exactly 2,330 kg/m^3. **(a)** What is the oscillation frequency f for your cantilever? (Give a result with eight significant figures, assuming all the numbers used as inputs are exact.) **(b)** You now coat your cantilever with a layer of receptor molecules, which can bind the pathogen. Each receptor molecule has a molecular weight of 278.1 u, where 1 u $= 1.66053886 \times 10^{-27}$ kg. The receptor molecules coat all the exposed surfaces of your cantilever, each occupying an average area of exactly $(5 \text{ nm})^2$. Assume that this distributed mass counts toward the effective mass in the same way that the mass of the silicon itself counts. What is the new oscillation frequency? **(c)** Each of the pathogens you must detect has a mass of 20,000 u. Assuming that 1% of the receptor molecules bind a pathogen, what is the new oscillation frequency? (Again, assume that the distributed mass of the bound pathogens counts the same way that the mass of the silicon itself counts.) **(d)** What is the minimum time for which you must make a frequency measurement to detect the difference between the frequencies in parts (b) and (c)?

3 Damped Oscillations

Decay is inherent in all component things.
Work out your salvation with diligence.
> — The next-to-last thing said by the Buddha

3.1 Damped mechanical oscillators

Any real macroscopic oscillator that is displaced from equilibrium and released does not actually show the pure sinusoidal oscillations described in chapter 1; instead, the oscillations decay over time, as shown in figure 3.1.1. Fundamentally, this damping occurs because a macroscopic oscillator is always coupled to its surroundings, even if weakly, and the energy initially present in the oscillation leaks away through these couplings. For example, the string of a violin is coupled to the body of the violin. If the string is set into vibration, it causes the body to vibrate, which in turn sets the air vibrating and broadcasts sound waves, which carry away energy. The violin body also transmits vibrations into the violinist, providing another channel for energy to leak away.

Following the approach of physics, we start by studying the simplest possible form of energy leakage: a drag force, also known as a "viscous damping force." If you stick your hand out of the window of a moving car, you feel the force of the air pushing against you. The force is opposite to the velocity of the car and increases the faster you go. Thus, we might reasonably expect that the force of the air is

$$F_{\text{drag}} = -bv, \tag{3.1.1}$$

where b is a constant, and the minus sign shows that the force is in the opposite direction to the velocity.

Quantitative experiments with various objects show that equation (3.1.1) is correct, but only if the object is moving slowly enough that the airflow around it isn't turbulent. The transition from nonturbulent or "laminar" flow at low speeds to turbulent flow at higher speeds can be visualized in a wind tunnel. In wind tunnel tests, it is easier to hold the object fixed and blow the air past it, as shown in figure 3.1.2a. At low to moderate wind speeds, the air simply moves around the object, as shown in figure 3.1.2b. At higher speeds, turbulent flow sets in, as shown in figure 3.1.2c, and the drag force is no longer described by equation (3.1.1). We will discuss the turbulent regime a little more in section 3.6.

Figure 3.1.1 Decaying oscillations.

Figure 3.1.2 a: A stream of smoke is used to test the aerodynamics of a car in a wind tunnel. The car is stationary, and large fans (not shown) force a flow of air past it. The smoke flows smoothly over the car without turbulence, indicating laminar flow. This is desirable because it reduces air drag, improving fuel efficiency. (Image courtesy of NASA.) b: A sphere in a wind tunnel. At low speeds, the airflow is laminar. c: At higher speeds, the flow is turbulent.

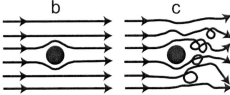

For now, we focus on the laminar flow regime. It is important in its own right, and mathematically convenient. Further, many of the conclusions we will draw apply qualitatively to other energy leakage mechanisms. In this regime, the drag force is given by equation (3.1.1).

Now, we apply the same procedure as in chapter 1 to find the motion of a mass experiencing both a restoring force from a spring and a viscous damping force.

1. Write Newton's second law for each object in the system

There is only one object in this system, so we have

$$m\ddot{x} = F_{\text{Total}} \Rightarrow m\ddot{x} = \underbrace{-kx}_{\substack{\text{spring} \\ \text{force}}} \underbrace{-bv}_{\substack{\text{viscous} \\ \text{damping}}}$$

that is,

$$m\ddot{x} + b\dot{x} + kx = 0 \tag{3.1.2}$$

or

$$\ddot{x} + \gamma\dot{x} + \omega_0^2 x = 0, \tag{3.1.3}$$

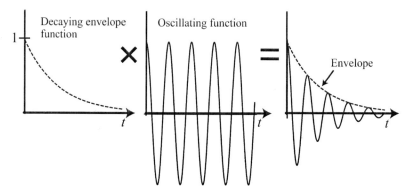

Figure 3.1.3 (decaying envelope) × (oscillating function) = decaying oscillation.

where

$$\gamma \equiv \frac{b}{m} \tag{3.1.4}$$

characterizes the damping, and $\omega_0 \equiv \sqrt{\frac{k}{m}}$ is the angular frequency of oscillations *in the absence of damping* (i.e., in the absence of a drag force).

Following the lead of chapter 1, we now cast the problem into complex form. Equation (3.1.3) is the real part of

$$\ddot{z} + \gamma \dot{z} + \omega_0^2 z = 0, \tag{3.1.5}$$

where z may be complex. If we can find a solution to equation (3.1.5), we automatically get a solution to equation (3.1.3) by setting $x = \mathrm{Re}\, z$ (since the operation of taking the real part commutes with taking derivatives).

2. Use physical and mathematical intuition to guess a solution

As suggested in figure 3.1.3, the solution *might* be of the form

$$x = [\text{decaying envelope function}] \times A_0 \cos(\omega_v t + \varphi),$$

where ω_v is the angular frequency in the presence of viscous damping (to be determined). It is not clear what the mathematical form of the decaying envelope function should be. However, we can start by trying a decaying exponential, since this is easy to handle mathematically.[1] So, we'll try

$$[\text{decaying envelope function}] \stackrel{?}{=} \mathrm{e}^{-\sigma t}, \tag{3.1.6}$$

---

1. We could instead try to guess the envelope function from energy considerations. Since Work = Force × distance, Power = Force × velocity. Therefore, the power dissipated by the damping force is $P = F_{\text{damp}} v = -bv^2$. The average power over a cycle is then given by $P_{\text{average}} = \langle P \rangle = \left\langle \frac{dE}{dt} \right\rangle = -b \langle v^2 \rangle$, where the angle brackets indicate the average over a cycle. Let us assume for now that the damping is relatively light. Then, the velocity will vary approximately sinusoidally over a single cycle. The average value of the square of a sinusoid is exactly half the maximum of the square, as you can show by integration in problem 3.4. Therefore, $\langle v^2 \rangle = \frac{1}{2} v_{\text{max}}^2 = E/m$

where the constant σ is to be determined. So, our complete guess is

$$x \overset{?}{=} e^{-\sigma t} A_0 \cos\left(\omega_v t + \varphi\right) = \text{Re}\,(z),$$

$$\text{where } z = A_0 e^{-\sigma t} e^{i(\omega_v t + \varphi)} \tag{3.1.7}$$

3. Plug the guess back into the differential equation to see if it is indeed a solution, and whether there are restrictions on the parameters.

Let's prepare to plug equation (3.1.7) into (3.1.5) by computing the derivatives:

$$z = A_0 e^{i\varphi} e^{(-\sigma + i\omega_v)t},$$

$$\dot{z} = \left(-\sigma + i\omega_v\right) z,$$

$$\ddot{z} = \left(-\sigma + i\omega_v\right)^2 z.$$

Plugging into equation (3.1.5), $\ddot{z} + \gamma \dot{z} + \omega_0^2 z = 0$, gives

$$\left(-\sigma + i\omega_v\right)^2 z + \gamma \left(-\sigma + i\omega_v\right) z + \omega_0^2 z = 0.$$

Cancelling the z's gives

$$\left(-\sigma + i\omega_v\right)^2 + \gamma \left(-\sigma + i\omega_v\right) + \omega_0^2 = 0,$$

$$\Rightarrow \sigma^2 - i2\sigma\omega_v - \omega_v^2 - \gamma\sigma + i\gamma\omega_v + \omega_0^2 = 0.$$

This is actually two equations; the real part of the left side must equal the real part of the right side, and also the imaginary part of the left side must equal the imaginary part of the right side. So,

$$\text{imaginary part:} -2\sigma\omega_v + \gamma\omega_v = 0 \Rightarrow \sigma = \frac{\gamma}{2}$$

(since $E = \frac{1}{2}mv_{max}^2$). Plugging this in yields

$$\left\langle \frac{dE}{dt} \right\rangle = -b\frac{E}{m} \Rightarrow$$

$$\left\langle \frac{dE}{dt} \right\rangle = -\gamma E. \tag{3.1.8}$$

If we restrict ourselves to timescales much longer than one period, we can write this as

$$\frac{dE}{dt} = -\gamma E \Rightarrow \frac{dE}{E} = -\gamma\, dt$$

Integrating both sides gives $\ln E = -\gamma t + \text{const.} \Rightarrow E = e^{-\gamma t + \text{const.}} = e^{\text{const.}} e^{-\gamma t}$. Setting $E\,(t = 0) \equiv E_0$, we can identify the constant, so that

$$E = E_0\, e^{-\gamma t}. \tag{3.1.9}$$

The damping force does not affect the potential energy, so we still have $U = \frac{1}{2}kx^2 \Rightarrow E = \frac{1}{2}kx_{max}^2$. Combining this with equation (3.1.9) gives $x_{max} = \sqrt{\frac{2E_0}{k}}\, e^{-\frac{\gamma}{2}t}$. This is of the same form as equation (3.1.6).

(in agreement with footnote 1 about guessing the envelope function from energy considerations), and

$$\text{real part: } \sigma^2 - \omega_v^2 - \gamma\sigma + \omega_0^2 = 0.$$

Substituting for σ gives

$$\frac{\gamma^2}{4} - \omega_v^2 - \frac{\gamma^2}{2} + \omega_0^2 = 0 \Rightarrow$$

$$\omega_v = \sqrt{\omega_0^2 - \frac{\gamma^2}{4}}. \tag{3.1.10}$$

Thus, our guess

$$z = A_0 e^{-\frac{\gamma}{2}t} e^{i(\omega_v t + \varphi)}. \tag{3.1.11}$$

is indeed a solution, but only if equation (3.1.10) is satisfied. This means that x is given by

$$x = \text{Re}(z) = A_0 e^{-\frac{\gamma}{2}t} \cos(\omega_v t + \varphi). \tag{3.1.12}$$

Note that ω_v is smaller than ω_0, as shown by equation (3.1.10). This is perhaps intuitively reasonable, since the viscous damping slows down the oscillating mass. However, for most cases of practical interest, the difference between ω_v and ω_0 is small, as we'll explore further in section 3.4. The adjustable constants A_0 and φ are determined by the initial position and velocity, as you can explore in problem 3.7.

3.2   Damped electrical oscillators

In section 1.5, we analyzed an LC oscillator. Now, we add a damping element to the circuit: a resistor, as shown in figure 3.2.1. Air drag converts mechanical energy into thermal energy; a resistor converts electrical energy into thermal energy. The voltages across the capacitor and inductor are

$$V_C = \frac{q}{C}(1.5.1) \quad \text{and} \quad V_L = L\ddot{q}. \quad (1.5.6).$$

The magnitude of the voltage across the resistor is given by Ohm's Law: $\left|V_R\right| = IR$. We go clockwise around the loop when evaluating voltage changes, and we also define

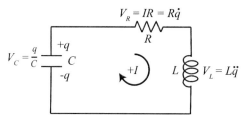

Figure 3.2.1 Damped electrical oscillator.

positive current to be clockwise. Since the voltage drops across the resistor, $V_R = -IR$. As in section 1.5, positive I decreases the charge on the capacitor, so $I = -\dot{q}$. Therefore,

$$V_R = -IR = \dot{q}R.$$

Kirchhoff's loop rule (which says that the sum of the voltage changes around a loop must be zero) then gives

$$V_L + V_R + V_C = 0 \Rightarrow$$

$$L\ddot{q} + R\dot{q} + \frac{1}{C}q = 0. \tag{3.2.1}$$

This is isomorphic to (i.e., identical to except for a change of symbols) (3.1.2):

$$m\ddot{x} + b\dot{x} + kx = 0.$$

Therefore, we simply change symbols in the solution (3.1.12):

$$q = A_0 e^{-\frac{\gamma}{2}t} \cos(\omega_v t + \varphi), \tag{3.2.2}$$

where, as before, $\omega_v = \sqrt{\omega_0^2 - \frac{\gamma^2}{4}}$.

Your turn (answer[2] below): Explain why, for the electrical oscillator, $\gamma \equiv \dfrac{R}{L}$ and $\omega_0 \equiv \sqrt{\dfrac{1}{LC}}$.

3.3 Exponential decay of energy

The solution for the damped mass/spring is

$$(3.1.12): x = A_0 e^{-\frac{\gamma}{2}t} \cos(\omega_v t + \varphi).$$

If the damping is relatively light, then γ is small, and the amplitude of oscillations changes only slowly. In this limit, we can say that the amplitude of oscillation is

$$A = A_0 e^{-\frac{\gamma}{2}t}, \tag{3.3.1}$$

so that, within each cycle the potential energy reaches a maximum value of

$$U_{\max} = E_{tot} = \tfrac{1}{2}kA^2 = \tfrac{1}{2}kA_0^2 e^{-\gamma t}$$

$$\boxed{\Rightarrow E_{tot} = E_0 e^{-\gamma t},} \tag{3.3.2}$$

where the initial energy is $E_0 = \tfrac{1}{2}kA_0^2$. This gives us a more direct interpretation of γ: in a time γ^{-1}, the energy is reduced by a factor $1/e$.

---

2. Comparing equation (3.1.2) with (3.2.1), we see that m becomes L, b becomes R, and k becomes $1/C$. Making these substitutions into $\gamma \equiv b/m$ and $\omega_0 \equiv \sqrt{k/m}$ gives $\gamma \equiv \frac{R}{L}$ and $\omega_0 \equiv \sqrt{\frac{1}{LC}}$.

Concept test (answer[3] below): Although the energy decays by a factor $1/e$ in a time γ^{-1}, according to equation (3.3.1) it takes a time $2\gamma^{-1}$ for the amplitude to decay by a factor $1/e$. How can the energy decay more quickly than the amplitude?

Self-test (answer[4] below): How long does it take for the energy to decay to 1% of its initial value? Express your answer in terms of γ^{-1}.

3.4   The quality factor

In virtually all circumstances, whether in science or engineering, people don't specify the degree of damping by providing a value for the b that appears in $F_{\text{damp}} = -bv$, or for $\gamma \equiv b/m$. Instead, they quote the quality factor Q:

$$Q \equiv \frac{\omega_0}{\gamma}. \tag{3.4.1}$$

Since ω_0 and γ both have units of s^{-1}, this is a dimensionless number. It is a ratio of the rate of oscillations to the rate of energy loss through damping. A system with low damping has a low value of γ, and therefore a high Q.

$$\boxed{\text{High } Q \Leftrightarrow \text{low damping}}$$

For many applications, a large Q is desirable, because one often wants the oscillation to persist for a long time. Also, a large Q means that the decay of oscillations is slow, so that the actual waveform is close to a perfect sinusoid, allowing for the most accurate timing. In a good watch, the time is determined by counting the vibrations of a quartz crystal having a Q of about 10^5. (The crystal is often in the shape of a tuning fork, as shown in figure 3.4.1. The usual oscillation frequency is 32,768 Hz $= 2^{15}$ Hz. It is easy in digital circuitry to divide a frequency into half. Doing this 15 times results in a frequency of 1 Hz, which can be used to drive the second hand of the watch.)

However, in other applications, a low Q is preferable. For example, when the suspension of a car is set into oscillation by driving over a pothole, the occupants would prefer that it not keep oscillating for a long time! The suspension of a typical car has a Q of about 1.

---

3. Since $E = \frac{1}{2}kA^2$, the amplitude need only decrease by a factor of $\sqrt{2}$ in order for the energy to decrease by a factor of 2.

4. $\ln(100)\,\gamma^{-1} \cong 4.6\gamma^{-1}$.

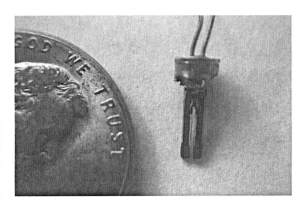

Figure 3.4.1 The quartz crystal oscillator from a watch must have a high Q, so that its oscillation frequency is very well-defined. (Image courtesy of and © Dr. Erhard Schreck.)

We can re-express the angular frequency in the presence of damping using Q. From equation (3.1.10), we have

$$\omega_v = \sqrt{\omega_0^2 - \frac{\gamma^2}{4}} = \omega_0 \sqrt{1 - \frac{\gamma^2}{4\omega_0^2}}$$

$$\Rightarrow \omega_v = \omega_0 \sqrt{1 - \frac{1}{4Q^2}}. \qquad (3.4.2)$$

Thus, we can see that if Q is large, then $\omega_v \cong \omega_0$. To make a more quantitative statement about this, you must first show the following:

Your turn: For a system with light damping (for which $E = E_0 e^{-\gamma t}$ and $\omega_v \cong \omega_0$), show that

$$E = E_0 e^{-n2\pi/Q}, \qquad (3.4.3)$$

where $n = t/T$ is the number of oscillation cycles in time t, and the period is $T = 2\pi/\omega_0$.

So, if $Q = 2\pi$, then the energy decays by about a factor of e per cycle of oscillation. This is fairly heavy damping by most people's standards, and yet equation (3.4.2) tells us that ω_v is only about 0.3% smaller than ω_0 for this case! So, it many circumstances, one can ignore the difference between ω_v and ω_0.

Connection to current research: There are some applications that depend on detecting the small difference between ω_v and ω_0. For example, the micromachined cantilever shown in figure 3.4.2 can be used to detect the pressure of gas or the molecular weight of the gas in its surroundings. Higher gas pressure, and gases with higher molecular weight, produce more damping and so a greater shift in the oscillation frequency. (A maximum frequency shift of 7% was observed for one atmospheric pressure of Argon gas, as compared to the frequency in vacuum.)

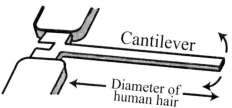

Figure 3.4.1 A micromachined silicon cantilever used to detect the pressure or molecular weight of gas via the change in the oscillation frequency due to damping. The cantilever is produced by a combination of photolithography and etching ("micromachining") and is suspended a few μm above the substrate, so that it can oscillate in a plane parallel to the substrate, as shown in the schematic diagram. Top image reprinted with permission from Y. Xu, J. –T. Lin, B. W. Alphenaar, and R. S. Keynton, *Appl. Phys. Lett.* **88**, 143513 (2006). Copyright 2006, American Institute of Physics.

3.5   Underdamped, overdamped, and critically damped behavior

> **Concept test:** (Please cover the bottom part of the page, below this box, so that you don't see the answer right away.) What is disturbing about equation (3.4.2): $\omega_{\mathrm{v}} = \omega_0 \sqrt{1 - \dfrac{1}{4Q^2}}$ if you consider the behavior at large damping?

What's disturbing is that, if Q is less than $\frac{1}{2}$, then ω_{v} is imaginary. Since low Q corresponds to heavy damping, we can definitely make Q less than $\frac{1}{2}$. This limit is called "overdamping." (Damping with $Q > \frac{1}{2}$, which is of the most interest to us, and corresponds to the damped oscillations discussed previously in this chapter, is called "underdamping.") The math that we went through in section 3.1 works fine even if ω_{v} is imaginary, so the solution for the overdamped case is still $x = \operatorname{Re} z$, where

$$(3.1.11)\text{: } z = A_0 e^{-\frac{\gamma}{2}t} e^{\mathrm{i}(\omega_{\mathrm{v}} t + \varphi)}.$$

It is revealing to rewrite things so that the imaginary nature of ω_{v} (for $Q < \frac{1}{2}$) is shown explicitly:

$$\omega_{\mathrm{v}} = \omega_0 \sqrt{1 - \frac{1}{4Q^2}} = \omega_0 \sqrt{(-1)\left(\frac{1}{4Q^2} - 1\right)} = \pm \mathrm{i}\omega_0 \sqrt{\frac{1}{4Q^2} - 1}.$$

We define

$$\beta \equiv \omega_0 \sqrt{\frac{1}{4Q^2} - 1}, \tag{3.5.1}$$

which is real if $Q < \frac{1}{2}$. So, $\omega_{\mathrm{v}} = \pm \mathrm{i}\beta$. Because of the $\pm$, plugging this into equation (3.1.11) above gives two possible solutions:

$$z_1 = A_1 e^{-\frac{\gamma}{2}t} e^{\mathrm{i}(+\mathrm{i}\beta t + \varphi_1)} = A_1 e^{\mathrm{i}\varphi_1} e^{-\frac{\gamma}{2}t} e^{-\beta t} \text{ and } z_2 = A_2 e^{\mathrm{i}\varphi_2} e^{-\frac{\gamma}{2}t} e^{+\beta t}.$$

As discussed in footnote 3 of section 1.3, the general solution of a linear, homogeneous second order differential equation, such as the differential equation (3.1.5): $\ddot{z} + \gamma \dot{z} + \omega_0^2 z = 0$ that describes the damped oscillator, is the sum of two independent solutions, such as z_1 and z_2 above. Therefore, the most general solution is:

$$z = z_1 + z_2 = e^{-\frac{\gamma}{2}t}\left[A_1 e^{i\varphi_1} e^{-\beta t} + A_2 e^{i\varphi_2} e^{+\beta t}\right].$$

The actual behavior is

$$x = \mathrm{Re}\, z = e^{-\frac{\gamma}{2}t}\left[A_1 \cos\varphi_1 e^{-\beta t} + A_2 \cos\varphi_2 e^{+\beta t}\right].$$

We define $B_1 \equiv A_1 \cos\varphi_1$ and $B_2 \equiv A_2 \cos\varphi_2$, so that finally

$$x = e^{-\frac{\gamma}{2}t}\left[B_1 e^{-\beta t} + B_2 e^{+\beta t}\right]. \tag{3.5.2}$$

Overdamped behavior (i.e., behavior for $Q < 1/2$)

The constants B_1 and B_2 are determined by the values of $x\,(t=0)$ and $\dot{x}\,(t=0)$. Note that there is no oscillation involved. Therefore, the system can cross the equilibrium point at most once, as shown in figure 3.5.1a. (You can show this rigorously in problem 3.14.)

The case $Q = 1/2$ is called "critical damping." It is only of mathematical interest, since in any real system Q would never exactly equal $1/2$, but would always be at least slightly greater or slightly less. However, for thoroughness, and because the name "critical damping" makes it sound as though it ought to be important, we discuss it briefly.

Again, the math from section 3.1 is unchanged, so that

$$x = \mathrm{Re}\left[A_0 e^{-\frac{\gamma}{2}t} e^{i\left(\omega_v t + \varphi\right)}\right].$$

However, now $\omega_v = \omega_0 \sqrt{1 - \frac{1}{4Q^2}} = 0$, so that

$$x = \mathrm{Re}\left[A_0 e^{-\frac{\gamma}{2}t} e^{i\varphi}\right] = A_0 \cos\varphi\, e^{-\frac{\gamma}{2}t}.$$

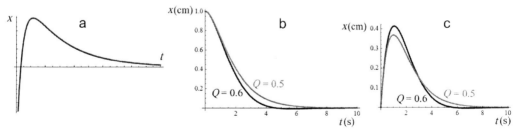

Figure 3.5.1 a: In the overdamped case, there is at most one crossing of the equilibrium point; even this can only occur if the initial velocity is high and the initial position is close to $x = 0$. b and c: Comparison of underdamped and critically damped systems. In part b, the initial velocity is zero but the initial position is nonzero. In part c, the initial position is zero but the initial velocity is non-zero. In both cases, the underdamped system ($Q = 0.6$) returns to equilibrium a little more quickly than the critically damped system.

We define $B = A_0 \cos \varphi$, so that

$$x = B \, e^{-\frac{\gamma}{2}t},\tag{3.5.3}$$

Where B is determined by the values of $x(t = 0)$ and $\dot{x}(t = 0)$.

Concept test: (Please cover the bottom part of the page, below this box, so that you don't see the answer right away.) Something about that last phrase should trouble you. What is troubling about it?

As you've probably realized, we can't get the desired values of both $x(t = 0)$ and $\dot{x}(t = 0)$ with just the one adjustable constant B. Therefore, equation (3.5.3) can't be the complete solution for the critically damped case. In problem 3.15, you can show that the complete solution is

$$x = B \, e^{-\frac{\gamma}{2}t} + Dt \, e^{-\frac{\gamma}{2}t}.\tag{3.5.4}$$

Critically damped behavior (i.e., behavior for $Q = 1/2$)

It is sometimes claimed that, if an engineer wants to design a system that moves from a displaced value to equilibrium as quickly as possible, then s/he should make it critically damped. This is incorrect, although it is not far from the truth, as shown in figure 3.5.1b and c—an underdamped system with a Q a little above the critically damped limit returns more quickly to equilibrium.

3.6   Types of damping

As we discussed in section 3.1, for low speeds the drag force is given by

$$(3.1.1) : F_{\text{drag}} = -bv,$$

where v is the speed. For a sphere of radius r,

$$b_{\text{sphere}} = 6\pi \, r\mu,\tag{3.6.1}$$

Stokes' Law

where μ is the viscosity of the surrounding fluid. This relation was derived by George Stokes in 1851.[5]

Recall, as discussed briefly in section 3.1, that the flow of the fluid around a moving object is laminar at low speeds, and turbulent at higher speeds, as shown in figure 3.6.1.

---

5. Stokes made important contributions in mathematics and physics, including central results in fluid dynamics, the first description of fluorescence, and early ideas that were close to the mark in correctly explaining spectroscopic lines. Stokes' theorem, a central result in vector calculus, is named for him, although it was derived by Lord Kelvin. (It was associated with Stokes because during oral exams he would ask students to derive it.) However, Stokes was himself generous in sharing credit with Kelvin on other fronts, so perhaps this misnomer reflects his good karma.

Table 3.6.1. Viscosities of common fluids

Fluid	μ (kg m^{-1} s^{-1})
Air	$1.784 . \times 10^{-5}$
Water	$1.004 . \times 10^{-3}$
Light motor oil	~ 0.1
Honey	~ 10
Peanut butter	~ 250

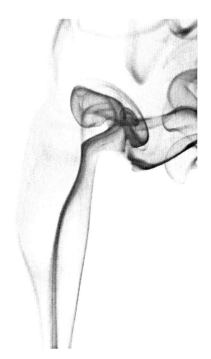

Figure 3.6.1 Smoke from an incense stick initially shows laminar flow, but when it has accelerated to a high enough velocity as it rises, it transitions to turbulent flow. (Image © Matt Trommer/Dreamstime.com)

The conditions for this transition are determined by the Reynolds number:

$$\text{Re} \equiv \frac{\rho \, \dot{v} L}{\mu}, \tag{3.6.2}$$

where L is the characteristic length and ρ is the density of the fluid. (For flow in a pipe, L is the pipe diameter. For a sphere moving through a liquid, L is the diameter of the sphere.) Re is a dimensionless number.[6] For Re $< 2{,}000$, the flow is laminar, while for

---

6. The number is named for Osborne Reynolds (1842–1912). There is a wonderful anecdote told by Sir J. J. Thomson (who discovered the electron), a student of Reynolds at Manchester University: "Occasionally in the higher classes he would forget all about having to lecture and, after waiting for ten minutes or so, we sent the janitor to tell him that the class was waiting. He would come rushing into the room pulling on his gown as he came through the door, take a volume of Rankine [a standard textbook of the time] from the table, open it apparently at random, see some formula or other and say it was wrong. He then went up to the blackboard to prove this. He wrote on

Re > 4,000 the flow is turbulent. For a baseball moving through air, the transition thus occurs at about 0.6 m/s. Be careful not to confuse the notation Re for Reynolds number with Re meaning "real part of." Usually the correct interpretation will be obvious from context.

In turbulent flow, $F_{drag} \propto v^2$, as you can show in problem 3.21. The fact that the force depends on the *square* of the velocity makes the differential equation describing the motion of an oscillator experiencing such a force impossible to solve analytically, although the motion can be solved by numerical methods. To be specific, in that problem, you can show that in turbulent flow

$$F_{drag} = \frac{C_d}{2} A v^2 \rho, \tag{3.6.3}$$

where A is the cross-sectional area. The "drag coefficient" C_d varies from about 0.03 for a high-performance jet airplane to about 0.3 for an automobile to about 1.1 for a person standing up. (In fact, the drag coefficient is not really constant, and depends somewhat on the velocity.)

There are a number of other mechanisms which can suck energy out of an oscillation, and so provide damping. A good example is friction, which provides a force essentially independent of velocity. This can be addressed analytically, but it is surprisingly messy, and is beyond the scope of this text. Another type of damping comes from atomic-scale motions within the oscillator, the details of which are a subject of current research. Because of these motions, if you put a mass/spring system in a vacuum and set it oscillating, the motion eventually damps out, although all fluid damping has been eliminated. Although important aspects of the behavior of damped oscillators vary from one type of damping to another, we can still use the results obtained for viscous damping in this chapter and the next as a qualitative guide.

Concept and skill inventory for chapter 3

After reading this chapter, you should fully understand the following terms:

Viscous drag (3.1, 3.6)
Laminar flow (3.1, 3.6)
Turbulence (3.1, 3.6)
Envelope function (3.1)

---

the board with his back to us, talking to himself, and every now and then rubbed it all out and said it was wrong. He would then start afresh on a new line, and so on. Generally, towards the end of the lecture, he would finish one which he did not rub out, and say this proved Rankine was right after all. This, though it did not increase our knowledge of facts, was interesting, for it showed the workings of a very acute mind grappling with a new problem." (From *Recollections and Reflections*, by Sir J. J. Thomson, Macmillan, New York, 1937, p. 15, cited in *Eurekas and Euphorias: The Oxford Book of Scientific Anecdotes*, by Walter Gratzer, Oxford University Press, Oxford, England, 2004, p. 333.)

Exponential decay (3.1)
Quality factor (3.4)
Underdamped, overdamped, critically damped (3.5)
Reynolds number (3.6)
Drag coefficient (3.6)

You should understand the following connections:

Damping and coupling to other parts of the universe (3.1)
Resistors and air resistance (3.2)
Damped electrical oscillators and damped mechanical oscillators (3.2)
Q and damping (3.4)
Laminar *versus* turbulent flow and v *versus* v^2 damping (3.1, 3.6)

You should be familiar with the following additional concepts:

A single complex equation contains two equations, one for the real part and one for the imaginary part (3.1)

You should be able to:

Find the amplitude or energy of a damped oscillator as a function of time (3.1–3.2)
Calculate the speed at which the drag force changes from being proportional to v to being proportional to v^2

In addition to all of the above, you should be able to combine the concepts you've learned to address new situations.

Problems

Note: Additional problems are available on the website for this text.

Instructors: Difficulty ratings for the problems, full solutions, and important additional support materials are available on the website.

3.1 Consider two identical oscillators of the type shown in figure 1.4.1a. For one of the oscillators, the surface is frictionless, but there is viscous damping $F_{\text{drag}} = -bv$. For the other, the surface has coefficient of kinetic friction μ_k, but there is no viscous damping. The masses are both pulled away from equilibrium the same distance to the right and then released. When they first reach the equilibrium position, the magnitude of the viscous damping force for the first oscillator is equal to the magnitude of the frictional force for the second oscillator. Which oscillator damps down to one tenth the initial amplitude more quickly? Explain your reasoning.

3.2 Eddy current damping. A changing magnetic field $\mathbf{B}$ creates an electromotive force $\varepsilon = -\dfrac{d\phi_B}{dt}$, where $\phi_B \equiv \int \mathbf{B} \cdot \hat{\mathbf{n}}\, dA$ is the magnetic flux. If a changing field is applied to a bulk piece of metal, the resulting current pattern is complicated, since the current isn't confined to simple geometries by wires. Such currents are called "eddy currents," and are used to provide a damping force for vibration isolation in sensitive scientific equipment. They are also used to provide a braking force for large trains, resulting in less wear on the main brakes. When the train engineer throws the appropriate lever, permanent magnets are brought close to the rails, without touching them. The motion of the train then causes a $\dfrac{d\phi_B}{dt}$ through any loop within the rail. Thus, $\dfrac{d\phi_B}{dt}$ is proportional to the train speed; we will take the proportionality constant to be α, so that $\dfrac{d\phi_B}{dt} = \alpha v$. The current that flows within the rail is proportional to $\varepsilon = -\dfrac{d\phi_B}{dt}$; we will take the proportionality constant for the average current I to be R: $\varepsilon = IR$. What is the damping constant b in terms of α and R?

3.3 Exponentials versus Power law. Using a suitable computer program, superpose plots of the functions $e^{-t/\tau}$ and t^{-n}. Choose whatever value you like for the constants τ and n. (Suggestion: the plots are easier to make if you choose n to be no greater than 20.) Try to make the power law (t^{-n}) decay faster than the exponential. Make two plots: in the first, let the vertical axis range from 0 to 2. In the second plot, choose the scales for the axes to show that, even though the power law may initially decay faster than the exponential, the exponential eventually always catches up and falls below the power law.

3.4 Show that the average value of the square of a sinusoid (averaged over one cycle) is exactly half the maximum value (of the squared sinusoid).

3.5 Time dependence of the dissipated power. For a damped harmonic oscillator, use a suitable computer program to superpose plots of $x(t)$ and the power dissipated as a function of time. (Choose whatever values you like for the relevant parameters, and make sure your plot covers at least one full period.) Comment qualitatively on why your plot looks the way it does.

3.6 A bank account earns interest according to $\dfrac{dB}{dt} = gB$, where B is the account balance and g is a constant.

    (a) Write an equation for $B(t)$ (i.e., B as a function of time) in terms of B_0 (the value of B at $t = 0$) and g.

    (b) If $B_0 = \$100.00$, and after 1 year B has increased by 5.00%, what is the value of g? *Hint: quote g in units of year* $^{-1}$. In financial terms, g is the "continually compounded interest rate," while 5% is the "Annual Percentage Rate" or APR.

3.7 We found in section 3.1 that the position of a damped oscillator is given by $x = A_0 e^{-\frac{\gamma}{2}t} \cos(\omega_v t + \varphi)$. Without using a symbolic algebra program or calculator, find A_0 and φ in terms of the initial position x_0, the initial

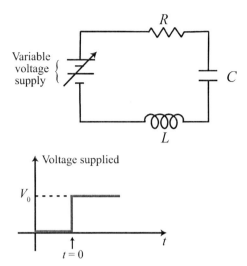

Figure 3.P.1 An RLC circuit suddenly has a nonzero voltage applied to it.

velocity $\dot{x}_0$, ω_v, and γ. *Hint: You should be able to show that* $\tan \varphi = \dfrac{B}{D}$, *where B and D are constants involving* x_0, $\dot{x}_0$, ω_v, *and* γ. *To find* A_0, *you will need to find* $\cos \varphi$. *To do this, draw a right triangle with B and D as the two legs.*

3.8 For the circuit shown in figure 3.P.1, at times $t < 0$ the output of the voltage supply is set to 0 V, there is no current flowing, and there is no charge on the capacitor. At $t = 0$, the voltage supply is suddenly changed to a voltage V_0, as shown, and remains constant thereafter.

    (a) Show that, for $t > 0$, the charge q on the capacitor is described by the following differential equation:

$$\ddot{q} + \frac{R}{L}\dot{q} + \frac{1}{LC}q = \frac{V_0}{L}.$$

    (b) Show that

$$q = A + Be^{-\alpha t}\cos(\omega t + \varphi)$$

is the solution to the above differential equation, and determine the values of A, α, and ω in terms of V_0, R, L, and C.

Hints:

    1. Cast the problem into complex form.

    2. If f(t) is a function of time, and $f(t) + G = H$, where G and H are constants, then we must have $G = H$ and $f(t) = 0$. (Otherwise the equation $f(t) + G = H$ could not be satisfied for all times.)

    3. You may, if desired, use exact analogies if they are appropriate in solving this problem. It is not necessary for you to use analogies to get the answers, and the analogy approach might not give answers for all three quantities A, α, and ω. If you choose to use the analogy approach, you should explain very clearly the analogy you're using.

(c) Because the inductor will not allow the current to change discontinuously, we must have $\dot{q}_0 = 0$. Use this initial condition to show that $\varphi = \tan^{-1}\left(\dfrac{-R}{2\omega L}\right)$.

(d) It is clear from the above that $\varphi \to 0$ or π when the damping is light. Let's choose $\varphi \to 0$. In this light damping limit, use the fact that $q_0 = 0$ to show that $B \to -CV_0$.

(e) Sketch $q(t)$ for fairly light, but nonzero damping. Your sketch can be qualitative.

3.9 Show that, perhaps surprisingly, the time between displacement maxima for a damped oscillator is exactly $\dfrac{2\pi}{\omega_v}$, independent of the degree of damping (so long as the underdamped limit applies, that is, so long as equation (3.1.12) describes the solution).

3.10 At $t = 0$, a particle of mass m attached to a spring of spring constant k is at rest a distance $x = A_0$ away from its equilibrium position. It is released, and begins oscillating. The system is immersed in fluid which leads to a damping force of the form $F_{\text{damping}} = -b\dot{x}$. You may assume the damping is light. Find the time for the *envelope* of the oscillations to drop to $A_0/10$.

3.11 Thermal vibrations.
The "Equipartition Theorem" of statistical mechanics tells us that thermal fluctuations impart an average potential energy of $\frac{1}{2}k_BT$ to a harmonic oscillator and on top of this also impart an average kinetic energy of $\frac{1}{2}k_BT$, giving a total thermal energy of k_BT, where $k_B = 1.38 \times 10^{-23}$ J/K is Boltzmann's constant, and T is the absolute temperature of the oscillator. A damped harmonic oscillator has mass m, spring constant k, and quality factor Q, which is greater than $\frac{1}{2}$. **(a)** Explain how, by measuring the amplitude of these thermal vibrations and the temperature, one can determine the spring constant of a mass/spring system for which both the mass and the spring constant are unknown. Make your explanation as quantitative as possible, remembering the word "average" which appears above. This method is used in atomic force microscopy to determine the spring constant of microfabricated cantilevers. These cantilevers can then be used to quantitatively measure the forces of interaction between individual molecules. **(b)** A damped harmonic oscillator has mass m, spring constant k, and quality factor Q, which is greater than $\frac{1}{2}$. It is set into motion with an initial amplitude A_0. In principle, the mass crosses the equilibrium point an infinite number of times. However, the damping eventually causes the motion

to become smaller than random thermal motions. In terms of m, k, Q, A_0, k_B, and T *about* how long does this take?

3.12 For a damped harmonic oscillator, show that if $F_{damp} = -b\dot{x}^\beta$, then the energy as measured on timescales longer than an oscillation period only decays exponentially if $\beta = 1$. Assume light damping. You may also assume that $P_{average} = GP_{max}$, where P is the dissipated power and G is a constant. *Hint: If you assume that E decays exponentially, what differential equation involving $P_{av} = \dot{E}$ must be obeyed? Then, in this equation, write E in terms of the maximum kinetic energy. Finally, develop a second equation for $\dot{E}$ using $F_{damp} = -b\dot{x}^\beta$.*

3.13 **Damping by radiation**. From classical electromagnetism, one can show that the instantaneous power of electromagnetic radiation that is emitted by a point charge q undergoing acceleration a is $P = \dfrac{q^2 a^2}{6\pi \, \varepsilon_0 \, c^3}$, where $c = 2.9979 \times 10^8$ m/s is the speed of light and $\varepsilon_0 = 8.8542 \times 10^{-12}$ C^2/Nm^2 is the permittivity of free space. Consider an electron that oscillates in harmonic motion about a stable position with frequency f and amplitude A. It is held close to this position by forces from other charges. As usual, we assume that the net restoring force has a magnitude proportional to displacement away from equilibrium. In this problem, you'll show that the energy radiated away provides an effective viscous damping force. Assume that the energy radiated by the electron in one cycle is much less than the energy in the oscillator, so that over one cycle the motion is approximately sinusoidal. **(a)** Show that the energy radiated in one cycle is $E_{cycle} = \dfrac{4\pi^3 e^2 A^2 f^3}{3\varepsilon_0 c^3}$. **(b)** Show that therefore $\dfrac{\Delta E}{\Delta t} \propto E$, where ΔE is the energy lost in one cycle and Δt equals one period. **(c)** By comparison with equation (3.1.8), which appears in footnote 1, show that the energy decays exponentially, just as with a viscous damping force, and find the time for the energy to decrease by a factor of e. **(d)** Show that $Q \cong \dfrac{3\varepsilon_0 mc^3}{e^2 f}$, where m is the mass of the electron.

3.14 Show that an overdamped oscillator can cross the equilibrium point one time at most.

3.15 Show that $x = Dte^{-\frac{\gamma}{2}t}$ is a solution to equation (3.1.3) for the case of critical damping. (In section 3. 5, we showed that $x = Be^{-\frac{\gamma}{2}t}$ is also a solution, so the general solution is $x = Be^{-\frac{\gamma}{2}t} + Dte^{-\frac{\gamma}{2}t}$.)

3.16 For a given object moving through air or water, is the speed needed to produce turbulence higher in air or water? Explain.

3.17 Consider a round rock of radius r in a shallow stream. For about what velocity of the stream should the flow around the rock become turbulent, according to the ideas discussed in this chapter? Is this velocity consistent with your experience?

3.18 You are designing a chemical processing plant, in which water from a pipe is used to fill a spherical reaction vessel that is 2 m in diameter. If there is turbulence inside the pipe, then the required pressure to force the water through is significantly higher, so this is to be avoided. If the reaction vessel must be filled in less than 100 min, what is the minimum pipe diameter?

3.19 Calibration of laser tweezers. As you may recall from an electricity and magnetism course, matter is attracted to regions of high electric field. (The atoms in the matter are polarized by the field, so that each atom forms a dipole. The two ends of the dipole experience fields of different strength, leading to a net force.) You may also recall that light consists of rapidly oscillating electric and magnetic fields. Combining these two ideas, we see that there is an attractive force which draws matter into a region with the most intense light. For example, if you shine a laser beam into a drop of water in which there are small plastic spheres suspended, the spheres are attracted into the beam. This is called a "laser tweezers apparatus." By moving the beam around, you can drag one of these spheres wherever you like. You can also grab a living cell that is floating around in the water, and drag it around without harming it. By combining two laser tweezers with a more powerful laser that can cut through the cell wall ("laser scissors"), you can fuse two cells together. Going back to the plastic beads, you can attach one sort of molecule to the surface of a bead, allow it to bind to another sort of molecule on a different bead or on a solid surface, and measure the strength of the binding by tugging with the laser tweezers. In order to perform this experiment, you must calibrate the strength of the force applied to the bead by the laser tweezers. Since the tweezers hold the bead in a position of stable equilibrium, the force applied to the bead by the laser can be approximated by Hooke's law, $F = -kx$, where x is the displacement of the center of the bead away from the center of the laser beam. One way to find the spring constant x is to flow water (or another fluid) past the bead at a known speed v_0 and measure the resulting displacement x_0 of the bead. Find an expression for k in terms of x_0, v_0, the viscosity μ of the surrounding fluid, the density ρ of the fluid, and the radius r of the bead, assuming the flow remains laminar.

3.20 (a) Explain how our guess for the complex version of the solution of the damped oscillator, (3.1.7): $z = A_0 e^{-\sigma t} e^{i(\omega_v t + \varphi)}$ could be written in the form $z = C e^{\alpha t}$, where C and α are both complex. **(b)** Show that $z = C e^{\alpha t}$ is not a solution for the damped oscillator in the turbulent regime, where (3.6.3):
$$F_{\text{drag}} = \frac{C_{\text{d}}}{2} A v^2 \rho.$$

3.21 Drag force for turbulent flow. (a) It is easier to think about drag forces in a reference frame in which the object is stationary and the fluid medium (air, water, etc.) is moving past it. In laminar flow, the fluid moves around the object with relatively little net disruption to its velocity. However, in turbulent flow, the velocity of the fluid is changed violently. Consider the volume of fluid that will impinge on the object in a time Δt, as shown in figure 3.P.2. This has a length $v \Delta t$, and a cross-sectional area A equal to that of the object. Assume that a fraction $C_{\text{d}}/2$ of the momentum of this volume of fluid is transferred to the object. (C_{d} is called the "drag coefficient.") Show

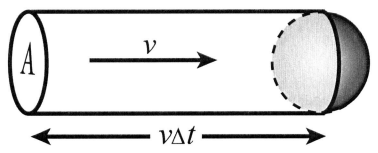

Figure 3.P.2 In time Δt, all the air in the cylinder will hit the front of the spherical object.

that the resulting force on the object has a magnitude $F_{\text{drag}} = \dfrac{C_d}{2}Av^2\rho$, where ρ is the density of the fluid. **(b)** For an airplane, $C_d \approx 0.03$. What power is required from the engines for an airplane with $A = 15$ m^2 to fly at a speed of 150 kph near sea level, where the density of air is 1.2 kg/m^3? **(c)** Repeat part (b) for a speed of 300 kph.

4 Driven Oscillations and Resonance

When you drive an oscillator at its resonant frequency
Then the amplitude of the oscillation will become huge.
In the equation above, it becomes infinite,
But in practice there will be some damping
That prevents that.

You have known this since your childhood,
This is how you swing on a swing.
If you live in a snowy climate, you know
(or at least should know) that a trick
To get your car out of a snow bank is
To rock it back and forth—

If you get the frequency right you will make the car oscillate
With a large amplitude
And dislodge it.

The electrical analogue is used to tune a radio.

From "The Driven Oscillator" (not originally intended to be a poem)
by Professor B. Paul Padley, Rice University

4.1 Resonance

As we saw in chapter 3, the oscillations of a macroscopic oscillator decay over time because the energy leaks out into the surroundings. For an oscillation to be sustained, this energy loss must be balanced by the energy added to the oscillator. We see examples of this all the time. A child on a swing uses her legs to pump the amplitude higher and higher. The seat in a bus vibrates, because it is driven by the engine which is shaking in a regular way. A washing machine shakes, especially in the spin cycle; in this case, the energy is provided by the motor which spins the tub with the clothes in it. The current inside a radio receiver oscillates, driven by the radio waves broadcast by the transmitter. The straps on your backpack, and your hair (if it's long enough) swing back and forth as you walk, driven by the energy you put into moving your legs (which also moves your body up and down a little with each step).

The behavior of oscillators that are driven, including the above examples, is perhaps the single most important idea in all of physics. Essentially every area of physics has strong connections with driven oscillators, and many seemingly unrelated phenomena can be understood qualitatively by analogy with driven oscillators.

You have probably noticed that if an oscillator is driven with a periodic force of just the right frequency, the motion of the oscillator becomes very large, much larger than for other frequencies of drive force. For example, if you have an old car that rattles sometimes, you may have noticed that the rattle is worst when the engine is running at a particular speed—for lower speeds *or* for higher speeds the rattle is less noticeable. This phenomenon of a strong response at a certain frequency is called "resonance." It can be annoying and even destructive, leading to failure of mechanical components, but it can also be used to almost magical effect in the design of ultrasensitive detectors, and in medical imaging, as shown in figure 4.1.1.

Following the approach of physics, we begin with the simplest possible driven oscillator: a mass on a spring. However, by now you recognize that this can represent a vast array of physical systems, including electronic circuits (which we'll explore later in this chapter). We will also assume to start that the energy injected into the oscillator comes from the simplest possible periodic source: a sinusoidal driving force. However, this is actually no restriction at all, since we will show that *any* periodic driving force can be represented as a sum of sinusoids, and that the response of the linear

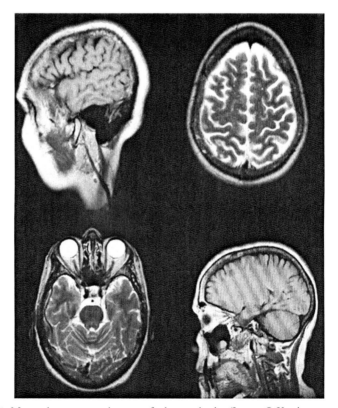

Figure 4.1.1 Magnetic resonance images of a human brain. (Image © Katrina Brown/Dreamstime.com)

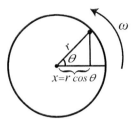

Figure 4.1.2 An unbalanced washing machine (top view).

oscillators that are our main focus is simply the sum of the responses to the individual sinusoids.

Furthermore, many real-world driving forces *do* have a simple sinusoidal form. Many driving forces arise from a circular motion. For example, consider again the tub of a washing machine, in which the clothes are unevenly distributed. The position of the heaviest part of the clothes is shown by the dot in figure 4.1.2, and the angle of a line to this point relative to horizontal is θ. If the tub rotates at constant angular velocity ω, then $\theta = \omega t$. The horizontal position of the bunch of clothes is then $x = r \cos \theta = r \cos \omega t$, so as the tub rotates, this creates a sinusoidal driving force in the x-direction.

For a mass hanging on a spring (figure 4.1.3), one way to apply the driving force is to move the support point for the spring sinusoidally. The force from the spring is clearly related to the motion of x *relative to* x_c, i.e.,

$$F_{\text{spring}} = -k\left(x - x_c\right)$$

(e.g., if x and x_c are both shifted upward by the same amount, the spring force should be zero.) We move x_c sinusoidally with amplitude A_d and angular frequency ω_d, where subscript "d" indicates "drive." For example, if $x_c = A_d \cos \omega_d t$, then we get a sinusoidal driving force:

$$F_{\text{spring}} = -k\left(x - x_c\right) = -kx + kA_d \cos \omega_d t = \underbrace{-kx}_{\substack{\text{usual} \\ \text{spring} \\ \text{force}}} + \underbrace{F_0 \cos \omega_d t}_{\substack{\text{sinusoidal} \\ \text{driving} \\ \text{force}}}$$

with

$$F_0 = kA_d. \tag{4.1.1}$$

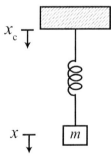

Figure 4.1.3 A sinusoidal driving force can be applied by moving the support point x_c sinusoidally. The position x of the mass is defined relative to the equilibrium position, as indicated by the short horizontal line next to x. Similarly the position of the support point, x_c, is defined relative to its own equilibrium position, as indicated by the short horizontal line next to x_c.

Before starting our quantitative analysis, let's qualitatively consider what to expect. If the frequency of the drive is very low, then the mass should simply move up and down together with the support point, that is, $x = x_c = A_d \cos \omega_d t$. On the other hand, if ω_d is very high, then the sign of the drive force keeps changing at a high frequency. So, the drive force doesn't have much time to accelerate the mass before the sign of the drive force changes. So, for large ω_d, we expect a small motion of the mass. Somewhere between low ω_d and high ω_d we expect to find a resonant response, that is, a special frequency of drive force for which the mass oscillates with large amplitude.

The damped oscillator without a drive force is just a special case of the damped driven oscillator, with zero drive amplitude, that is, $F_0 = 0$. Thus, the *full* solution for the oscillator with both driving and damping should include the possibility of an oscillation at angular frequency ω_v that decays away exponentially and has an amplitude and phase that depend on the initial position and velocity. However, for times well beyond $t = 0$, it is reasonable to expect that the effect of these initial conditions will decay away, and that we will see a "steady-state" oscillation in which the energy per cycle delivered to the system by the drive equals the energy per cycle that leaks away from the system because of the damping. This "steady-state solution" is actually the one of most interest, since it persists indefinitely.

We consider a general sinusoidal drive force

$$F_{\text{drive}} = F_0 \cos \omega_d t, \tag{4.1.2}$$

which might be applied by a motion of the support point (equation (4.1.1), or might be applied in some other way.

To find the behavior of the system, $x(t)$, we again follow our three-step procedure:

1. Write Newton's second law for each object in the system:

$$m\ddot{x} = -kx - b\dot{x} + F_0 \cos \omega_d t \iff$$

$$m\ddot{x} + b\dot{x} + kx = F_0 \cos \omega_d t \iff \tag{4.1.3a}$$

$$\ddot{x} + \gamma \dot{x} + \omega_0^2 x = \frac{F_0}{m} \cos \omega_d t. \tag{4.1.3b}$$

1b. Cast the DEQ into complex form

Following the lead given by previous chapters, we write the simplest complex differential equation for which equation (4.1.3b) is the real part:

$$\ddot{z} + \gamma \dot{z} + \omega_0^2 z = \frac{F_0}{m} e^{i\omega_d t} \tag{4.1.4}$$

with $x = \text{Re}(z)$.

2. Guess a solution, based on physical and mathematical intuition

Driven oscillators appear to show a regular motion, at least once the steady-state discussed earlier has been achieved. It is not obvious whether the angular frequency of this motion would be that of oscillation without damping or driving (ω_0) or that of oscillation with damping but no driving (ω_v), or that of the drive (ω_d), or some combination of these. So, for now, we write the angular frequency of the motion in steady-state as ω_s. To be as general as possible, we include a phase factor, and of course an amplitude. Thus, our guess is

$$z \stackrel{?}{=} Ae^{i(\omega_s t - \delta)}.$$

Here, the phase factor is written as $-\delta$, rather than, for example, $+\varphi$. We do this because it will turn out that δ as defined this way is always positive, though this is not yet obvious. The above guess can be rewritten

$$z \stackrel{?}{=} Ae^{-i\delta}e^{i\omega_s t}. \tag{4.1.5}$$

3. Substitute the guess into the DEQ to see if it is a solution, and if there are restrictions on the parameters

Your turn: Plug our guess (4.1.5) into (4.1.4) to show that

$$(-\omega_s^2 A + i\gamma \omega_s A + \omega_0^2 A)e^{-i\delta}e^{i\omega_s t} \stackrel{?}{=} \frac{F_0}{m}e^{i\omega_d t}. \tag{4.1.6}$$

We see that the left side oscillates at an angular frequency ω_s, while the right side oscillates at ω_d. So, if they are to be equal, we must have $\omega_s = \omega_d$. In other words (assuming other aspects of our guess turn out to be correct):

> In the steady-state, the oscillator moves with the same angular frequency as the drive.

This is perhaps an unexpected result. In the steady-state, the system does *not* move with its "natural" angular frequency ω_0 or at ω_v, but rather it moves at ω_d.

So, our guess now becomes

$$z \stackrel{?}{=} Ae^{-i\delta}e^{i\omega_d t}, \tag{4.1.7}$$

and is shown graphically in figure 4.1.4. But we haven't yet shown that our guess really works. Since $\omega_s = \omega_d$, equation (4.1.6) becomes

$$(-\omega_d^2 A + i\gamma \omega_d A + \omega_0^2 A)e^{-i\delta}e^{i\omega_d t} \stackrel{?}{=} \frac{F_0}{m}e^{i\omega_d t}.$$

$$\Rightarrow (-\omega_d^2 A + i\gamma \omega_d A + \omega_0^2 A)e^{-i\delta} \stackrel{?}{=} \frac{F_0}{m}.$$

$$\Leftrightarrow (\omega_0^2 - \omega_d^2)A + i\gamma \omega_d A \stackrel{?}{=} \frac{F_0}{m}e^{i\delta}$$

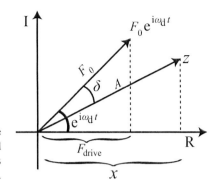

Figure 4.1.4 Our guess represented in the complex plane. Note that, since the units of z and F_0 are different, the length of the two vectors cannot be compared.

For this equation to hold, the real part of the left side must equal the real part of the right side, and also the imaginary part of the left side must equal the imaginary part of the right side:

$$\text{Real: } (\omega_0^2 - \omega_d^2)A \stackrel{?}{=} \frac{F_0}{m}\cos\delta. \tag{4.1.8}$$

$$\text{Imaginary: } \gamma\,\omega_d A \stackrel{?}{=} \frac{F_0}{m}\sin\delta. \tag{4.1.9}$$

To make these equations true, we will need particular values of A and δ. To isolate A, we square these two equations and add them, giving

$$A = A\left(\omega_d\right) = \frac{F_0/m}{\sqrt{(\omega_0^2 - \omega_d^2)^2 + (\gamma\,\omega_d)^2}}, \tag{4.1.10}$$

in which we emphasize that A is a function of ω_d. To isolate δ, we instead divide equation (4.1.9) by (4.18), giving

$$\tan\delta\left(\omega_d\right) = \frac{\gamma\,\omega_d}{\omega_0^2 - \omega_d^2}, \tag{4.1.11}$$

again emphasizing that δ is a function of ω_d.[1]

So, our guess (4.1.7) is indeed a solution of the differential equation (4.1.4)! (But only if A and δ are as given above.) Of course,

$$x = \text{Re}\,(z) = A\cos\left(\omega_d t - \delta\right). \tag{4.1.12}$$

We can see from equation (4.1.10) that the amplitude depends on the angular frequency of the drive, ω_d. It might appear from the equation that the maximum amplitude occurs

---

1. It is correct to write $\delta\left(\omega_d\right) = \tan^{-1}\dfrac{\gamma\,\omega_d}{\omega_0^2 - \omega_d^2}$, however recall that one must be careful with the $\tan^{-1}$ function. For a negative argument, your calculator (or Mathematica or other equivalent program) returns a negative value for the $\tan^{-1}$. Since we want δ to be positive, we must add π to such a result.

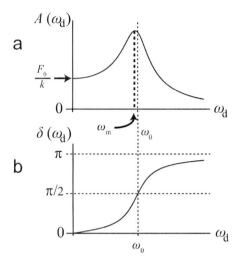

Figure 4.1.5 Amplitude and phase of a damped driven oscillator as a function of the angular frequency of the drive.

at $\omega_d = \omega_0$; this is almost right, but the actual maximum is at a slightly lower value[2] of ω_d. However, the difference is quite small, except for heavy damping, and is usually unimportant. (We shall discuss this in more detail in section 4.2.) This maximum amplitude is the resonance discussed earlier in this section.

We also see from equation (4.1.11) that the phase δ by which the oscillator's response lags behind the drive force also depends on ω_d. When $\omega_d = \omega_0$, equation (4.1.11) becomes $\tan \delta \to \infty$, so that $\delta = \pi/2$. The dependencies of A and δ on ω_d are shown in figure 4.1.5.

We see from equation (4.1.10) that the response amplitude at high frequencies approaches zero, as anticipated in our initial qualitative discussion. In the opposite limit, $\omega_d \to 0$, equation (4.1.10) reduces to

$$A\left(\omega_d \to 0\right) = \frac{F_0/m}{\omega_0^2} = \frac{F_0}{k}.$$

If the drive force is applied by moving the support point, then, using equation (4.1.1): $F_0 = kA_d$, this becomes

$$A\left(\omega_d \to 0\right) = A_d, \tag{4.1.13}$$

which is also as we anticipated.

_____

2. To see this, we rewrite equation (4.1.10):

$$A = \frac{F_0/m}{\sqrt{(\omega_0^2 - \omega_d^2)^2 + (\gamma \omega_d)^2}} = \frac{1}{\omega_d} \underbrace{\frac{F_0/m}{\sqrt{\frac{(\omega_0^2 - \omega_d^2)^2}{\omega_d^2} + \gamma^2}}}_{\text{peaks at } \omega_d = \omega_0}.$$

The second part of this has a peak at exactly $\omega_d = \omega_0$. However, it is multiplied by the factor $1/\omega_d$, which increases as ω_d decreases, shifting the peak to a slightly lower value of ω_d.

Effects of damping

The shapes of the curves in figure 4.1.5 are profoundly important for applications of resonance, both those applications in which we want to maximize resonance effects (such as in radio receivers) and applications in which we want to minimize them (such as in building designs). These curves are strongly affected by the degree of damping, as we'll explore in this section.

Often, it is revealing to re-express functions in terms of their dependence on dimensionless variables; this frequently reveals a universal behavior that was obscured in the original form of the function. In our case, we will try re-expressing the steady-state amplitude A and the phase shift δ of the response relative to the drive force. In equations (4.1.10) and (4.1.11), these two functions are expressed in terms of the angular frequency of the drive, ω_d and the factor $\gamma \equiv b/m$ which characterizes the damping. We will now re-express them in terms of $\omega_\mathrm{d}/\omega_0$ and $Q \equiv \omega_0/\gamma$, both of which are dimensionless.

Your turn: Starting from equation (4.1.10), $A = \dfrac{F_0/m}{\sqrt{(\omega_0^2 - \omega_\mathrm{d}^2)^2 + (\gamma\,\omega_\mathrm{d})^2}}$, and using

$\gamma = \omega_0/Q$, show that $A = \dfrac{F_0/m}{\omega_0\,\omega_\mathrm{d}\sqrt{\left(\dfrac{\omega_0}{\omega_\mathrm{d}} - \dfrac{\omega_\mathrm{d}}{\omega_0}\right)^2 + \dfrac{1}{Q^2}}}$.

We can rewrite this result as $A = \dfrac{F_0}{m\omega_0^2}\dfrac{\omega_0/\omega_\mathrm{d}}{\sqrt{\left(\dfrac{\omega_0}{\omega_\mathrm{d}} - \dfrac{\omega_\mathrm{d}}{\omega_0}\right)^2 + \dfrac{1}{Q^2}}}$. Since $\omega_0 \equiv \sqrt{\dfrac{k}{m}}$, this

becomes

$$A = \frac{F_0}{k}\frac{\omega_0/\omega_\mathrm{d}}{\sqrt{\left(\dfrac{\omega_0}{\omega_\mathrm{d}} - \dfrac{\omega_\mathrm{d}}{\omega_0}\right)^2 + \dfrac{1}{Q^2}}}. \tag{4.2.1}$$

We can now see the universal behavior that we hoped would arise; the particular values of ω_d and ω_0 are not really central – what really matters is the ratio $\omega_\mathrm{d}/\omega_0$.

If the drive force is applied by moving the support point, then equation (4.1.1): $F_0 = kA_\mathrm{d}$, so that

$$A = A_\mathrm{d}\frac{\omega_0/\omega_\mathrm{d}}{\sqrt{\left(\dfrac{\omega_0}{\omega_\mathrm{d}} - \dfrac{\omega_\mathrm{d}}{\omega_0}\right)^2 + \dfrac{1}{Q^2}}}. \tag{4.2.2}$$

This relation is graphed in figure 4.2.1. Several different curves are shown for different values of Q. The most important effect of increasing the Q (i.e., lowering the damping) is to make the peak higher and sharper.

A more subtle effect is that the peak, which is always at an angular frequency close to ω_0, moves even closer to ω_0 as Q increases. You can show in problem 4.8 that the

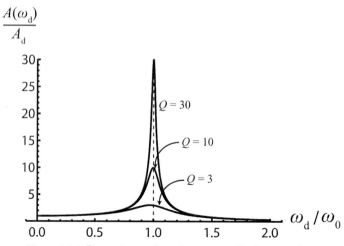

Figure 4.2.1 Dependence of steady state amplitude on ω_d / ω_0.

peak occurs at

$$\omega_{d, peak} = \omega_0 \sqrt{1 - \frac{1}{2Q^2}}. \qquad (4.2.3)$$

Recall from chapter 3 that $Q = 2\pi$ corresponds to reasonably heavy damping. (Without a driving force, the energy decays by a factor of $1/e$ per cycle, and the amplitude decays by a factor of $1/\sqrt{e} \cong 0.61$ per cycle.) Substituting $Q = 2\pi$ into equation (4.2.3) gives $\omega_{d,peak} = 0.994\,\omega_0$. For light damping, $\omega_{d,peak}$ is even closer to ω_0, so for most purposes the difference can be ignored.

For moderate to light damping, for which the peak amplitude occurs at $\omega_d \cong \omega_0$, we can obtain a useful result for the height of the peak. Substituting $\omega_d = \omega_0$ into equation (4.2.2) gives

$$A\left(\omega_d = \omega_0\right) = QA_d. \qquad (4.2.4)$$

In other words,

> For light to moderate damping:
> (peak response amplitude) $\cong Q \times$ (drive amplitude).

Since the low-frequency response amplitude is equal to the drive amplitude, we could also say that the peak response amplitude is Q times the low-frequency response amplitude.

For $\omega_d \gg \omega_0$, the response becomes

$$A = A_d \frac{\omega_0/\omega_d}{\sqrt{\left(\frac{\omega_0}{\omega_d} - \frac{\omega_d}{\omega_0}\right)^2 + \frac{1}{Q^2}}} \xrightarrow{\lim \omega_d \gg \omega_0} A_d \left(\frac{\omega_0}{\omega_d}\right)^2, \qquad (4.2.5)$$

so that, at high frequencies, the response is universal, and does not even depend on the degree of damping.

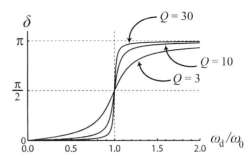

Figure 4.2.2 Phase shift δ of the steady-state response.

Now, we will consider how the graph of the phase shift δ is affected by damping. Again, we begin by expressing δ in terms of the dimensionless quantities ω_d/ω_0 and Q.

Your turn: Starting from equation (4.1.11), $\tan\delta = \dfrac{\gamma\,\omega_d}{\omega_0^2 - \omega_d^2}$, and using $\gamma = \omega_0/Q$, show that

$$\tan\delta = \frac{1/Q}{\dfrac{\omega_0}{\omega_d} - \dfrac{\omega_d}{\omega_0}}. \tag{4.2.6}$$

This relation is graphed in figure 4.2.2.[3] At low ω_d, the phase shift is zero, meaning (as we anticipated) that the oscillator moves in phase with the drive. At high ω_d, the phase shift is 180°; the oscillator is exactly out of phase with the drive. At $\omega_d = \omega_0$, the phase shift is exactly 90°. The effect of increasing the Q (i.e., lowering the damping) is to sharpen the transition from $\delta = 0$ at low drive frequency to $\delta = \pi$ at high-drive frequency.

Connection to current research: Tapping Mode Atomic Force Microscopy
The Atomic Force Microscope (AFM), invented in 1986 by Gerd Binnig, Calvin Quate, and Christoph Gerber, images a sample by touching it very lightly with a sharp tip mounted on the end of a cantilever (figure 4.2.3). The upward force exerted on the tip by the sample causes a bend in the cantilever, so that this force can be measured quantitatively. In the original mode of operation ("contact mode"), the tip is first lowered into contact with the sample until a pre-set force is achieved (typically about 100 nN). Then, the tip is moved laterally across the sample. As it moves, the force is monitored, and the base of the cantilever is moved up or down as needed to keep the force constant at the pre-set level. In this way, the microscope "feels" the shape

continued

---

3. We can rewrite equation (4.2.6) as $\delta = \tan^{-1}\dfrac{1/Q}{\dfrac{\omega_0}{\omega_d} - \dfrac{\omega_d}{\omega_0}}$. Recall that, in this case if the argument of the $\tan^{-1}$ function is negative, that is, if $\omega_d > \omega_0$, we must add π to the result given by a calculator or a program like Mathematica in order to get the correct value of δ. (See footnote 1 after equation (4.1.11).)

of the surface, while only applying a light force. This works well for semiconductor and metal samples, but the lateral motion of the tip is very destructive for the soft samples which are of most interest in biology and in molecular electronics.

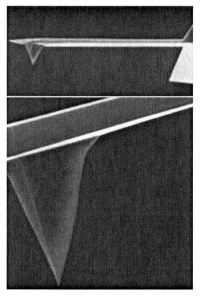

Figure 4.2.3 The cantilever for an atomic force microscope (top image) must have a moderately high Q, so that a small amplitude vibration applied at the base of the cantilever results in a vibration amplitude at the tip of about 100 nm. The length of the cantilever is 125 μm, about the same as the diameter of a human hair. The pyramid-shaped tip located at the end of the cantilever (bottom image) must be very sharp to obtain high resolution images. (Image courtesy of Veeco Instruments Inc.)

In 1993, Zhong, Inniss, Kjoller, and Elings introduced the Tapping mode™ AFM. (The trademark belongs to Digital Instruments (now part of Veeco Corporation), which was founded by Virgil Elings and is the leading manufacturer of AFMs.) In this mode, the base of the cantilever is vibrated vertically by means of a piezoelectric crystal. The frequency of vibration is chosen to match the resonant frequency of the cantilever, so that the tip vibrates with an amplitude of about 100 nm. The vibration amplitude is measured, and the base of the cantilever is slowly lowered toward the sample. When the tip begins to tap the sample, this contact reduces the oscillation amplitude. The base is lowered further until the amplitude reaches a pre-set value. Then, as in contact mode, the base of the cantilever is moved laterally across the sample. As it moves, the amplitude of the tip's oscillation is monitored, and the base of the cantilever is moved up or down as needed to keep the amplitude constant. Because the tip only touches the sample briefly during each cycle of oscillation, the lateral forces applied to soft features on the sample are minimized, so that they can be imaged without damage.

The Q of the cantilever is typically about 100. Therefore, since $A\left(\omega_{d} = \omega_{0}\right) = QA_{d}$, the drive amplitude A_{d} applied by the piezoelectric crystal to the base of the cantilever needs to be only about 1 nm in order to produce the desired tip vibration amplitude A of 100 nm. This large ratio is essential, because if one had to vibrate the base by the full 100 nm, the entire AFM would vibrate, dramatically degrading image quality.

4.3   **Energy flow**

Energy is perhaps the most fundamental idea in physics. In many situations, including many parts of quantum mechanics, finding the energy is the central problem. We will find that, for damped driven oscillators, a careful consideration of the energy for the steady-state behavior provides insights that can be very broadly applied.

The power (energy per time) supplied to the oscillator by the drive force is

$$P_{drive} = F_{drive}v.$$

In the steady-state, $x = A \cos (\omega_d t - \delta)$, and so

$$v = \dot{x} = -A\omega_d \sin (\omega_d t - \delta) = A\omega_d \sin (\delta - \omega_d t) = A\omega_d \cos (\delta - \pi/2 - \omega_d t)$$

$$= A\omega_d \cos (\omega_d t + \pi/2 - \delta)$$

Therefore,

$$P_{drive} = \underbrace{(F_0 \cos \omega_d t)}_{F_{drive}} \underbrace{(A\omega_d \cos (\omega_d t + \pi/2 - \delta))}_{v}. \tag{4.3.1}$$

Concept test (answer[4] below): For what value of ω_d/ω_0 are the oscillations of F_{drive} in phase with the oscillations in v? *Hint: review figure 4.2.2.*

Since, for the value you just found, the oscillations are always in phase, P_{drive} is always positive, that is, the drive force always supplies power *to* the oscillator. For any other value of ω_d/ω_0, the oscillations of F_{drive} are not in phase with the oscillations in v; therefore, for some parts of the cycle P_{drive} is positive (the drive force supplies power to the oscillator), and for some parts of the cycle P_{drive} is negative (the oscillator supplies power to the entity providing the drive force). An example is shown in figure 4.3.1a. We can define $P_{drive, av}$ to be the power supplied by the drive averaged over a cycle. For the example shown in figure 4.3.1a, P_{drive} is positive for most of the cycle, so $P_{drive, av} > 0$.

As $\omega_d \to 0$, $\delta \to 0$, so that F_{drive} and v are out of phase by $\pi/2$. For this condition, the net energy flowing from the drive to the oscillator is zero (i.e., $P_{drive, av} = 0$), as suggested in figure 4.3.1b. (You can show rigorously that the net energy flow is zero in problem 4.11.) As $\omega_d \to \infty$, $\delta \to \pi$, so that F_{drive} and v are out of phase by $\pi/2$ in the other direction; again the net flow of energy from the drive over a cycle is zero. Putting all this together, we expect that the graph of $P_{drive, av}$ as a function of ω_d/ω_0 must start at zero, reach a peak (probably at $\omega_d/\omega_0 = 1$), and then go back to zero.

Let's examine this quantitatively. The average power, $P_{drive, av}$, can be computed using the average value theorem from calculus:

$$P_{drive,av} = \frac{1}{T} \int_0^T P_{drive} dt,$$

---

4. For the oscillations to be in phase, we need $\delta = \pi/2$, which occurs when $\omega_d/\omega_0 = 1$.

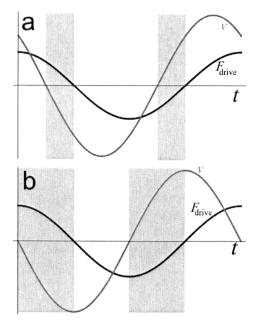

Figure 4.3.1 a: F_{drive} and v for the case $\delta = 45°$. In the shaded regions, the two have opposite sign, so P_{drive} is negative, meaning that the oscillator supplies power to the drive. In the other regions, F_{drive} and v have the same sign, so P_{drive} is positive, and the drive supplies power to the oscillator. (Since F_{drive} and v have different units, the vertical scales cannot be compared.) b: F_{drive} and v for $\omega_{\text{d}} \to 0$; in this limit $\delta \to 0$, so F_{drive} and v are out of phase by $\pi/2$. In the shaded regions, the two have opposite sign, so P_{drive} is negative, meaning that the oscillator supplies power to the drive. In the other regions, F_{drive} and v have the same sign, so P_{drive} is positive, and the drive supplies power to the oscillator. (The velocity scale for this plot is greatly magnified compared to part a; for small ω_{d}, the velocity is also small.)

where T is the period of the steady-state motion. Plugging in our expression from equation (4.3.1), we get

$$P_{\text{drive}} = \frac{1}{T} \int_0^T \left(F_0 \cos \omega_{\text{d}}t\right) A\omega_{\text{d}} \cos\left(\omega_{\text{d}}t + \pi/2 - \delta\right) \, dt$$

$$= \frac{F_0 A\omega_{\text{d}}}{T} \int_0^T \cos \omega_{\text{d}}t \, \cos\left(\omega_{\text{d}}t + \pi/2 - \delta\right) \, dt.$$

This integral can be evaluated using Mathematica, or a similar symbolic algebra program or calculator; it can also be done "by hand" as shown in the footnote.[5] Bearing

---

5. First, we re-express the second term in the integral: $\cos\left(\omega_{\text{d}}t + \pi/2 - \delta\right) = \cos\left(\delta - \pi/2 - \omega_{\text{d}}t\right) = \sin\left(\delta - \omega_{\text{d}}t\right)$. Therefore, $P_{\text{drive}} = \frac{F_0 A\omega_{\text{d}}}{T} \int_0^T \cos \omega_{\text{d}}t \, \sin\left(\delta - \omega_{\text{d}}t\right) \, dt$. Next, we use $\sin\left(A + B\right) = \sin A \cos B + \cos A \sin B$, so that $P_{\text{drive}} = $

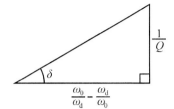

Figure 4.3.2 Geometric representation of equation (4.2.6).

in mind that $T = 2\pi/\omega_{\rm d}$, the result is

$$P_{\rm drive,\,av} = \frac{F_0 A \omega_{\rm d}}{2} \sin \delta. \tag{4.3.2}$$

To evaluate $\sin \delta$, we can represent equation (4.2.6) geometrically, as shown in figure 4.3.2. From this, we see that

$$\sin \delta = \frac{1/Q}{\sqrt{\left(\dfrac{\omega_0}{\omega_{\rm d}} - \dfrac{\omega_{\rm d}}{\omega_0}\right)^2 + \dfrac{1}{Q^2}}}.$$

Your turn: Substitute the above, and also equation (4.2.1): $A = \dfrac{F_0}{k} \dfrac{\omega_0/\omega_{\rm d}}{\sqrt{\left(\dfrac{\omega_0}{\omega_{\rm d}} - \dfrac{\omega_{\rm d}}{\omega_0}\right)^2 + \dfrac{1}{Q^2}}}$

into equation (4.3.2), and show that the result is

$$P_{\rm drive,\,av} = \frac{F_0^2 \omega_0}{2kQ} \frac{1}{\left(\dfrac{\omega_0}{\omega_{\rm d}} - \dfrac{\omega_{\rm d}}{\omega_0}\right)^2 + \dfrac{1}{Q^2}}. \tag{4.3.3}$$

$\dfrac{F_0 A \omega_{\rm d}}{T} \left[\sin \delta \displaystyle\int_0^T \cos \omega_{\rm d} t \, \cos\left(-\omega_{\rm d} t\right) \, {\rm d}t \; + \; \cos \delta \int_0^T \cos \omega_{\rm d} t \, \sin\left(-\omega_{\rm d} t\right) \, {\rm d}t \right]$. This simplifies

to $P_{\rm drive} = \dfrac{F_0 A \omega_{\rm d}}{T} \left[\sin \delta \displaystyle\int_0^T \cos^2 \omega_{\rm d} t \, {\rm d}t \; - \; \cos \delta \int_0^T \cos \omega_{\rm d} t \, \sin \omega_{\rm d} t \, {\rm d}t \right]$. These are both standard integrals which can be looked up in a table. The first one is especially important: $\displaystyle\int_0^T \cos^2 \omega_{\rm d} t \, {\rm d}t = \left[\dfrac{t}{2} + \dfrac{1}{4\omega_{\rm d}} \sin 2\omega_{\rm d} t \right]_0^T$. Since $T = 2\pi/\omega_{\rm d}$, the second term evaluates to zero at both limits, so that the integral is just $T/2$. Since we integrated over a whole period, we can see that **the average value of a squared sinusoid over one period is 1/2.** The second integral can also be looked up in a table (or you can integrate it by parts): $\displaystyle\int_0^T \cos \omega_{\rm d} t \, \sin \omega_{\rm d} t \, {\rm d}t = \left[\dfrac{1}{2\omega_{\rm d}} \sin^2 \omega_{\rm d} t \right]_0^T$. Since $T = 2\pi/\omega_{\rm d}$, this evaluates to zero at both limits. Putting this all together gives $P_{\rm drive} = \dfrac{F_0 A \omega_{\rm d}}{T} \dfrac{T}{2} \sin \delta = \dfrac{F_0 A \omega_{\rm d}}{2} \sin \delta$.

This relation is plotted in figure 4.3.3a. Inspection of the equation shows that the peak is at exactly $\omega_d/\omega_0 = 1$. As for the graph of A *versus* ω_d (figure 4.2.1), larger Q (less damping) leads to a higher and sharper peak.

The width of the peak is quite important for a variety of applications. One common way to define the width is the "full width at half maximum," or FWHM, which is the width at half of the peak height, as shown in figure 4.3.3b. Let us calculate the FWHM exactly. Referring to equation (4.3.3), we see that the values of ω_d corresponding to half the maximum height are determined by

$$\left(\frac{\omega_0}{\omega_\pm} - \frac{\omega_\pm}{\omega_0}\right)^2 = \frac{1}{Q^2}, \tag{4.3.4}$$

where $\omega_\pm$ is either ω_+ or ω_-. First, we consider ω_-. Since $\omega_0 > \omega_-$, equation (4.3.4) gives

$$\omega_0^2 - \omega_-^2 = \frac{\omega_0 \omega_-}{Q} \Leftrightarrow \omega_-^2 + \frac{\omega_0 \omega_-}{Q} - \omega_0^2 = 0 \Rightarrow \omega_- = \frac{-\frac{\omega_0}{Q} \pm \sqrt{\frac{\omega_0^2}{Q^2} + 4\omega_0^2}}{2}.$$

Since ω_- must be greater than 0, we choose the positive square root, giving

$$\omega_- = -\frac{\omega_0}{2Q} + \frac{1}{2}\sqrt{\frac{\omega_0^2}{Q^2} + 4\omega_0^2}.$$

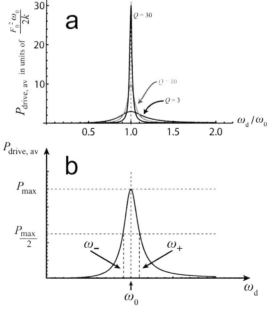

Figure 4.3.3 a: Power delivered by drive force to oscillator (averaged over a cycle) as a function of ω_d. This is called the power resonance curve. b: Definition of FWHM.

Next, we consider ω_+. Since $\omega_+ > \omega_0$, equation (4.3.4) gives

$$\omega_+^2 - \omega_0^2 = \frac{\omega_0 \omega_+}{Q} \Leftrightarrow \omega_+^2 - \frac{\omega_0 \omega_+}{Q} - \omega_0^2 = 0$$

$$\Rightarrow \omega_+ = \frac{\frac{\omega_0}{Q} \pm \sqrt{\frac{\omega_0^2}{Q^2} + 4\omega_0^2}}{2}.$$

Since ω_+ must be greater than 0, we choose the positive square root, giving

$$\omega_+ = \frac{\omega_0}{2Q} + \frac{1}{2}\sqrt{\frac{\omega_0^2}{Q^2} + 4\omega_0^4}.$$

The FWHM is given by $\omega_+ - \omega_-$, so that

$$\boxed{\text{FWHM} = \frac{\omega_0}{Q} = \gamma.} \tag{4.3.5}$$

Full Width at Half Maximum for the graph of $P_{\text{drive, av}}$ *versus* ω_{d}.

This important equation tells us that systems with more damping have a broader power resonance peak. This is a very universal behavior, which is observed even for systems with turbulence or frictional damping.

4.4 Linear differential equations, the superposition principle for driven systems, and the response to multiple drive forces

What happens if we apply a nonsinusoidal drive force, or if we apply two or three different sinusoids at different frequencies and amplitudes? We will show in chapter 8 that *any* function can be represented as a sum of sinusoids. Therefore, if we can figure out how a damped driven oscillator responds to multiple sinusoidal drives, it will be relatively easy, after understanding the contents of chapter 8, to determine the response to any driving force.

We will make use of an important property of the differential equation governing a damped driven oscillator:

$$m\ddot{x} = -kx - b\dot{x} + F_0 \cos\omega_{\text{d}}t \quad \Rightarrow \quad \ddot{x} + \gamma\dot{x} + \omega_0^2 x = \frac{F_0}{m}\cos\omega_{\text{d}}t.$$

This differential equation is linear, meaning that it has no terms proportional to x^2, or to $\dot{x}^2$, or to $x\dot{x}$, etc. That means, for example, that if we multiply x by two, then the entire left side of the above equation becomes twice as big. (This would not be true, for example, if the left side were x^2.) It also means, as we're about to show, that we can simply add the steady-state solutions for each of the individual driving forces to get the steady-state solution for the combined driving force. The term on the right side, which does not depend on x, and which comes from the drive (also called the excitation) is called the "source term."

We'll be able to see the pattern for the response to any number of sinusoidal driving forces by considering the case of just two:

$$F_{\text{drive}} = F_1 \cos \omega_1 t + F_2 \cos \omega_2 t.$$

To find the behavior of the system, $x(t)$, we begin to follow our three-step procedure, though this time it will be surprisingly easy so that we don't have to go through the whole process.

1. **Write Newton's second law for each object in the system:**

$$m\ddot{x} = -kx - b\dot{x} + F_1 \cos \omega_1 t + F_2 \cos \omega_2 t \quad \Rightarrow$$

$$\ddot{x} + \gamma \dot{x} + \omega_0^2 x = \frac{F_1}{m} \cos \omega_1 t + \frac{F_2}{m} \cos \omega_2 t. \tag{4.4.1}$$

If the drive force is simply $F_{\text{drive}} = F_1 \cos \omega_1 t$, then we know that the steady-state response is $x_1(t) = A_1 \cos(\omega_1 t - \delta_1)$, where A_1 and δ_1 are as given by equations (4.1.10) and (4.1.11), simply replacing F_0 by F_1 and ω_d by ω_1:

$$A_1 = \frac{F_1/m}{\sqrt{(\omega_0^2 - \omega_1^2)^2 + (\gamma \omega_1)^2}} \quad \text{and} \quad \tan \delta_1 = \frac{\gamma \omega_1}{\omega_0^2 - \omega_1^2}. \tag{4.4.2}$$

We know that, for this simple drive force, we have

$$\ddot{x}_1 + \gamma \dot{x}_1 + \omega_0^2 x_1 = \frac{F_1}{m} \cos \omega_1 t. \tag{4.4.3}$$

Similarly, if the drive force is simply $F_{\text{drive}} = F_2 \cos \omega_2 t$, then the steady-state response is $x_2(t) = A_2 \cos(\omega_2 t - \delta_2)$, where A_2 and δ_2 are defined analogously to equation (4.4.2). For this simple drive force, we have

$$\ddot{x}_2 + \gamma \dot{x}_2 + \omega_0^2 x_2 = \frac{F_2}{m} \cos \omega_2 t. \tag{4.4.4}$$

Adding equation (4.4.4) to (4.4.3) gives

$$\ddot{x}_1 + \gamma \dot{x}_1 + \omega_0^2 x_1 + \ddot{x}_2 + \gamma \dot{x}_2 + \omega_0^2 x_2 = \frac{F_1}{m} \cos \omega_1 t + \frac{F_2}{m} \cos \omega_2 t$$

$$\Leftrightarrow (\ddot{x}_1 + \ddot{x}_2) + \gamma (\dot{x}_1 + \dot{x}_2) + \omega_0^2 (x_1 + x_2) = \frac{F_1}{m} \cos \omega_1 t + \frac{F_2}{m} \cos \omega_2 t.$$

$$\Rightarrow \ddot{x}_{\text{tot}} + \gamma \dot{x}_{\text{tot}} + \omega_0^2 x_{\text{tot}} = \frac{F_1}{m} \cos \omega_1 t + \frac{F_2}{m} \cos \omega_2 t, \tag{4.4.5}$$

where $x_{\text{tot}}(t) \equiv x_1(t) + x_2(t) = A_1 \cos(\omega_1 t - \delta_1) + A_2 \cos(\omega_2 t - \delta_2)$. We see that equation (4.4.5) is the same as equation (4.4.1). Therefore,

> The steady-state response of a damped driven oscillator to two sinusoidal drive forces is simply the sum of the steady state responses to each force by itself.

It is easy to see that if we have seven sinusoidal drive forces, or seventy times seven sinusoidal drive forces, then the steady-state response is simply the sum of the individual responses.

Figure 4.4.1 Waves on a water surface are governed by a linear differential equation (if they are small in amplitude), and so different wave patterns simply add, as shown here. Image © Andrew Davidhazy, Rochester Institute of Technology.

This principle, that the total solution is simply the sum of the individual solutions, is called the "superposition principle" and only works for linear differential equations. Luckily, many important differential equations in physics are linear, so we can use this principle in many contexts. (See, e.g., figure 4.4.1.)

> **The superposition principle for driven systems: For a system governed by a linear differential equation, the total response to multiple excitations is the sum of the response for each excitation applied individually.**

4.5 Transients

Recall that the goal of section 4.1 was to find the steady-state behavior, that is, the behavior after any effects of initial conditions have decayed away because of the damping. Indeed, examination of the solution we found there shows that there are no constants that can be adjusted to take into account variations of the initial position or velocity:

$$(4.1.12): x = A \, \cos(\omega_d t - \delta), \quad \text{where}$$

$$(4.1.10): A = \frac{F_0/m}{\sqrt{(\omega_0^2 - \omega_d^2)^2 + (\gamma \, \omega_d)^2}} \quad \text{and}$$

$$(4.1.11): \tan \delta \left(\omega_d\right) = \frac{\gamma \, \omega_d}{\omega_0^2 - \omega_d^2}.$$

In almost every case, the steady-state behavior is the only thing of interest. But, there are a few circumstances in which the initial behavior, which *does* depend on x_0 and $\dot{x}_0$, is important. (See, e.g., figure 4.5.1.)

The solution $x = A \cos (\omega_d t - \delta)$ is *a* solution to our differential equation,

$$(4.1.3b): \ddot{x} + \gamma \, \dot{x} + \omega_0^2 x = \frac{F_0}{m} \cos \omega_d t,$$

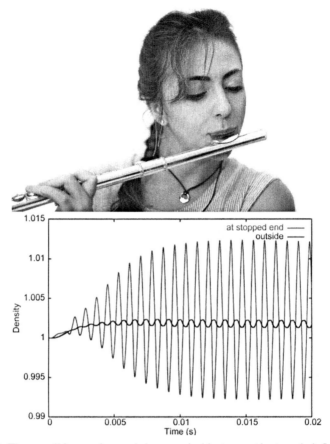

Figure 4.5.1 The overall impression made by a musical instrument is strongly influenced by the "attack," that is, the way a note starts up from silence. This is an example of transient behavior, and the transition to steady state. The graph shows a simulation of the attack for a flute; the larger amplitude waveform shows air density variations inside the flute, while the smaller waveform shows the variation outside near the mouthpiece. Top image © Galina Barskaya | Dreamstime.com. Bottom image © and courtesy of Dr. Helmut Kuehnelt.

but it cannot be the *general* solution; we know the general solution must include two adjustable constants, which can be changed to reflect the effects of x_0 and $\dot{x}_0$. So, we can anticipate that the general solution must be the sum of $x = A \cos(\omega_d t - \delta)$ with something that decays away over time and that depends on x_0 and $\dot{x}_0$. It is actually quite easy to find the general solution, using the work we have already done.

Consider the special case $F_0 = 0$; this is one of the cases that must be described by the general solution. The differential equation for this special case is

$$\ddot{x} + \gamma \dot{x} + \omega_0^2 x = 0, \tag{4.5.1}$$

which describes the damped oscillator without driving. We already found the general solution for this case in section 3.1:

$$(3.1.12): x = A_0 e^{-\frac{\gamma}{2} t} \cos\left(\omega_v t + \varphi\right),$$

where A_0 and φ are determined by the initial conditions x_0 and $\dot{x}_0$. Since this is the solution to equation (4.1.3b): $\ddot{x} + \gamma \dot{x} + \omega_0^2 x = \frac{F_0}{m} \cos \omega_\mathrm{d} t$ for the special case of $F_0 \to 0$, it is reasonable to guess that the general solution to equation (4.1.3b) is the sum of this and equation (4.1.12):

$$x_\mathrm{G} \stackrel{?}{=} \underbrace{A_0 e^{-\frac{\gamma}{2}t} \cos\left(\omega_\mathrm{v} t + \varphi\right)}_{\text{transient behavior}} + \underbrace{A \cos\left(\omega_\mathrm{d} t - \delta\right)}_{\text{steady-state behavior}} . \tag{4.5.2}$$

We can see that this works for the special case $F_0 = 0$ (since, according to equation (4.1.10), $A = 0$ for this case), and also that it works for the steady-state, since the first term decays away over time. It has two adjustable constants A_0 and φ, as we know the general solution must. Is it in fact a solution of our differential equation (4.1.3b) above?

To facilitate our checking, we introduce some nomenclature: equation (4.5.1) is called the "homogeneous" version of equation (4.1.3b), because it has zero on the right side. So, equation (3.1.12) is the general solution to the homogeneous version of the differential equation, and we'll call it x_H:

$$x_\mathrm{H} = A_0 e^{-\frac{\gamma}{2}t} \cos\left(\omega_\mathrm{v} t + \varphi\right). \tag{4.5.3}$$

The steady-state solution (4.1.12) is a "particular solution" to the full version (the inhomogeneous version) of the differential equation (4.1.3b), so we call it x_P:

$$x_\mathrm{P} = A \cos\left(\omega_\mathrm{d} t - \delta\right). \tag{4.5.4}$$

Now, we are ready to check whether our guess $x_\mathrm{G} \stackrel{?}{=} x_\mathrm{H} + x_\mathrm{P}$ is a solution to equation (4.1.3b):

$$\ddot{x}_\mathrm{G} + \gamma \dot{x}_\mathrm{G} + \omega_0^2 x_\mathrm{G} \stackrel{?}{=} \frac{F_0}{m} \cos \omega_\mathrm{d} t,$$

$$\Leftrightarrow \frac{d^2}{dt^2}\left(x_\mathrm{H} + x_\mathrm{P}\right) + \gamma \frac{d}{dt}\left(x_\mathrm{H} + x_\mathrm{P}\right) + \omega_0^2 \left(x_\mathrm{H} + x_\mathrm{P}\right) \stackrel{?}{=} \frac{F_0}{m} \cos \omega_\mathrm{d} t$$

$$\Leftrightarrow \left(\ddot{x}_\mathrm{H} + \gamma \dot{x}_\mathrm{H} + \omega_0^2 x_\mathrm{H}\right) + \left(\ddot{x}_\mathrm{P} + \gamma \dot{x}_\mathrm{P} + \omega_0^2 x_\mathrm{P}\right) \stackrel{?}{=} \frac{F_0}{m} \cos \omega_\mathrm{d} t.$$

We know that x_H is a solution of equation (4.5.1): $\ddot{x} + \gamma \dot{x} + \omega_0^2 x = 0$, so the first term in parentheses equals zero. We know that x_P is a solution of equation (4.1.3b), so the second term in parentheses equals $\frac{F_0}{m} \cos \omega_\mathrm{d} t$. Thus, the equation is indeed satisfied, and the general solution for the damped driven harmonic oscillator is

$$x_\mathrm{G} = \underbrace{A_0 e^{-\frac{\gamma}{2}t} \cos\left(\omega_\mathrm{v} t + \varphi\right)}_{\text{transient behavior}} + \underbrace{A \cos\left(\omega_\mathrm{d} t - \delta\right)}_{\text{steady-state behavior}}, \tag{4.5.5}$$

where A_0 and φ are determined by the initial conditions x_0 and $\dot{x}_0$. A typical example is shown in figure 4.5.2.

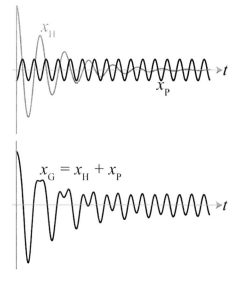

Figure 4.5.2 The full solution (bottom trace) for the damped driven harmonic oscillator is the sum of the general solution for the damped oscillator x_H (top gray trace) and the steady-state solution x_P (top black trace).

4.6   Electrical resonance

In section 3.2, we discussed the RLC electrical oscillator, shown in the top part of figure 4.6.1. We applied Kirchhoff's loop rule to obtain

$$V_L + V_R + V_C = 0 \Rightarrow$$

$$L\ddot{q} + R\dot{q} + \frac{q}{C} = 0.$$

To add driving, we open the loop and apply a drive voltage, as shown in the bottom part of figure 4.6.1. Now, instead of the voltage changes around the loop summing to zero, they must sum to the applied voltage, so that

$$L\ddot{q} + R\dot{q} + \frac{q}{C} = V_0 \cos \omega_d t. \tag{4.6.1}$$

This is isomorphic to the differential equation for the damped driven mass/spring system, (4.1.3a):

$$m\ddot{x} + b\dot{x} + kx = F_0 \cos \omega_d t.$$

So, the full isomorphism is as shown in table 4.6.1.

Concept test (answer below[6]): In terms of R, L, and C, what is Q for the circuit shown in the bottom part of figure 4.6.1?

---

6. $Q = \omega_0/\gamma$. For the mass/spring, $\omega_0 = \sqrt{k/m}$, which translates for the electrical oscillator into $\omega_0 = \sqrt{1/LC}$. For the mechanical oscillator, $\gamma = b/m$, which translates into $\gamma = R/L$. Combining gives $Q = \sqrt{\dfrac{1}{LC}} \dfrac{L}{R} = \dfrac{1}{R}\sqrt{\dfrac{L}{C}}$.

Table 4.6.1. Isomorphism between mechanical and electrical oscillators

Mass and Spring	Electrical Oscillator
Position relative to equilibrium x	Charge q on capacitor
Mass m	Inductance L
Spring constant k	Inverse of capacitance: $1/C$
Damping constant b	Resistance R
Drive force amplitude F_0	Drive voltage amplitude V_0

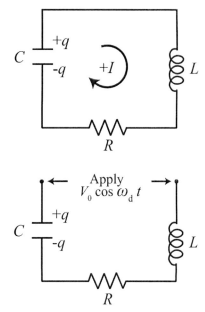

Figure 4.6.1 Top: RLC Oscillator. Bottom: Driven RLC oscillator.

The series RLC oscillator can be used to make a simple radio receiver, as shown in the top part of figure 4.6.2. The voltage from the antenna ΔV_{IN} is used to drive a series RLC circuit. The inductor is adjustable, so that the resonant angular frequency of the circuit, $\omega_0 = \sqrt{1/LC}$ can be tuned to match the angular frequency of the radio station. (For AM radio, this is in the range of 2π times 520–1,610 kHz.) Typically, there are many radio stations within range, but each broadcasts at a different frequency. By tuning the resonant frequency to match the broadcast frequency of one of the stations, that signal is amplified by the factor Q (which can be very high), while other stations that do not match the resonant frequency are amplified by a much smaller factor. The output could be taken across the capacitor (as shown), or across the resistor, or across the inductor, since at resonance q, $\dot{q}$, and $\ddot{q}$ all oscillate with large amplitude. Practical radio receivers have more complex input circuits than shown here, but the circuit still includes resonance in a circuit containing an inductor and a capacitor.

Example: Two adjacent radio stations. In the United States, the minimum frequency separation between AM radio stations is 20.4 kHz. (Stations in a particular broadcast area usually have a much greater separation than this.) In this example, station WINE broadcasts at 1,210 kHz, while WART broadcasts at 1,230.4 kHz. A circuit such as that

continued

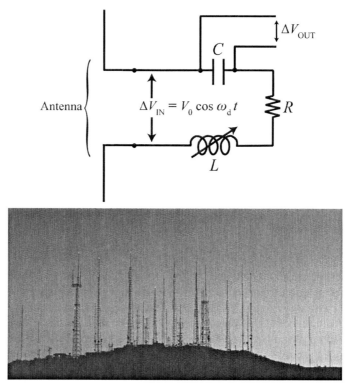

Figure 4.6.2 Top: The front end of a simple radio receiver. The arrow on the inductor indicates that it has a variable inductance, allowing the resonance frequency of the LRC circuit to be tuned to match the frequency of the desired radio station. Bottom: Radio broadcast towers are often grouped together to take advantage of a good site. As we'll see in chapter 9, the differential equation governing radio waves is linear, so the different signals simply add. A resonant RLC circuit inside a receiver, tuned to the broadcast frequency of one of the stations, is used as part of the circuitry that selects one of the stations out of the many signals that are received. Image © Tose | Dreamstime.com.

shown in figure 4.6.2 is tuned to a resonant angular frequency of $2\pi \cdot 1, 210$ kHz. What is the required Q if the power dissipated in the resistor due to the WART signal is to be half the power dissipated due to the WINE signal?

Solution: Because the circuit is tuned to the angular frequency of WINE, the center angular frequency ω_0 for the power resonance curve (see figure 4.3.3b) is that of WINE. In steady state, the power dissipated is equal to the power provided by the drive. We need the power dissipated by the WART signal to be half that dissipated by the WINE signal, meaning that the angular frequency of the WART signal should correspond to the half maximum point of the power resonance curve (marked ω_+ in figure 4.3.3b). We know from equation (4.3.5) that the full width at half maximum of the power resonance curve is given by FWHM $= \frac{\omega_0}{Q} \Leftrightarrow Q = \frac{\omega_0}{\text{FWHM}}$. We need half the FWHM to equal the difference in angular frequency between WART and WINE:

$$\frac{\text{FWHM}}{2} = 2\pi \times 20.4 \text{ kHz} \Leftrightarrow \text{FWHM} = 4\pi \times 20.4 \text{ kHz}.$$

Recall that the resonant angular frequency of the circuit is given as $\omega_0 = 2\pi \times 1, 210$ kHz. So, we have $Q = \dfrac{2\pi \times 1, 210 \text{ kHz}}{4\pi \times 20.4 \text{ kHz}} = 29.7.$

4.7   Other examples of resonance: MRI and other spectroscopies

Of course, every oscillator described in chapter 2 can exhibit resonance when driven. However, the idea of resonance, meaning a strong response of system at or near a particular excitation frequency, appears in many other situations as well. In this section, we briefly explore two of these.

Semiclassical description of Magnetic Resonance Imaging (MRI). MRI is based on the phenomenon of Nuclear Magnetic Resonance (NMR). The nuclei in the atoms of the body are comprised of protons and neutrons. We will treat these using a "semiclassical" description, meaning a hybrid of classical and quantum mechanical ideas. Although this description does not capture the full quantum mechanical truth, it does allow us to get a feel of what is going on, and it also allows fully quantitative predictions for the results of experiments. In this description, we visualize the protons as tiny spinning spheres. Because the protons carry charge, the spin creates a circulating electrical current. As you'll recall from a course in electricity and magnetism, a circulating current creates a magnetic field. Therefore, as shown in figure 4.7.1, each proton acts as a tiny bar magnet, albeit one with some unusual properties. Because the magnetic moment $\boldsymbol{\mu}$ of the proton is due to the spin, it is proportional to the spin angular momentum $\mathbf{J}$:

$$\boldsymbol{\mu} = \gamma \mathbf{J}, \tag{4.7.1}$$

where the proportionality constant γ is called the gyromagnetic ratio, and depends on the type of the nucleus. (In quantum mechanics, one often uses $\mathbf{J}$ to represent the angular momentum, rather than $\mathbf{L}$.)

Now, we apply an external magnetic field, $\mathbf{B}_{\text{applied}}$. (In an MRI instrument, this is usually produced by current flowing through a large coil that surrounds the patient.) From introductory electricity and magnetism, this field creates a torque on any magnetic

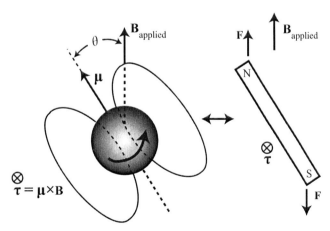

Figure 4.7.1 Semiclassical model for the magnetic moment of a proton. The proton is pictured as a spinning sphere, which creates a magnetic field similar to that of a bar magnet. Thus, the proton has a magnetic moment μ. An applied external magnetic field exerts a torque τ on this magnetic moment; the torque points into the page.

moment (including that of the nucleus):

$$\boldsymbol{\tau} = \boldsymbol{\mu} \times \mathbf{B}_{\text{applied}}. \tag{4.7.2}$$

The torque tends to align the magnetic moment along the direction of $\mathbf{B}_{\text{applied}}$.

However, quantum mechanics tells us that, surprisingly, μ doesn't actually fully line up with $\mathbf{B}_{\text{applied}}$; instead, there is a well-defined angle θ between these two vectors. One of the most important types of nuclei used for MRI is hydrogen; this is abundant in virtually all the molecules that make up the body. For hydrogen, $\theta = 54.7°$.

Because μ is not fully lined up with $\mathbf{B}_{\text{applied}}$, we have a body with a significant angular momentum (the nucleus) that experiences a torque. As you may recall from a demonstration in an earlier course involving a weighted bicycle wheel or a gyroscope, surprising things happen in this circumstance. Let's take the example of a child's top that is tilted, as shown in figure 4.7.2a. In the side view, we can calculate the torque due to gravity around the point at the bottom of the top. The weight can be taken as applied at the center of mass, and therefore creates a torque

$$\boldsymbol{\tau} = \mathbf{r} \times \mathbf{F}.$$

Using the right-hand rule for the cross-product, this torque points out of the page. Another way to find the direction of the torque: If the top weren't spinning, it would fall down, rotating counterclockwise about the point at the bottom. Curling the fingers of your right hand in the direction of rotation that the torque is "trying" to produce, your thumb points out of the page, in the direction of $\boldsymbol{\tau}$. However, because the top already has angular momentum, this torque does not cause it simply to fall over. Recall that

$$\boldsymbol{\tau} = d\mathbf{L}/dt \Leftrightarrow d\mathbf{L} = \boldsymbol{\tau} dt,$$

where $\mathbf{L}$ is the angular momentum of the top. Therefore, the change in the angular momentum is in the direction of $\boldsymbol{\tau}$, which is perpendicular to $\mathbf{L}$. So, the length of $\mathbf{L}$ doesn't change, but its direction does: in time dt, the the tip of the $\mathbf{L}$ vector moves an infinitesimal amount out of the page. Since $\mathbf{L}$ is along the axis of rotation of the top, this means that the axis of rotation moves as well. This doesn't fundamentally change the situation, so the direction of the torque $\boldsymbol{\tau}$ changes by an infinitesimal amount as well, and is still perpendicular to $\mathbf{L}$. Thus, the tip of $\mathbf{L}$ will trace out a circle with a "radius" (in units of angular momentum) equal to the x-component of $\mathbf{L}$, as shown in figure 4.7.2c:

$$\text{"radius" of circle} = L_x = L \sin \theta.$$

This motion of the axis of rotation is called "precession." If there were no friction to slow down the spinning of the top, it would continue to precess in this way without ever falling down.

The "speed" (in units of angular momentum per time) of the tip of $\mathbf{L}$ as it moves along this circle is $\tau = d L/dt$. Therefore, the time it takes for one full circle is

$$\text{time} = \frac{\text{"distance"}}{\text{"speed"}} = \frac{2\pi \text{ "radius"}}{\tau} = \frac{2\pi L \sin \theta}{\tau}.$$

The frequency of the precession is the inverse of this time:

$$f = \frac{\tau}{2\pi L \sin \theta}. \tag{4.7.3}$$

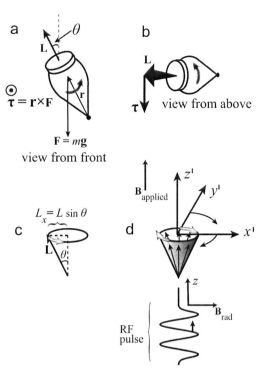

Figure 4.7.2 a and b: A child's top that is tilted experiences a torque which is perpendicular to the angular momentum **L**. c: The torque causes **L** to precess around a circle. d: We define a reference frame that rotates at the precession rate. In MRI, a pulse of RF radiation is applied, with the magnetic field along the x-axis (in the nonrotating frame).

Applying these ideas to the nucleus with the applied external magnetic field, the tip of μ precesses in the same way as does the axis of rotation of the top, moving in a circle around the direction of $\mathbf{B}_{\text{applied}}$. As the tip moves in this circle, the vector μ sweeps through a cone, as shown in figure 4.7.2d.[7] Combining equations (4.7.2) and (4.7.3), and using J to denote the magnitude of the spin angular momentum (rather than L), we find that the precession frequency is

$$f = \frac{\tau}{2\pi J \sin\theta} = \frac{\mu B_{\text{applied}} \sin\theta}{2\pi J \sin\theta} = \frac{\mu B_{\text{applied}}}{2\pi J}.$$

---

7. Recall that this is a semi-classical treatment. In a fully quantum mechanical treatment, the direction of μ cannot be fully determined, due to an uncertainty principle similar to those discussed in section 1.12. We discussed there how the momentum and the position of a particle cannot both be precisely defined simultaneously. Similarly, it turns out that the components of μ along the x-, y-, and z-axes cannot all be precisely defined simultaneously. In our example, the length of μ is precisely defined, and the component along the direction of $\mathbf{B}_{\text{applied}}$ is precisely defined, but the components in the other directions are completely undefined. We sometimes say that μ is "delocalized" around the cone. Careful consideration of these effects produce the same results as our semi-classical argument.

Substituting for μ from equation (4.7.1) gives

$$f = \frac{\gamma J B_{\text{applied}}}{2\pi J} \Rightarrow .$$

$$\boxed{f_{\text{Larmor}} = \frac{\gamma}{2\pi} B_{\text{applied}},}$$
(4.7.4)

where this precession frequency is named the "Larmor frequency" after Joseph Larmor, an Irish physicist who occupied the same professorship at Cambridge University as did Isaac Newton. Because there is no dissipation in the system, the magnetic moment μ of the nucleus precesses around the direction of $\mathbf{B}_{\text{applied}}$ forever, unless something else happens.

For MRI, the "something else" happens when we apply a pulse of radio frequency (RF) electromagnetic radiation, with a frequency equal to f_{Larmor}. Let us define the z-axis to lie along $\mathbf{B}_{\text{applied}}$. The MRI machine is arranged so that this RF radiation travels in the z-direction. As you may recall from a course in electricity and magnetism, electromagnetic radiation consists of waves of perpendicular electric and magnetic fields, both of which are perpendicular to the direction of travel. (We shall explore these in more detail in chapter 9.) Therefore, the magnetic field of the RF radiation, $\mathbf{B}_{\text{rad}}$ could oscillate along any direction in the x–y plane. For convenience, we take it to oscillate along the x-axis. This oscillation at frequency f_{Larmor} and angular frequency $\omega_{\text{Larmor}} = 2\pi f_{\text{Larmor}}$ can be considered to be the sum of two counter-rotating magnetic field vectors, each of which has angular velocity ω_{Larmor}, as shown in figure 4.7.3a. We call the clockwise-rotating vector $\mathbf{B}_{\text{rad, cl}}$; it rotates at the same rate and in the same direction as does the precessing magnetic moment μ of the nucleus.

We now consider what things look like in a rotating reference frame x', y', z' that rotates clockwise along with μ, with z' parallel to z. Since $\mathbf{B}_{\text{rad, cl}}$ rotates at the same rate as μ, it points in a constant direction in this frame. Let us define this direction to be x', as shown in figure 4.7.3b.

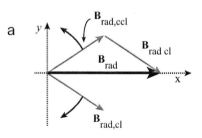

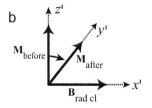

Figure 4.7.3 a: An oscillating magnetic field along the x-axis equals the sum of two counter-rotating magnetic fields. If the angular frequency of the RF pulse equals the Larmor frequency, then $\mathbf{B}_{\text{rad, cl}}$ rotates at the precession rate, and so is stationary in the rotating reference frame. b: Taking $\mathbf{B}_{\text{rad, cl}}$ to lie along x', it causes a precession of the magnetization around the x' axis. If the pulse length is chosen suitably, the magnetization can be rotated by 90°.

At this point, we take a step back from thinking about the individual nuclei, and instead consider the net magnetization **M** of a small region of the patient inside the MRI machine. This magnetization is due to the average of the magnetic moments of the nuclei, and so points along z. We have seen that a magnetic moment precesses about the direction of an applied magnetic field. Because the magnetization is due to the average of many magnetic moments, it also precesses about any applied field. Before the application of $\mathbf{B}_{rad}$, **M** is parallel to $\mathbf{B}_{applied}$, and so is stationary. When $\mathbf{B}_{rad}$ is applied, $\mathbf{B}_{rad,\,cl}$ points in a constant direction in the rotating reference frame, and **M** precesses about it. (In the rotating reference frame, $\mathbf{B}_{rad,\,ccl}$ rotates at angular frequency $2\omega_{Larmor}$, and so has no average effect.)

We can control how far **M** precesses by the duration of the RF radiation pulse. Because **M** results from the average of all the individual magnetic moments of the atoms, the precession of **M** around the magnetic field due to the RF pulse, as seen in the rotating frame, is governed by the same physics leading to equation (4.7.4), so that the angular rate of rotation is $\omega = \gamma B_{rad,\,cl}$. In the simplest version of MRI, the pulse length is chosen so that **M** precesses by 90° in the rotating reference frame, so that after this precession it lies along the y' axis, as shown in figure 4.7.3b. A pulse of this length is called a 90° or $\pi/2$ pulse.

In the stationary reference frame, the y' axis rotates clockwise at angular frequency ω_{Larmor} around the z-axis. Therefore, after a 90° pulse has been applied so that **M** lies along the y' axis, it also rotates clockwise at ω_{Larmor}. This rotating magnetization broadcasts electromagnetic radiation, which can be detected by the MRI machine. This whole mechanism only works when the frequency of the radiation in the RF pulse matches f_{Larmor}. Any other frequency of radiation would have a magnetic field that is not stationary in the rotating frame, and so produces no average effect.

Several different techniques are used to get contrast between different body tissues in MRI. The simplest is associated with the effect of nearby electrons on the $\mathbf{B}_{applied}$ that is felt by a particular nucleus. The externally applied magnetic field affects the motion of these electrons, and since moving electrons create a magnetic field, the total magnetic field experienced by the nucleus depends on the local density of electrons. Nuclei in different types of molecules, and in different parts of the same molecule, are surrounded by different densities of electrons. From equation (4.7.4), f_{Larmor} depends on the total $\mathbf{B}_{applied}$. Therefore, these nuclei respond to different frequencies for the RF radiation pulse, giving contrast.

The rotation of **M** away from $\mathbf{B}_{applied}$ increases the potential energy of the system; this energy comes from the RF radiation pulse. We have seen that this transfer of energy from the radiation to the nucleus can only occur if the radiation has the correct frequency, f_{Larmor}.

There is a rather different way of understanding the transfer of energy from the radiation to the nucleus, which generalizes to many other situations, but is less connected to classical physics. Even in a rather strong applied magnetic field, the effect of the field on the nuclei is small compared to that of random thermal fluctuations. This fact does not affect the above arguments, which are based on the average magnetization. Let us consider again the particular case of hydrogen nuclei. If one measures μ_z (the component of μ in the direction of $\mathbf{B}_{applied}$), then surprisingly one finds that the result is always one of two possibilities. Slightly more than half the time, one gets the result for μ_z that one would expect from figure 4.7.1, with $\theta = 54.7°$; this is called "spin up," because the direction of the nuclear spin axis is as close to the direction of $\mathbf{B}_{\text{applied}}$ as it ever gets. Slightly less than half the time one gets the result, one would expect for $\theta = 180° - 54.7° = 125.3°$; this is called "spin down." This surprising finding that only these two results are measured and nothing in between, is a purely quantum mechanical effect with no classical analog. You might ask, "Weren't we just talking about the magnetization precessing by 90°? Isn't that an in-between result?" However, recall that that part of our discussion was centered on the magnetization of a small region inside the patient, which is the combined result of the magnetic moments of many individual spins. Each of the individual spins is either "spin up" (meaning that it points somewhere along the upward-pointing cone, with an angle of 54.7° to $\mathbf{B}_{\text{applied}}$) or it is "spin down." The combined effect of all these spins is the magnetization of the small region, which can change direction in an essentially continuous way.

The spin angular momentum of the hydrogen nucleus, which is simply a proton, is a fixed quantity. It is a property of the particle, much as the charge is. The magnitude of this angular momentum is

$$J = \frac{\sqrt{3}}{2}\hbar,$$

where $\hbar = \frac{h}{2\pi} = 1.05457 \times 10^{-34}$ Js and $h = 6.6260693 \times 10^{-34}$ Js are both called "Planck's constant." To avoid confusion to the extent possible, $\hbar$ is usually just called "h-bar." So, for the spin-up result, the z-component of the spin angular momentum is

$$j_{z\uparrow} = \frac{\sqrt{3}}{2}\hbar \cos 54.7° = \frac{\hbar}{2},$$

where the up arrow indicates "spin up." Using equation (4.7.1), this means that

$$\mu_{z\uparrow} = \gamma\frac{\hbar}{2}.$$

Similarly, for the "spin down" possibility, we have

$$\mu_{z\downarrow} = -\gamma\frac{\hbar}{2}.$$

From introductory electricity and magnetism, the potential energy for a magnetic dipole in an applied field is

$$U = -\boldsymbol{\mu} \cdot \mathbf{B}_{\text{applied}}.$$

In our case, $\mathbf{B}_{\text{applied}}$ is in the z-direction, so

$$U = -\mu_z B_{\text{applied}}.$$

Therefore, the potential energies for spin up and spin down are

$$U_\uparrow = -\gamma\frac{\hbar}{2}B_{\text{applied}} \quad \text{and} \quad U_\downarrow = \gamma\frac{\hbar}{2}B_{\text{applied}}.$$

The difference between these is

$$\Delta U = U_\downarrow - U_\uparrow = \gamma\hbar B_{\text{applied}}.$$

You may recall from section 1.12 that there is a connection between oscillation frequency and energy for an electron:

$$(1.12.5): E = hf.$$

It turns out that the same relation applies to light. Experimentally, we find that, whenever energy is absorbed from light, it is always absorbed in discrete packets called "photons," each of which has energy given by equation (1.12.5), with f equal to the oscillation frequency of the light. To "promote" the hydrogen nucleus from the lower energy spin up state to the higher energy spin down state, and to conserve energy, the system must absorb a photon from the RF radiation pulse that has energy

$$E = hf = \Delta U = \gamma \hbar B_{\text{applied}} \Rightarrow$$

$$f = \frac{\gamma \hbar B_{\text{applied}}}{h} = \frac{\gamma B_{\text{applied}}}{2\pi}.$$

At this point, there may be chills running up and down your spine, because this frequency, which we have arrived at using purely quantum mechanical methods, is exactly the same as the Larmor frequency (4.7.4), which we derived using semi-classical methods![8]

Other types of spectroscopy. By either line of reasoning, only radiation of frequency very close to f_{Larmor} can transfer energy to the system of nuclear magnetic moments. The quantum version of the argument generalizes perfectly to a large variety of circumstances. For example, in ultraviolet/visible spectroscopy, light is passed through a solution containing molecules of interest. Within the molecules, the electrons occupy quantum mechanical states with well-defined energies. It is possible to promote an electron from a low-energy state to a higher-energy state by the absorption of a photon with a frequency f such that hf equals the energy difference between the two states. This is essentially the same process that we described earlier, in which the absorption of a photon from the RF radiation pulse promotes the hydrogen nucleus from the spin up state to the spin down state.

In a more sophisticated version of the argument, we would see that the frequency match need not be exact, although the energy transfer is greatest when there is an exact match. This is closely analogous to the resonant response of a damped, driven mass/spring system; the system absorbs the most energy from the drive force when the frequency of the drive matches the resonant frequency of the system, but energy can also be absorbed from drives with frequencies that do not match the resonance exactly. Recall that the width of the power resonance curve for a mass/spring system increases when the damping is increased. Also recall that the damping represents a coupling of the energy in the oscillator to other parts of the universe, such as the air surrounding the mass. In just the same way, the range of radiation frequencies that can be absorbed by a molecule increases when the molecule is coupled to other things

---

8. For more detailed treatments, see *Introduction to Physics in Modern Medicine, 2nd Ed.*, by Suzanne Amador Kane, CRC Press, Boca Raton FL, 2009, and *Physical Methods for Chemists, 2nd Ed.*, by Russell S. Drago, Surfside Scientific Publishers, Gainesville, FL, 1992.

in the universe. Thus, the frequency range is very small when the molecule is in a dilute gas phase, becomes broader when the gas molecules collide more frequently ("pressure broadening"), and becomes much broader when it sticks via multiple strong bonds to the surface of a solid. (Note that there are additional mechanisms that broaden the frequency range in actual spectroscopic experiments, such as local variations in the environments felt by different copies of the same molecule.)

4.8   Nonlinear oscillators and chaos

In all real systems, Hooke's Law $F = -kx$ is only valid for a small enough displacement away from equilibrium. Beyond this limit, the restoring force is no longer proportional to displacement. This nonlinearity leads to a host of fascinating effects, many of which are of great practical importance. Unfortunately, even the simplest modification to $F = -kx$ leads to a differential equation with no analytic solutions, meaning that no combination of exponentials, trigonometric functions, polynomials, etc., gives an exact solution. Therefore, scientists and engineers approximate actual systems with a linear restoring force whenever possible. Nonlinear systems *can* be studied quantitatively, using a combination of sophisticated approximation techniques and numerical (i.e., computer-based) methods, but these techniques are beyond the scope of this text.[9] In this section, we use rough and mostly qualitative arguments to explore a few of the most important phenomena.

Harmonic generation[10]

The potential energy for Hooke's Law is $U = \frac{1}{2}kx^2$. It will make the math easier when we consider a nonlinear oscillator to maintain the symmetry of the potential energy around $x = 0$. (Later, we will consider nonsymmetrical potentials qualitatively.) The simplest modification that preserves the symmetry is the addition of a term proportional to x^4:

$$U = \tfrac{1}{2}kx^2 \left(1 + \alpha x^2\right), \tag{4.8.1}$$

where α is a constant. We assume that the nonlinearity is small, meaning that

$$\alpha A^2 \ll 1,$$

where A is the maximum magnitude of x. This potential energy is shown in figure 4.8.1a, and compared to the harmonic potential energy $U = \frac{1}{2}kx^2$. (You can see that, for small x the harmonic potential is a good approximation.) Instead of only thinking about a mass on a non-Hookian spring, we can extend our thinking and consider any particle

---

9. It is not too difficult to use a computer for numerical integration; you might inquire with your instructor about doing an independent project in this area.

10. This discussion is adapted from that in *Vibrations and Waves in Physics, 3rd Ed.*, by Ian G. Main, Cambridge University Press, Cambridge, 1993, pp. 93–97.

which experiences this potential. For example, the particle might be an atom which experiences a potential due to chemical bonds with its neighbors.

If the particle has total energy E, then it can move between points 1 and 2 marked in figure 4.8.1a. When released from point 1, the particle oscillates between the two points. The condition $\alpha A^2 \ll 1$ is equivalent to $\left(\frac{1}{2}kA^2\right) \alpha A^2 \ll \left(\frac{1}{2}kA^2\right)$. You can see from figure 4.8.1a that this is not obeyed for the energy E chosen for our example, although we do have $\left(\frac{1}{2}kA^2\right) \alpha A^2 < \left(\frac{1}{2}kA^2\right)$. (For our example, $\alpha A^2 = 0.6$.) We use this example to highlight the changes that occur when the potential energy is not harmonic. Because our example does not obey $\alpha A^2 \ll 1$, the results obtained below are only qualitatively correct for our example. However, they are quantitatively correct for oscillations with lower E, corresponding to lower oscillation amplitude. (For an energy corresponding to half the amplitude of our example, $\alpha A^2 = 0.15$, which is small enough compared to 1 that the results below would be fairly good approximations.)

Because the potential energy is not harmonic, we expect that $x(t)$ will not be a simple cosine function, that is, we will not simply have $x = A \cos \omega t$ as we did for the simple harmonic oscillator. In fact, figure 4.8.1b shows the actual behavior that we will derive below, for the example energy shown in figure 4.8.1a. Also shown in figure 4.8.1b is a cosine (dashed line) with the same amplitude and angular frequency as the actual $x(t)$. You can see that the two curves are almost identical, but not quite. Figure 4.8.1d shows the slopes (i.e., $\dot{x}$) for these two curves, which shows the difference a little more clearly.

> **Concept test (answer below[11]):** In the regions near $x = 0$, the shape of the $\dot{x}$ curve corresponding to $x(t)$ is a little flatter than the $\dot{x}$ curve corresponding to the cosine. Why is this to be expected, based on the shape of $U(x)$?

However, even though $x(t)$ is not sinusoidal, it still must be a periodic function (if we ignore damping). We will show in chapters 7 and 8 that any periodic function can be expressed as a weighted sum of the sinusoids with the same periodicity; this is the process of Fourier synthesis. In our case, the period for oscillation from 1 to 2 and back to 1 is T, as shown in figure 4.8.1b. We define $\omega \equiv 2\pi/T$. Of course, the function $\cos \omega t$ has periodicity T, meaning that it has the same shape from $t = 0$ to $t = T$ as it does from $t = T$ to $t = 2T$. However, the function $\cos 2\omega t$ also has periodicity T; it completes two full cycles during the time T, and in the interval from $t = T$ to $t = 2T$ it again completes two full cycles. In fact any function $\cos n\omega t$, where n is an integer, has periodicity T, as do the functions $\sin n\omega t$. Even the integer $n = 0$ works; this just gives a constant for either the sin or the cos. Therefore, we must be able to write

$$x(t) = \text{const.} + a_1 \cos \omega t + b_1 \sin \omega t + a_2 \cos 2\omega t + b_2 \sin 2\omega t$$
$$+ a_3 \cos 3\omega t + b_3 \sin 3\omega t + \cdots \tag{4.8.2}$$

where the a's and b's are constants.

---

11. Because the potential energy curve $U(x)$ is flatter near $x = 0$ than a harmonic potential energy $\frac{1}{2}kx^2$, the velocity for the $U(x)$ case does not change as much near $x = 0$.

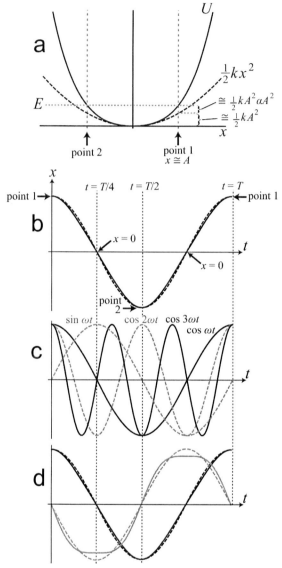

Figure 4.8.1 a: A nonharmonic potential energy $U(x)$ is compared with the harmonic potential energy $\frac{1}{2}kx^2$. A particle is released at rest from point 1, so that it has total energy E. Thereafter, it oscillates between points 1 and 2. At point 1, x is only approximately equal to A, because of the added $\cos 3\omega t$ term in equation (4.8.7). b: The full solution $x(t)$ from equation (4.8.7) is plotted as a solid line, and compared with a cosine of the same amplitude and angular frequency (dashed line). Although they are almost the same, $x(t)$ is slightly below the cosine from $t = 0$ to $t = T/4$, and slightly above the cosine from $t = T/4$ to $t = T/2$. c:Comparison of three cosines and one sine. Sines have odd symmetry about $t = T/2$, and so cannot contribute to $x(t)$. Cosines of even multiples of ωt, such as $\cos 2\omega t$ have even symmetry about $t = T/4$, and so cannot contribute to $x(t)$. Only cosines of odd multiples of ωt, such as the two shown in black, have the correct symmetries. d: The black curves are the same as in part b – $x(t)$ is shown as a solid black line, and a cosine with the same amplitude and angular frequency is shown as a dashed black line. The grey curves show the time derivatives, that is, $\dot{x}$ for each of the black curves, so that the solid gray curve is the time derivative of $x(t)$ and the dashed gray curve is the time derivative of the cosine.

However, because our potential energy function is symmetrical, we know that the shape of $x(t)$ from point 1 to 2 must be the mirror image of the shape from point 2 to 1, that is, $x(t)$ must have even symmetry about the time $T/2$, as shown in figure 4.8.1b. The sines have odd symmetry about $T/2$ as shown in figure 4.8.1c, therefore all the b's must equal zero. Furthermore, again because of the symmetry of the potential energy, the shape of $x(t)$ as the particle moves from point 1 (at $t = 0$) to $x = 0$ (at $t = T/4$) should be the time reversed version of the shape as the particle moves from $x = 0$ (at $t = T/4$) to point 2 (at $t = T/2$), that is, $x(t)$ must have odd symmetry about the time $T/4$, as shown in figure 4.8.1b. The odd cosines (cos ωt, cos $3\omega t$, etc.) have this property, but the even cosines don't, as shown in figure 4.8.1c. Therefore, the a's with even subscripts must all be zero. Finally, because of the symmetry of the potential energy function, the average position of the particle must be zero, so that the first term in equation (4.8.2) ("const.") must be zero. Therefore, we have

$$x(t) = a_1 \cos \omega t + a_3 \cos 3\omega t + a_5 \cos 5\omega t + \cdots .$$

Because we have assumed that the nonlinearity is small, it is reasonable to expect that the departure from harmonic behavior represented by the cos $3\omega t$ and higher terms is also small. Therefore, we can write

$$x(t) \cong A \left(\cos \omega t + \varepsilon_3 \cos 3\omega t + \varepsilon_5 \cos 5\omega t + \cdots \right), \qquad (4.8.3)$$

where $A = a_1$, $\varepsilon_3 = a_3/A$, etc., and ε_3, ε_5, etc., are much less than 1.

The restoring force associated with the potential energy (4.8.1) is

$$F_r = -\frac{dU}{dx} = -\frac{d}{dx} \left(\tfrac{1}{2}kx^2 + \tfrac{1}{2}k\alpha x^4 \right) = -kx - 2k\alpha x^3 .$$

In the absence of damping, this is the only force that acts on the mass. Therefore, we have

$$F_r = m\ddot{x} \Rightarrow \ddot{x} - \frac{F_r}{m} = 0 \Rightarrow$$

$$\ddot{x} + \omega_0^2 x + 2\omega_0^2 \alpha x^3 = 0, \qquad (4.8.4)$$

where $\omega_0 \equiv \sqrt{k/m}$. To determine the coefficients in equation (4.8.3), we need to plug it into equation (4.8.4). To prepare for this, we first evaluate the messiest term:

$$2\omega_0^2 \alpha x^3 = 2\omega_0^2 \alpha A^2 A \left(\cos \omega t + \varepsilon_3 \cos 3\omega t + \varepsilon_5 \cos 5\omega t + \cdots \right)^3 .$$

Unlike the other terms in equation (4.8.4), this has the overall multiplicative factor of αA^2, which is much smaller than 1. Therefore, since the ε's are small, we ignore all terms proportional to ε_3 or ε_5, etc. (Such terms would be proportional, for example, to $\varepsilon_3 \alpha A^2$, and so would be utterly negligible compared to the terms in equation (4.8.4) that are not proportional to an ε or to αA^2.) This leaves us with

$$2\omega_0^2 \alpha x^3 \cong 2\omega_0^2 \alpha A^2 A \cos^3 \omega t.$$

You can show in problem 4.21 that $\cos^3 \theta = \tfrac{1}{4} \cos 3\theta + \tfrac{3}{4} \cos \theta$. Applying this to the above gives

$$2\omega_0^2 \alpha x^3 \cong \frac{\omega_0^2 \alpha A^2}{2} A(\cos 3\omega t + 3 \cos \omega t).$$

Substituting this and equation (4.8.3) into (4.8.4) and cancelling the common factor of A gives

$$
\begin{aligned}
&- \omega^2 \left(\cos \omega t + 9\varepsilon_3 \cos 3\omega t + 25\varepsilon_5 \cos 5\omega t + \cdots \right) \\
&+ \omega_0^2 \left(\cos \omega t + \varepsilon_3 \cos 3\omega t + \varepsilon_5 \cos 5\omega t + \cdots \right) \\
&+ \frac{\omega_0^2 \alpha A^2}{2} \left(\cos 3\omega t + 3 \cos \omega t \right) \cong 0.
\end{aligned}
\tag{4.8.5}
$$

For this to be valid at all times, the coefficients of $\cos \omega t$ must sum to zero, as must the coefficients of $\cos 3\omega t$ and the coefficients of $\cos 5\omega t$, etc. Setting the sum of the coefficients of $\cos \omega t$ to zero we get

$$
- \omega^2 + \omega_0^2 + \frac{3\omega_0^2 \alpha A^2}{2} \cong 0 \Rightarrow
$$

$$
\omega \cong \omega_0 \sqrt{1 + \frac{3\alpha A^2}{2}}.
\tag{4.8.6}
$$

Thus, depending on the sign of α, the angular frequency will be shifted up or down relative to ω_0, by an amount that depends on the amplitude A. Thus, one of the hallmarks of harmonic oscillation, that the frequency is independent of amplitude, is removed when we consider a non-linear restoring force. In our example, $\omega = 1.38 \, \omega_0$, a quite substantial shift.

Returning to equation (4.8.5), setting the sum of the coefficients of $\cos 3\omega t$ to zero gives

$$
-9\varepsilon_3 \omega^2 + \varepsilon_3 \omega_0^2 + \frac{\omega_0^2 \alpha A^2}{2} \cong 0.
$$

Plugging in equation (4.8.6) yields

$$
-9\varepsilon_3 \omega_0^2 \left(1 + \frac{3\alpha A^2}{2} \right) + \varepsilon_3 \omega_0^2 + \frac{\omega_0^2 \alpha A^2}{2} \cong 0 \Rightarrow -8\varepsilon_3 \omega_0^2 - \frac{27}{2} \varepsilon_3 \omega_0^2 \alpha A^2 + \frac{1}{2} \omega_0^2 \alpha A^2 \cong 0.
$$

The middle term is proportional to both ε_3 and αA^2, and so is negligible compared to the others. This leaves us with

$$
\varepsilon_3 \cong \frac{1}{16} \alpha A^2.
$$

Thus, as we had anticipated, the departure from harmonic behavior in $x(t)$ is small, and proportional to the departure from linearity in the restoring force.

Finally, in equation (4.8.5) setting the sum of coefficients of $\cos 5\omega t$ to zero gives

$$
-25\varepsilon_5 \omega^2 + \omega_0^2 \varepsilon_5 \cong 0.
$$

Plugging in equation (4.8.6) yields

$$
-25\varepsilon_5 \omega_0^2 \left(1 + \frac{3\alpha A^2}{2} \right) + \omega_0^2 \varepsilon_5 \cong 0 \Rightarrow -24\varepsilon_5 \omega_0^2 \left(1 + \frac{3\alpha A^2}{2} \right) \cong 0.
$$

Since $\alpha A^2 \ll 1$, this gives

$$
-24\varepsilon_5 \omega_0^2 \approx 0 \Rightarrow \varepsilon_5 \approx 0.
$$

Thus, at the level of approximation we are using, there is a negligible response at angular frequency $5\omega t$. (Similar arguments would hold for the higher harmonics, such as $7\omega t$.) Plugging this result and $\varepsilon_3 \cong \frac{1}{16}\alpha A^2$ into equation (4.8.3) gives

$$x \cong A \left(\cos \omega t + \frac{\alpha A^2}{16} \cos 3\omega t \right). \qquad (4.8.7)$$

This relation is plotted for our example as the solid line in figure 4.8.1b. (If the departure from linearity is large enough, the higher harmonics such as $5\omega t$ become important.)

If we had considered an asymmetric potential energy, then we would not have been able to discard the even harmonics in equation (4.8.2), that is, terms such as $\cos 2\omega t$. In analogy with equation (4.8.7), we would find that the amplitude of the $\cos 2\omega t$ term is proportional to the departure from nonlinearity in the restoring force. Thus, we reach the important conclusion:

> If the restoring force is nonlinear but with symmetric magnitude about the equilibrium point, then we observe oscillations with angular frequency $\omega \approx \omega_0$ superposed with oscillations with angular frequency 3ω. If the force is nonlinear and asymmetric, we also observe oscillations with angular frequency 2ω. Thus, harmonics of the fundamental oscillation ω are generated. For highly nonlinear restoring forces, higher harmonics can also be generated.

Harmonic generation is important in applications of lasers. The available laser wavelengths are limited to certain discrete values depending on the material from which the laser is made. However, if a laser of wavelength λ, corresponding to angular frequency ω_d, is focused at high intensity onto a suitable crystal, there is a response of the electrons in the crystal at $2\omega_d$, resulting in the production of light at angular frequency $2\omega_d$, corresponding to a wavelength $\lambda/2$, which might be exactly what is needed for a particular application.

Sub-harmonic resonances

Now, we consider a damped, driven nonlinear oscillator. Above, we showed that, in a circumstance where we would get a response at ω_0 for a linear system, we did get a response at an angular frequency near ω_0, but also a response near $3\omega_0$ and (for an asymmetric restoring force) near $2\omega_0$. For a damped driven linear oscillator, we get a steady-state response at ω_d. Therefore, given the above results, it may seem reasonable that in the driven case we get a steady-state response at ω_d, but also responses at $3\omega_d$ and (for an asymmetric restoring force) at $2\omega_d$. In other words, the response is

$$x(t) = A \cos \left(\omega_d t - \delta \right) + A_2 \cos \left(2\omega_d t - \delta_2 \right) + A_3 \cos \left(3\omega_d t - \delta_3 \right).$$

For a highly nonlinear system, we can also get responses at higher harmonics. The math needed to show this is more complicated than for the undriven case.

For the nondriven case, we saw that, for a small nonlinearity, the oscillation with angular frequency near $3\omega_0$ is much smaller than the oscillation with angular frequency near ω_0. Similarly, for the damped driven case, one can show that the response at $2\omega_d$ and $3\omega_d$ is ordinarily much smaller than the response at ω_d. However, if $\omega_d = \omega_0/2$, then the response at $2\omega_d = \omega_0$ is amplified by the resonance of the system. If Q is large, the response at $2\omega_d = \omega_0$ can thus be significantly *larger* than the response at ω_d. Similar comments apply for $\omega_d = \omega_0/3$. Thus, we can excite the resonance of a system with a drive force frequency *below* the resonant frequency.

Sub-harmonic resonances are often annoying. For example, parts of speakers in an audio system can be modeled as mass/spring oscillators. (Since all the parts are in equilibrium positions, we know that there are spring-like forces holding them there.) If the volume is turned up high, the sound drives the mechanical components of the speaker at high amplitude, so that the nonlinear region of the restoring force becomes important. When the frequency of the sound is half or one third of the resonant frequency of a part, it can be excited through the sub-harmonic resonance effect, resulting in an unpleasant buzz.

Mixing

A resistor has a linear current–voltage relationship:

$$V = IR \Rightarrow I = GV,$$

where $G \equiv 1/R$ is the conductance. If we apply the voltage

$$V = V_1 \cos\omega_1 t + V_2 \cos\omega_2 t \tag{4.8.8}$$

to a resistor, nothing very exciting happens. The resulting current is simply

$$I = V/R = GV_1 \cos\omega_1 t + GV_2 \cos\omega_2 t.$$

However, now let's consider a circuit element with a nonlinear current–voltage relationship, such as

$$I = GV(1 + \alpha V) = GV + \alpha G V^2,$$

where α is a constant. (Note: the current as a function of voltage for any device can be written as a Taylor series, so for anything other than a resistor there will be a term proportional to V^2 and/or V^3. Thus, we need not be talking about a very exotic component; something like a diode would exhibit the effects we discuss here). Applying the voltage (4.8.8) to this nonlinear device gives a current

$$I = GV_1 \cos\omega_1 t + GV_2 \cos\omega_2 t + \alpha G\left(V_1 \cos\omega_1 t + V_2 \cos\omega_2 t\right)^2. \tag{4.8.9}$$

It is the third term which is of most interest:

$$\alpha G \left(V_1 \cos\omega_1 t + V_2 \cos\omega_2 t\right)^2$$
$$= \alpha G \left(V_1^2 \cos^2\omega_1 t + 2V_1 V_2 \cos\omega_1 t \cos\omega_2 t + V_2^2 \cos\omega_2 t\right). \tag{4.8.10}$$

Now, we use two trigonometric identities:

$$\cos^2\theta = \tfrac{1}{2}(1 + \cos 2\theta) \quad \text{and} \quad \cos A \cos B = \tfrac{1}{2}\left[\cos(A+B) + \cos(A-B)\right].$$

Applying these to equation (4.8.10) gives

$$\alpha G \left(V_1 \cos \omega_1 t + V_2 \cos \omega_2 t\right)^2 = \frac{\alpha G}{2} \left[V_1^2 + V_2^2 + V_1^2 \cos 2\omega_1 t + V_2^2 \cos 2\omega_2 t\right.$$

$$\left. +2V_1 V_2 \cos \left(\omega_1 + \omega_2\right) t + 2V_1 V_2 \cos \left(\omega_1 - \omega_2\right) t\right].$$

Thus, the current has components that oscillate at ω_1 and ω_2 (as we can see from equation (4.8.9), but also components that oscillate at $2\omega_1, 2\omega_2, \omega_1 + \omega_2$, and $\omega_1 - \omega_2$.

It is these last two that are perhaps the most interesting. We could call them the sum angular frequency and the difference angular frequency. The difference angular frequency plays the key role in the technique of "heterodyning," which is used in virtually every radio receiver. In AM radio, the information to be transmitted is used as an envelope function for a "carrier wave." The carrier wave must have a much higher frequency (around 1 MHz) than the audio information (about 10 kHz). In FM radio, the information to be transmitted is instead used to control the frequency of the carrier wave; a variation in the audio wave causes a small change in the carrier frequency. In either case, inside the radio receiver, the signal from the antenna is mixed with the signal from a "local oscillator," which is simply a source of AC voltage at a frequency near that of the carrier. The mixing occurs by applying both signals to a nonlinear element, as described earlier. Because the frequency of the local oscillator is close to that of the carrier wave, the difference frequency generated by the mixing is a much lower frequency, and so this signal can more easily be processed by subsequent circuits in the receiver.

The same mixing effects occur in a nonlinear oscillator that is driven by two forces at angular frequencies ω_1 and ω_2; this results in a response that is the sum of functions that oscillate at $2\omega_1, 2\omega_2, \omega_1 + \omega_2$, and $\omega_1 - \omega_2$, with amplitudes proportional to the amplitudes of the drive forces. (The math required to show this for the case of a nonlinear oscillator is more complicated than for the nonlinear circuit element described earlier.)

Sensitivity to initial conditions

For some systems with nonlinear restoring forces, one observes (for appropriate choices of parameters) very complicated motion, even though the system itself is simple, and fully described by exact equations. In a system without a driving force, motion eventually ceases due to dissipation, but the final state of the system can be sensitive to the tiniest changes in the initial conditions.

One example is the "magnetic pendulum," consisting of a rigid pendulum arm ending in a small piece of iron, which hangs above a flat surface that has three identical magnets on it, as shown in the top part figure 4.8.2. One of the magnets is colored white, one black, and one gray. The piece of iron is attracted to each of the magnets, by a force proportional to $1/r^3$. Thus, the end of the pendulum moves in response to a nonlinear force. If the pendulum is started from rest near one of the magnets, it simply swings toward the magnet, and oscillates around it until damping brings it to rest close to that magnet. Each of these three possible final resting points is called an "attractor."

If the pendulum is released at rest at a point far away from any of the magnets, it follows a very complex trajectory. Eventually, it does come to rest at one of the

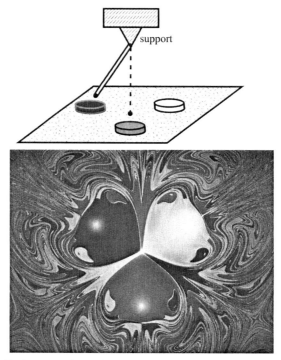

Figure 4.8.2 Top: In the magnetic pendulum, a rigid rod is free to swing from a top support. The bottom of the rod is attached to a small piece of iron, which is attracted to the three magnets on the surface below. Bottom: Basins of attraction for the three magnets, as simulated on a computer using an approximate model for the forces. If the pendulum is released from rest in any of the white regions, it eventually winds up pointing to the white magnet. If it is released from one of the black regions, it winds up at the black magnet, and similarly for the gray regions. The overall shading indicates the length of time required for the pendulum to settle to a final resting place, with darker shading indicating a longer time. (Lower image courtesy of and © Paul Nylander, www.bugman123.com)

three attractors. However, if we try to repeat the experiment by releasing the pendulum a second time from the same initial point, we find that it may end up at a different one of the three attractors. If we repeat the experiment over and over again, the final resting place appears to be random, even though the system is governed by nonrandom, deterministic equations.

This sensitivity can be demonstrated through a computer simulation, as shown in the bottom part of figure 4.8.2. If the pendulum is released in one of the white colored regions, it eventually winds up at the white magnet. Similarly, if released in a black region it eventually winds up at the black magnet, and if released in the gray regions it winds up at the gray magnet. Each colored region is called a "basin of attraction"; it is conceptually similar to a watershed (the area of land from which rainfall feeds into a particular river). The boundary between the three basins is extremely complex. In fact, if you look at it in higher magnification, it looks just as complex as it does at low magnification. This self-similarity at different scales is the hallmark of a "fractal." Many shapes in nature, such as coastlines and tree branches, have fractal character.

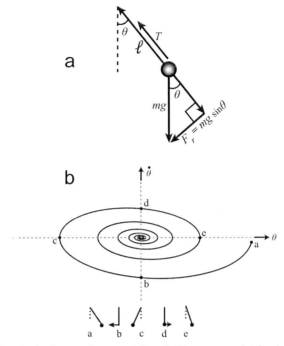

Figure 4.8.3 a: Free body diagram for a pendulum. b: Phase space plot for the damped, undriven pendulum.

Chaos

If a system with nonlinear restoring forces is also subjected to a periodic drive force, this can result in "chaos," in which the system is not only extremely sensitive to initial conditions, but also is sensitive to minute perturbations at any later time. Even a tiny perturbation, such as a gentle puff of air, leads to a huge change in the later behavior.[12] Furthermore, a chaotic system exhibits very complicated, ongoing nonperiodic motion, even though the drive force is periodic and the system is fully described by exact equations.

The damped, driven pendulum provides a good example of chaotic behavior. Let us find the exact version of the differential equation that governs this system. We showed in section 2.2 that, for small displacements, the restoring force for the pendulum is proportional to displacement. However, this is an approximation; as shown in figure 4.8.3a, the restoring force is $F_r = mg \sin \theta$. Therefore, the restoring torque is

$$\tau = -\ell mg \sin \theta.$$

12. This sensitivity to initial conditions and small perturbations is often referred to as the "butterfly effect." The name stems from a talk given in 1963 by one of the founding fathers of the field of chaos. During the talk, Edward Lorenz said of chaos theory, "One meteorologist remarked that if the theory were correct, one flap of a seagull's wings would be enough to alter the course of the weather forever." In later talks, the seagull evolved into a butterfly.

We include a viscous drag force

$$F_{\text{drag}} = -bv = -b\ell\dot{\theta},$$

which produces an associated torque

$$\tau_{\text{drag}} = -b\ell^2\dot{\theta}.$$

Finally, we include a periodic driving torque $\tau_0 \cos\omega_d t$. The angular version of $F_{\text{TOT}} = m\ddot{x}$ is $\tau_{\text{TOT}} = I\ddot{\theta}$, which gives us

$$-\ell mg \sin\theta - b\ell^2\dot{\theta} + \tau_0 \cos\omega_d t = I\ddot{\theta}.$$

For a simple pendulum, $I = m\ell^2$, so that

$$-\ell mg \sin\theta - b\ell^2\dot{\theta} + \tau_0 \cos\omega_d t = m\ell^2\ddot{\theta} \Leftrightarrow$$

$$\ddot{\theta} + \gamma\dot{\theta} + \omega_0^2 \sin\theta = \frac{\tau_0}{m\ell^2}\cos\omega_d t, \tag{4.8.11}$$

with $\gamma \equiv b/m$ and $\omega_0 \equiv \sqrt{g/\ell}$. This is a differential equation in the variable θ. Because of the $\sin\theta$ term, it is nonlinear, and therefore there is the possibility that, for appropriate choices of parameters, the system will show chaotic behavior.

To visualize the behavior of the system, we need a couple of new tools. A phase space plot is simply a plot with one axis for each variable that is time-dependent and that is needed to specify the state of the system. (Such quantities are called "dynamical variables.") For the case of zero driving torque, there are only two such variables: the angular position θ and the angular velocity $\dot{\theta}$. We can specify the state of the system entirely[13] by giving the values of θ and $\dot{\theta}$. Thus, if we make a plot with θ on the horizontal axis and $\dot{\theta}$ on the vertical axis, then each point in the plot specifies a state of the system. If we release the pendulum from rest at θ_0, with zero driving torque, it follows the phase space path shown in figure 4.8.3b, spiraling in toward rest at $\theta = 0$.

One can show that, for a system to exhibit chaos, there must be at least three dynamical variables. When we turn on the driving torque, we get the third variable: $\omega_d t$, which is the argument of the $\cos\omega_d t$ term in equation (4.8.11), and is therefore called the "drive phase." Now, in order to completely specify the state of the system, we need to specify θ, $\dot{\theta}$, and $\omega_d t$. You might ask, "Why not just use t as the dynamical variable instead of $\omega_d t$?" The answer is that we don't actually need to know the time in order to specify the state of the system, but we do need to know the argument of the $\cos\omega_d t$ term; we do need to know whether the drive torque is at a maximum or a minimum, or somewhere in between, and whether it is increasing or decreasing at the moment in question. Note that $\omega_d t$ is an angular variable, measured in radians; the drive torque has the same value at $\omega_d t = 0$ as it does at $\omega_d t = 2\pi$.

Since we now have three-dynamical variables, we need a three-dimensional phase space to represent the system. The axis for $\omega_d t$ need only run from 0 to 2π. Let's now

---

13. Note that we need not also specify $\ddot{\theta}$, since (for $\tau_0 = 0$) we can use equation (4.8.11) to calculate $\ddot{\theta}$ from θ and $\dot{\theta}$. Also note that specifying θ by itself is not enough to determine the state of the system. For example, if $\theta = 0$, the pendulum could be swinging to the right or to the left.

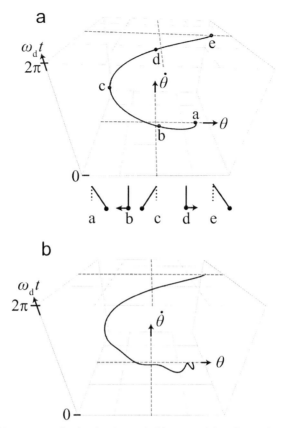

Figure 4.8.4 a: Phase space plot for the damped, driven pendulum in steady state. The angle θ is shown on the left-right axis, the angular velocity $\dot{\theta}$ is shown on the axis that is more-or-less perpendicular to the page, and the drive phase $\omega_d t$ is shown on the vertical axis. The system starts at a maximum value of θ, as shown by point a and by the small picture labeled a, then progresses through points b–e. b: Similar plot for the pendulum with arbitrary initial conditions, leading to transient behavior that decays over time.

drive the pendulum at low amplitude (low enough that the nonlinearities are not important). We know that, in steady state, it follows simple harmonic motion:

$$\theta = A \cos\left(\omega_d t - \delta\right) \Rightarrow \dot{\theta} = -\omega_d A \sin\left(\omega_d t - \delta\right).$$

As an example, we choose $\omega_d = \frac{2}{15}\omega_0$, which gives $\delta \cong 0$. This means that $\theta = A$ and $\dot{\theta} = 0$ at $t = 0$. The phase space plot for this steady-state behavior is shown in figure 4.8.4a; it is a helix that starts at the bottom. The phase space point representing the system moves up the helix, and reaches the top of the helix at $\omega_d t = 2\pi$, which is equivalent to $\omega_d t = 0$, so the phase space point jumps down to the bottom of the helix and starts up again.

   If we start the system with some arbitrary initial conditions, there will be some transient behavior near $t = 0$, which damps away in time, until the system reaches steady state, as we explored in section 4.5; this behavior is shown in figure 4.8.4b. Once the transients damp away, the system returns to the helix phase space path that represents the steady state. Therefore, this helix is the attractor for this system. It is a

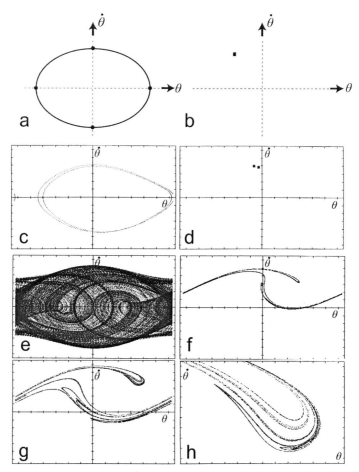

Figure 4.8.5 a: Trajectory in the $\theta - \dot{\theta}$ plane for pendulum driven at low amplitude. b: A Poincaré section of this motion. Parts c and f of the figure represent a higher amplitude motion of the pendulum, and so are at a larger scale than parts a and b. c: Trajectory in the $\theta - \dot{\theta}$ plane for higher drive torque, showing period doubling. d: A Poincaré section of this motion. e:Trajectory in the $\theta - \dot{\theta}$ plane for even higher drive torque, showing chaotic behavior. f: A Poincaré section of this motion. g: A Poincaré section for a pendulum with less damping than the one shown in parts a–f, showing chaotic behavior. (Note that, unlike the other figures in the left column, this is a Poincaré section rather than a phase space plot.) h: Zoom-in on the upper right part of the Poincaré section, showing that the complexity of the attractor does not diminish as we zoom in. Parts c–h used with permission from *Chaotic Dynamics: An Introduction, 2nd Ed.*, by G. L. Baker and J. P. Gollub, Cambridge University Press, Cambridge, 1996.

more complex attractor than the three points for the magnetic pendulum, and represents a dynamic behavior, but the idea is the same; it represents the behavior that the system eventually settles into.

Because it is complicated to look at the three-dimensional phase space plot, one often instead just shows the projection onto the plane of the θ and $\dot{\theta}$ axes (the $\theta - \dot{\theta}$ plane). For the driven pendulum at low-drive amplitude in steady state shown in figure 4.8.4a, this projection is an ellipse, as shown in figure 4.8.5a.

Now, imagine that we illuminate the pendulum with a strobe lamp, which flashes once per cycle of the drive torque. At each flash, we measure θ and $\dot{\theta}$, and plot this point in the $\theta - \dot{\theta}$ plane. The resulting plot is called a "Poincaré section." For the steady-state motion shown in figure 4.8.5a, the Poincaré section is simply a point, as shown in figure 4.8.5b. A single flash of the strobe is enough to create this particular Poincaré section, because the pendulum is in simple periodic motion. It is called a section because it is a slice through the three-dimensional phase space. We can choose to have the strobe flashes occur at $\omega_d t = 0, 2\pi, 4\pi$, etc., in which case we get a slice at the bottom of the cube shown in figure 4.8.4a. If instead we have the flashes occur at $\omega_d t = \pi, 3\pi, 5\pi$, etc., then we get a slice halfway up the cube. By varying the timing of the flashes, we can map out different slices through the attractor. (Again, for the case of low-drive amplitude, the attractor is the helix shown in figure 4.8.4a).

Next, we gradually increase the amplitude of the drive, τ_0, while keeping ω_d fixed. For our example, we now choose $\omega_d = \frac{2}{3}\omega_0$. We discussed earlier in this section that, for small nonlinearity and a symmetric potential energy function, there is a response not only at ω_d, but there is also a third harmonic response at $3\omega_d$, which in this case equals $2\omega_0$. Therefore, there can be an interplay between this response and the natural resonance angular frequency of the system, ω_0. The period of the drive torque is $T = \frac{2\pi}{\omega_d} = \frac{3}{2}\frac{2\pi}{\omega_0}$. In a time interval $2T$, the drive torque goes through exactly two cycles, the third harmonic response goes through exactly six cycles, while any oscillation at ω_0 goes through exactly three cycles. We can imagine therefore that the periodicity of the system could change, due to this interplay, from T to $2T$, and this "period doubling" is indeed observed experimentally. The projection onto the $\theta - \dot{\theta}$ plane of the phase-space trajectory is shown in figure 4.8.5c, and a Poincaré section in figure 4.8.5d. In this case, two successive flashes of the strobe are needed to complete the Poincaré section; in subsequent flashes, the point representing the system alternates between the first two points. As we increase the drive amplitude further, the system is driven to higher amplitudes, and samples the nonlinearity of the restoring force more. Additional complexities are added to the motion (such as the pendulum going "over the top"), and the period gets longer, but the motion is still periodic.

Then, with a small additional increase in the drive amplitude, the motion suddenly changes character. It is no longer periodic, as shown by the $\theta - \dot{\theta}$ plane plot in figure 4.8.5e, and the Poincaré section (figure 4.8.5f) becomes suddenly much more complex. In this case, many strobe flashes (in principle an infinite number) are needed to complete the Poincaré section, because the motion is not periodic. However, we see that the motion is not completely random; even with a large number of flashes, many parts of the Poincaré section remain blank. Again, the Poincaré section is a slice parallel to the $\theta - \dot{\theta}$ plane through the phase space attractor for the chaotic state (the equivalent of the helix of figure 4.8.4a). Just from this slice, we can tell that the attractor has a *much* more complicated shape, and so it is called a "strange attractor." If we attempt to examine the Poincaré section it in more detail, by expanding the plot in a small region, it still looks just as complex, as suggested in figures 4.8.5g and h. Thus, the strange attractor exhibits a fractal nature.

There are many plots, simulations, and animations relating to the chaotic pendulum available on the web; for a few suggestions, go to the web page for this text, and consult the part for this chapter section.

Chaos is important in a remarkably wide range of fields, including economics, medicine, meteorology, physics, and chemistry. We have only scratched the surface of the rich and active fields of chaos and nonlinear dynamics. For further study, you might begin with the very approachable text by Baker and Gollub.[14]

Concept and skill inventory for chapter 4

After reading this chapter, you should fully understand the following terms:

Resonance (4.1)
Steady state (4.1, 4.5)
Amplitude resonance curve (4.1–4.2)
Power resonance curve (4.3)
FWHM (4.3)
Superposition principle for driven systems (4.4)
Transient (4.5)
MRI (4.7)
NMR (4.7)
Magnetic moment (4.7)
Precession (4.7)
RF (4.7)
Gyromagnetic ratio (4.7)
Larmor frequency (4.7)
Rotating reference frame (4.7)
Spin up and spin down (4.7)
Nonlinear differential equation (4.8)
Harmonic generation (4.8)
Frequency doubling (4.8)
Subharmonic resonance (4.8)
Mixing (4.8)
Sensitivity to initial conditions (4.8)
Dynamical variable (4.8)
Basin of attraction (4.8)
Phase space (4.8)
Attractor (4.8)
Strange attractor (4.8)
Chaos (4.8)
Fractal (4.8)

---

14. *Chaotic Dynamics: An Introduction, 2nd Ed.*, by G. L. Baker and J. P. Gollub, Cambridge University Press, Cambridge, 1996.

You should understand the following connections:

Motion of the support point and resulting drive force (4.1)
Frequency of drive and frequency of steady-state response (4.1)
Phase of drive and phase of response (4.1–4.2)
Height of amplitude resonance curve and Q (4.2)
Width of power resonance curve and γ (4.3)
Height and width of power resonance curve (4.3)
Average power supplied and average power dissipated in steady state (4.3)
General solution for damped driven harmonic oscillator and solution for damped oscillator (4.5)
Damped driven mass/spring and driven RLC circuit (4.6)
Protons and a child's top (4.7)
Torque and precession (4.7)
Angular momentum and magnetic moment (4.7)
Symmetry of potential energy curve and which harmonics are generated (4.8)

You should be familiar with the following additional concepts:

Advantages of expressing things in terms of dimensionless variables (4.2)

You should be able to:

Find the steady-state response amplitude and phase given information about the oscillator, the damping, and the drive (4.1–4.2)
Find the response of a damped oscillator to superposed drive forces (4.4)
Given the gyromagnetic ratio and $\mathbf{B}_{applied}$, calculate the RF frequency that produces NMR (4.7)
Explain what a Poincaré section is (4.8)
Explain what an attractor is (4.8)

In addition to all of the above, you should be able to combine the concepts you've learned to address new situations.

Problems

Note: Additional problems are available on the website for this text.

Instructor: Ratings of problem difficulty, full solutions, and important additional support materials are available on the website.

    4.1 The following statement is incorrect. Provide a corrected version. "If you drive a damped oscillator at a frequency other than ω_0, the oscillator still oscillates at its resonance frequency ω_0. The amplitude of the response depends on how close the drive frequency is to the resonance frequency."

    4.2 **Connection to current research: Frequency Modulation AFM.** In Atomic Force Microscopy, a very sharp tip is used to "feel" the topography of the sample by moving the tip in an x–y pattern over the region to be imaged while monitoring the force between the tip and sample. For the highest resolution images, the tip is not allowed to "touch" the sample. Instead, it is brought close enough to feel an attractive interaction (due to Van der Waals and other forces). As the tip is moved laterally over the sample, a feedback circuit adjusts the z-position of the tip to keep this attractive force constant. The record of the adjustments needed in z then gives the topography of the sample. The tip is mounted at the end of a cantilever, which is set into oscillation, as shown schematically in figure 4.P.1a. (The actual amplitude of vibration is much less than that shown; for the image in figure 4.P.1b, the vibration amplitude was only 0.8 nm.) The attractive force between tip and sample is measured through its effect on the resonant frequency of the cantilever/tip system. Ordinarily, this technique is used in ultrahigh vacuum, and can give *sub-atomic* imaging resolution, as shown in figure 4.P.1b and c. For such an image, the measured forces are due to the interactions of *individual atomic orbitals*!

The force of interaction between the tip and sample is shown schematically in figure 4.P.1d. The position of the tip x is measured relative to the equilibrium position in the absence of this interaction; note that downward is defined as the positive direction for x. When the tip moves down (toward the sample), the force is initially attractive (positive). As the tip starts to "touch" the surface, the force becomes repulsive (negative). A typical point near which the AFM might be operated is shown as x_0. **(a)** Briefly explain why, near this point, we can model the force between tip and sample as $F \cong F_0 - k_{ts}\left(x - x_0\right)$, where $k_{ts} \equiv - \left.\dfrac{dF}{dx}\right|_{x_0}$. **(b)** Note that k_{ts} functions like a spring constant. It is negative because it acts oppositely to the spring of the cantilever; when the tip is closest to the sample (the point marked "A" in part a of the figure), the cantilever pulls it up whereas the tip–sample force is attractive and so pulls it down. Let the spring constant of the cantilever be k, and the effective mass of the cantilever and tip be m. Explain why, if $\left|k_{ts}\right| \ll k$, then the oscillation angular frequency is approximately $\omega \cong \omega_0 + \Delta\omega$, where $\omega_0 \equiv \sqrt{\dfrac{k}{m}}$ and $\Delta\omega \equiv \omega_0 \dfrac{k_{ts}}{2k}$. (Therefore, by measuring the frequency shift, one can measure the force of interaction between the tip and sample.)

4.3 It should be clear that, for a damped driven harmonic oscillator that has reached *steady state*, the *average* power supplied to the system by the driving forces is equal to the average power dissipated by the damping force. However, for most drive frequencies there are parts of the cycle in which power flows from the oscillator to the entity providing the driving force (the "driver"), and other parts of the cycle during which power flows from the driver into the oscillator. Thus, in general, the energy of the oscillator is not constant during the cycle, even though its value *averaged* over the whole cycle *is* constant. Assume that k, m, and b are known. For what finite, nonzero value of ω_d is the *instantaneous* power supplied by the driver exactly equal to the *instantaneous* power dissipated by the damping force at every instant in the cycle? Show your reasoning clearly. *Hints: This means that the energy*

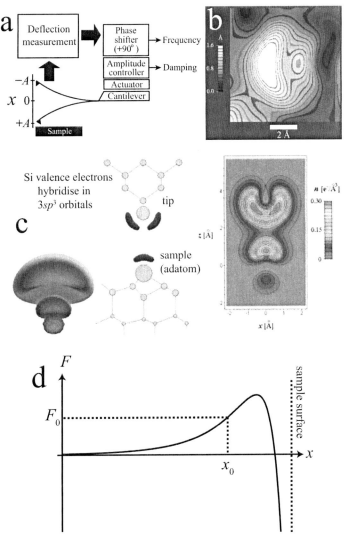

Figure 4.P.1 a: Schematic representation of frequency–modulation AFM. Note that a downward deflection of the tip is defined to be positive. b: Subatomic resolution image of an atom on the tip of the AFM. c: Explanation for the image. As the tip passes over a protruding atom on the sample ("adatom"), the adatom images the bottom-most atom of the tip, showing the two lobes of the sp^3 atomic orbitals. (The features shown in the image are actually determined both by the atom on the surface and by the atom on the tip.) d: Schematic diagram for the force of interaction between tip and sample. Parts a–c © and courtesy of Professors Jochen Mannhart and Franz Giessibl.

of the oscillator is constant throughout the cycle, so you might wish to start by writing an expression for the total energy as a function of time, and see if you can determine what value of ω_d would cause the total energy to be constant.

4.4 A harmonic oscillator has an undamped angular frequency ω_0. It is then put in a damping medium producing a damping characterized by Q. The oscillator is driven at a frequency such that, in the steady state, the response x lags behind F_{drive} by $45°$, meaning that the drive force reaches each peak a little earlier than x does. The drive force amplitude is F_0, and k is the effective spring constant for the oscillator.

(a) Show that the response amplitude is $A = \dfrac{\sqrt{2}\,F_0 Q^2}{k\left(\sqrt{1+4Q^2}-1\right)}$, and that

the dissipated power is $P_{\text{diss}} = \dfrac{\omega_0 Q F_0^2}{2k}\sin^2\left(\omega_d t - \delta\right)$.

(b) Sketch qualitative graphs of the drive, response, potential energy, velocity, and dissipated power over one complete steady-state cycle. Align the graphs in a vertical column, or superpose the graphs, so that the relationships between the five quantities are as clear as possible.

4.5 A mass is subjected to a spring force $F_{\text{spring}} = -kx$, a damping force $F_{\text{damp}} = -b\dot{x}$, an oscillating drive force $F_{\text{drive}} = F_0 \cos \omega_d t$, and an additional decaying force $F_{\text{extra}} = De^{-\beta t}$, where D and β are positive constants.

(a) Write the differential equation for x that describes this system.

(b) Write the simplest possible complex version of your DEQ from part a, that is, the simplest complex DEQ whose real part is your DEQ from part a.

(c) Show that $z = Ae^{i\varphi}e^{i\omega t}$ is *not* a solution for your DEQ from part b, no matter what the values of A, φ, and ω are.

4.6 Driving a pendulum with vertical motion. Make your own simple pendulum. (One easy way is to squeeze the tea out of a used teabag that has a string.) First, get your pendulum swinging by holding the end of the string and moving your hand laterally. Find the resonant frequency, and measure it roughly. With the pendulum still swinging, see if you can keep it going by moving your hand vertically instead of horizontally. If you experiment enough, you should be able to do this, though it only works if the pendulum is already swinging when you start moving your hand vertically. (a) What is the ratio of the frequency of your hand motion when you're moving your hand vertically divided by the frequency of your hand motion when you move your hand horizontally, assuming you drive the pendulum at resonance in both cases? (b) Explain your finding about the frequency ratio qualitatively.

4.7 Go to the website for this text, and under chapter 4 open the "Damped Driven Harmonic Oscillator" applet. Use it to answer the following questions:

(a) To use this animation, select a quality factor (Q) of 15 with the slider and then select a drive frequency by clicking on either of the two graphs. Try some different values of Q and drive frequency. At what drive frequency do you get the greatest amplitude?

(b) The phase graph shows the phase relationship between the motion of the drive and the motion of the oscillator. What is the phase relationship between the motion of the drive and the motion of the oscillator at the point that you just found of greatest amplitude? How about at the two extremes of very low drive frequency and very high drive frequencies?

(c) Now, click the check box labeled "Show Velocity" and observe the small velocity indicator to the right of the oscillator. What is its phase relationship to the drive force at resonance? Why does this make sense in terms of the power being delivered by the drive force to the oscillator?

(d) Generally speaking, how does varying Q affect the amplitude and phase diagrams?

(e) This animation ignores "transients." Because of this, what state of a damped-driven harmonic oscillator system are we seeing?

(f) How does the steady-state response frequency compare to the drive frequency at resonance? Well below resonance? Well above it?

4.8 Show that the peak of the amplitude resonance curve shown in figure 4.1.5 or 4.2.1 is at $\omega_d = \omega_0\sqrt{1 - \frac{1}{2Q^2}}$. Do not use a symbolic algebra program or calculator.

4.9 What, *if anything*, does the quality factor Q of a lightly damped mechanical oscillator have to do with each of the following? Be brief, but as quantitative as possible. **(a)** The rate at which the oscillator loses energy in the absence of a driving force. **(b)** The "amplification factor" at resonance, that is, the amplitude of the oscillator's motion when a driving force at the resonance frequency is applied, as compared to the amplitude of motion when the same amplitude driving force is applied at very low frequency. **(c)** The phase shift at resonance. (Assume the damping is light enough that amplitude resonance, velocity resonance, acceleration resonance, and power resonance all occur at essentially the same frequency.) **(d)** The full width at half-maximum of the "power resonance curve," that is, the graph of power absorbed as a function of drive frequency.

4.10 A simple pendulum is made by attaching a steel sphere ($\rho = 7,850$ kg/m^3) of radius 2.00 cm to the end of a thin string that is 1.00-m long. In this problem, take the resonant frequency to be $\sqrt{g/L}$, where L is the distance from the support point to the center of the sphere; assume that the damping is light enough that the differences between ω_0, ω_v, and the peak angular frequency for the amplitude response curve can be ignored. The support point (at the top of the string) is moved laterally (in the x-direction) in a sinusoidal pattern at the resonant frequency, with amplitude $1.00\ \mu\text{m} = 1.00 \times 10^{-6}$ m. What is the steady-state amplitude of the sphere's motion in the x-direction? Assume that the air provides viscous damping, that is, that the airflow is not turbulent.)

4.11 Using the results of section 4.1, show explicitly that, in steady state, the net energy provided by the drive force to a damped driven oscillator over one period equals the energy dissipated by the damping force.

4.12 **Steady-state response to a square wave.** In section 8.4, we will show that the square wave (shown in black in figure 4.P.2) can be represented as an

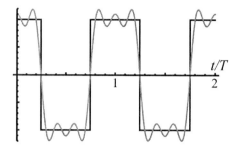

Figure 4.P.2 Synthesizing a square wave from sinusoids.

infinite sum of sinusoids:

$$\text{Square wave} = \cos \omega t - \frac{\cos 3\omega t}{3} + \frac{\cos 5\omega t}{5} - \cdots$$

The sum of the first three terms is shown in gray in the figure—you can see that it roughly approximates the square wave. The preciseness of the approximation improves as more terms are added. If a force with a square wave time dependence is used as the drive force for a damped driven harmonic oscillator, is the steady-state response a square wave? Explain your answer thoroughly.

4.13 The curve of power dissipated *versus* drive frequency for a damped driven oscillator is shown in figure 4.P.3. (The power dissipated averaged over a cycle in steady state is equal to the power absorbed from the drive force.)

(a) What is the Q for this system?

(b) The oscillator is driven at resonance until steady state is achieved. The drive force is then suddenly turned off. *About* how long does it take for the oscillations to die away? (Give your answer in seconds; all the numerical information you need is in the graph.)

(c) Now, instead, the oscillator is driven with $f_d = 850\text{Hz}$. The drive force is then turned off; thereafter, the oscillator continues to oscillate for a short time. At what frequency (approximately) does

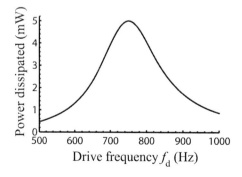

Figure 4.P.3 A power resonance curve.

it oscillate during this part of the experiment (after the drive force is turned off)? Explain briefly.

4.14 For the damped driven harmonic oscillator, find expressions for the adjustable constants A_0 and φ that appear in equation (4.5.2) in terms of the initial position x_0, the initial velocity $\dot{x}_0$, the steady-state amplitude A, the steady-state phase shift δ, ω_v, and ω_d. Do not use a symbolic algebra program or calculator. *Hints: Don't expect the final result to be "neat." You should be able to show that* $\tan \varphi = \frac{B}{D}$, *where B and D are constants involving* x_0, $\dot{x}_0$, ω_v, *and* γ. *To find* A_0, *you will need to find* $\cos \varphi$. *To do this, draw a right triangle with B and D as the two legs.*

4.15 A weasel holds an object of mass m at its equilibrium position $x = 0$. The mass hangs from a spring of constant k in a medium which provides $F_{damp} = -b\dot{x}$. The support point for the spring is moving, with position given by $x_C = A_d \cos \omega_d t$. The angular drive frequency ω_d happens to exactly equal $\sqrt{\frac{k}{m}}$. (Don't forget this, and its implications—otherwise, the math is a mess!)
    At $t = 0$, the weasel releases the mass, so that it can begin moving. Given these initial conditions, what is the *complete* solution $x(t)$? Everything in your solution should be expressed in terms of the symbols above only. However, you may make your life easier by defining symbols of your own in terms of those above, then expressing your solution using these new symbols.

4.16 An electrical engineer wishes to design an RLC oscillator, of the type shown in figure 4.6.1. The engineer wants the resonant response to be very strong, but also wants the system to respond strongly to a broad range of frequencies. Explain briefly why both goals can't be achieved.

4.17 Radio station WJJZ has hired you to help design radio receivers to go in the waiting rooms of doctors' offices. Each of these radios is to receive the WJJZ broadcast only—there will be no "tuning" knob. One of your fellow engineers hands you part of the schematic diagram for the radio, shown in the top part of figure 4.6.2. The input voltage to the LCR circuit from the antenna is $\Delta V_{IN} = V_0 \cos \omega_d t$, and the output voltage is taken across the capacitor (and then sent to the rest of the radio). "I already chose $L = 0.1\,\mu H$," he says. "You pick out the values for R and C." (Note: "μ" means "$\times 10^{-6}$".)

> **(a)** The broadcast frequency of WJJZ is $f = 101.9$ MHz. (Note: "M" means "$\times 10^6$"). What value of C should you pick so that the above circuit will resonate at this frequency? (Assume the circuit is very lightly damped.)
>
> **(b)** Rival station WART broadcasts at $f' = 101.5$ MHz. The steady-state oscillation amplitude of your circuit (i.e., the amplitude of ΔV_{OUT}) must be 100 times smaller at this frequency than at the WJJZ frequency, assuming equal drive (input) voltages for the two frequencies. What is the required value for R? Again, assume very light damping. Your answer need only be correct to within a few percent. *Hint: Don't forget that* $q = CV \Rightarrow \Delta V_{OUT} = \frac{q}{C}$.

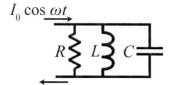

$I_0 \cos \omega t$

$R \quad L \quad C$

Figure 4.P.4 A parallel RLC oscillator.

  (c) Of course, the WJJZ management would be happy if the amplitude for WART is even less than 100 times below that for WJJZ. Bearing this in mind, is the R you just calculated the maximum allowed R or the minimum allowed R? Explain.

  (d) What is the Q of your circuit?

4.18 The parallel RLC oscillator. Consider the circuit shown in figure 4.P.4. This is driven by applying a current $I_0 \cos \omega t$, as shown. Assume the system is underdamped. **(a)** What is the resonance frequency ω_0? **(b)** What is the FWHM of the power resonance curve?

4.19 For this problem, the RLC oscillator shown in figure 4.6.1 is driven not simply by the voltage $V_0 \cos \omega t$, but instead by the more complicated voltage $V_1 \cos \omega_1 t + V_2 \sin \omega_2 t$. What is the charge on the capacitor, $q(t)$, once the system reaches the steady state?

4.20 A patient is undergoing MRI. A particular proton in her brain experiences a total applied magnetic field (including contributions from nearby electrons) of 1.50 T. The gyromagnetic ratio for a proton is $\gamma = 2\pi \cdot 42.576 \text{ MHz/T}$. **(a)** What frequency of RF radiation will this proton respond to most strongly? **(b)** For simplicity, in this part of the problem assume that the net magnetic moment and the net angular momentum of the protons in the local region of the brain can be treated by considering only their z-components. The main magnetic field in the MRI machine is applied in the z-direction. The magnetic field of the RF radiation is described by $\mathbf{B}_{\text{rad}} = B_0 \cos \omega t \, \hat{\mathbf{i}}$. If $B_0 = 30.0 \mu\text{T}$, how long of an RF pulse should be applied to rotate the magnetization from the z-direction into the x–y plane?

4.21 Show that $\cos^3 \theta = \frac{1}{4} \cos 3\theta + \frac{3}{4} \cos \theta$. (You will probably want to use a symbolic algebra program to do this.) This trigonometric identity is used in section 4.8.

4.22 Online exploration of a chaotic pendulum. Go to the entry for this problem on this book's website, and follow the instructions there.

5 Symmetric Coupled Oscillators and Hilbert Space

Gentlemen, I do not see that the sex of the candidate is an argument against her admission as a privatdozent. After all, the senate is not a bathhouse.
– David Hilbert, arguing in favor of allowing Emmy Noether a position at the University of Göttingen

5.1 Beats: An aside?

If you strike a tuning fork and listen, the resulting pressure variation at your ear is sinusoidal. We might write the *variation* in pressure (relative to the average background pressure) as

$$x_1 = \operatorname{Re} z_1, \quad \text{where } z_1 = Ae^{i\omega_1 t}.$$

Now, imagine that we strike two tuning forks of slightly different angular frequencies, ω_1 and ω_2. The resulting pressure variation at your ear is simply the sum of the variations from the two forks. $\boxed{\text{A/V:}}$ You can hear this right now by going to this book's web page and clicking on the "listen to beats" link under this chapter section. This sound is remarkable – you perceive it as a single note (i.e., a single frequency) with an oscillating loudness! We can see how the loudness variations come about graphically, as shown in figure 5.1.1a. This effect turns out to be of **tremendous** importance for our future studies, so let's see how it comes about mathematically.

To simplify the math, we'll look at the case where the two amplitudes are the same, that is,

$$z = z_1 + z_2 = A(e^{i\omega_1 t} + e^{i\omega_2 t}). \tag{5.1.1}$$

It will help to define

$$\omega_e \equiv \frac{\omega_1 - \omega_2}{2} \quad \text{and} \quad \omega_{av} \equiv \frac{\omega_1 + \omega_2}{2}. \tag{5.1.2}$$

(The subscript e is used for the first one because we will see that it is the angular frequency of an envelope function.) You should verify right now that

$$\omega_1 = \omega_{av} + \omega_e \quad \text{and} \quad \omega_2 = \omega_{av} - \omega_e.$$

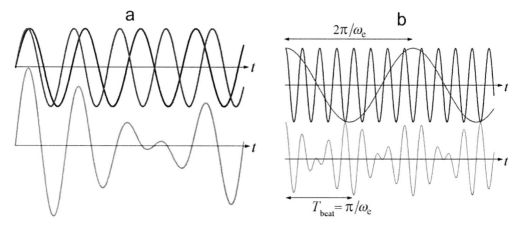

Figure 5.1.1 a: Two oscillations of slightly different frequencies (top) are added together (bottom). At first, the two waves are in phase, and so the sum has large amplitude. As time progresses, the two waves get out of phase, so the amplitude of the sum gets small. With further passage of time, the waves start to get back into phase. b: A rapidly oscillating function multiplied by a slowly varying envelope function of period $2\pi/\omega_e$ (top) produces beats (bottom).

Therefore,

$$z = A\left[e^{i(\omega_{av}+\omega_e)t} + e^{i(\omega_{av}-\omega_e)t}\right]$$

Your turn: Use the above and Euler's equation ($e^{i\theta} = \cos\theta + i\sin\theta$) to show that

$$z = 2Ae^{i\omega_{av}t}\cos\omega_e t$$

This means that

$$x = \mathrm{Re}\, z = \underbrace{2A}_{\substack{\text{constructive}\\ \text{interference}\\ \text{at maximum}}} \underbrace{\cos\omega_{av}t}_{\substack{\text{rapid}\\ \text{oscillation}}} \underbrace{\cos\omega_e t}_{\substack{\text{slow oscillation}\\ \text{due to the transition}\\ \text{from constructive to}\\ \text{destructive interference}\\ \text{(and back):}\\ \text{an envelope function}}}$$

We see that x is the product of three terms: an overall amplitude, a rapid oscillation at the average angular frequency, and a slow oscillation (corresponding to the oscillating loudness). The slow oscillation is called an "envelope function," because you can think of it as a time-dependent amplitude for the rapid oscillation, that is,

$$x = \underbrace{2A\cos\omega_e t}_{\substack{\text{time-dependent}\\ \text{amplitude}}}\; \underbrace{\cos\omega_{av}t}_{\substack{\text{rapid}\\ \text{oscillation}}} \qquad (5.1.3)$$

This multiplication is shown graphically in figure 5.1.1b. The slow variations in amplitude are called "beats," and you'll often hear people talking about one frequency "beating" against another, that is, the signals at two close frequencies combine to create beats. Following the usual convention, we define the beat period as the period between maximum amplitudes, as shown, that is,

$$T_{\text{beat}} \equiv \frac{\pi}{\omega_e} = \frac{2\pi}{\omega_1 - \omega_2} \tag{5.1.4}$$

Note that this is half the period of the envelope function. The beat frequency is then simply

$$f_{\text{beat}} = \frac{1}{T_{\text{beat}}} = \frac{\omega_1 - \omega_2}{2\pi} \Rightarrow$$

$$\boxed{f_{\text{beat}} = f_1 - f_2} \tag{5.1.5}$$

Self-test (answer below[1]): Serious musicians sometimes use this phenomenon for tuning their instruments. Two guitar players are trying to bring their guitars to the same pitch. They both pluck their lowest string, and hear beats. One of the players begins adjusting the pitch of her string, and the beat period gets shorter. Should she keep adjusting in this direction?

Applet: Click on the "beats applets" link under this chapter section on this book's web page to see other ways of understanding beats.

5.2   Two symmetric coupled oscillators: Equations of motion

You probably have seen much of the material up to this point in previous courses, though not at the same level of detail, nor with such a thrilling and masterful presentation.[2] Now, we begin the truly new (and really neat!) stuff. We will begin by considering two coupled oscillators, and this will lead directly to our treatment of waves (which we'll view as a large set of coupled oscillations), and to the most important ideas underlying quantum mechanics.

The simplest and easiest-to-draw example of coupled oscillators is a set of two pendula, connected by a spring, as shown in figure 5.2.1a. To start with, we use pendula with the same length and the same mass. Note that we define the position of each mass relative to its own equilibrium point, so that when the system is at equilibrium (both pendula hanging straight down), we have $x_1 = 0$ and $x_2 = 0$. Also, for simplicity we will ignore damping and driving forces for now.

If you displace mass 1 from equilibrium while holding mass 2 steady (at $x_2 = 0$), and then release both, you will observe a remarkable behavior. $\boxed{\text{A/V:}}$ You can see a

---

1. If the beat period is getting shorter, the beat frequency is getting larger. According to equation (5.1.5), this means the two frequencies are getting farther apart. Therefore, she should stop and adjust in the other direction instead. Perfectly matched pitches will result in a very long beat period (in principle infinitely long).

2. I asked my wife for ideas on how to punch up the humor of this sentence, but her suggestions were all too funny to include in a physics text.

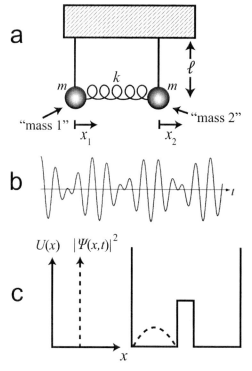

Figure 5.2.1 a: Coupled pendula. b: Position of mass 1 plotted as a function of time t. c: If the probability density $|\Psi(x, t)|^2$ of an electron is initially localized on the left of this double-well potential, it will oscillate to the right side, then back to the left, etc.

video clip of this by clicking on the "coupled pendula" video under this chapter section on this book's website. At first, mass 1 oscillates back and forth, and mass 2 hardly moves. Gradually the energy is transferred to mass 2; now mass 2 oscillates and mass 1 hardly moves. Gradually, the energy is transferred back to mass 1, and the process starts over. If you watch just one of the two masses, its behavior should remind you of something…something you've just been studying…what is it? Hmmm… Beats! If you plot the position of either mass as a function of time, it looks as if it's beating, as shown in figure 5.2.1b. Recall that beats result when you superpose two oscillations of slightly different frequencies. We will see that the best description of this motion of the coupled pendula is in terms of a superposition of two different oscillations. However, it is not at all obvious at this point why simply displacing one of the pendula from equilibrium should somehow excite two different oscillations.

One observes similar behavior for any set of two symmetric coupled oscillators. For the pendula, the coupling is obvious from just looking at the system, but there are other coupled oscillator systems for which the coupling is more subtle. For example, consider two cantilevers attached to the same support, with both cantilevers initially at rest. If one is displaced from equilibrium and then released, its oscillation causes a tiny amount of flexing in the support, which provides a coupling to the other cantilever. If the damping is small enough for the oscillations to persist for a long time, the oscillation energy starts to transfer to the other cantilever. (The transfer is slow, because the coupling is weak.)

Another example comes from quantum mechanics. If an electron is initially localized in the potential well on the left in figure 5.2.1c, its probability density[3] eventually moves entirely over to the right well, and then back to the left, just as the energy of the two coupled pendula moves back and forth.

Getting back to the example of the coupled pendula, our task (similar to those in the previous chapters) is to find $x_1(t)$ and $x_2(t)$ given the initial conditions. We begin our three-step procedure, through instead of steps 2 and 3 we will use a different method, in section 5.3.

1. Write Newton's second law for each object in the system:

> **Your turn:** Convince yourself that the force due to the spring on mass 1 is given by
>
> $$F_{\text{spring, 1}} = -k\left(x_1 - x_2\right).$$
>
> *Hint: First find the force when mass 1 is held at $x_1 = 0$ but mass 2 is allowed to move. Then find the force when mass 2 is held at $x_2 = 0$ but mass 1 is allowed to move.*

The above equation shows mathematically that the force on mass 1 depends not only on the position of mass 1, but also on the position of mass 2, so that the motion of the two masses is "coupled." The total force on mass 1 is the sum of the spring force and the "pendulum force" discussed in chapter 3. Recall that, for small displacements from equilibrium, the tension in the string and gravity combine to produce a spring-like force given by

$$F_{\text{pendulum}} = -\frac{mg}{\ell}x$$

Therefore, the total force on mass 1 is

$$F_1 = -\frac{mg}{\ell}x_1 - k\left(x_1 - x_2\right) = m\ddot{x}_1$$

$$\Leftrightarrow \ddot{x}_1 + \frac{g}{\ell}x_1 + \frac{k}{m}\left(x_1 - x_2\right) = 0. \tag{5.2.1a}$$

Similarly, the differential equation governing the motion of mass 2 is

$$\ddot{x}_2 + \frac{g}{\ell}x_2 + \frac{k}{m}\left(x_2 - x_1\right) = 0. \tag{5.2.1b}$$

These are a set of two DEQs, representing the motion of two objects. The equations are "coupled" because x_2 appears in the first equation and x_1 appears in the second equation. This mathematical coupling of the equations is a direct result of the physical coupling of the masses. The coupling makes the equations difficult to solve, because we must simultaneously find the solutions $x_1(t)$ and $x_2(t)$ for both equations. It would be much easier to find solutions if we could decouple the equations, that is, if we could find a way to write two other second-order differential equations that also completely describe the system, but which aren't coupled. There is a general recipe for doing this, and we *will* study it, but for this simple case we will use physical insight instead.

---

3. The concept of probability density was discussed in section 1.11.

5.3   Normal modes

If we could succeed in describing the system with two uncoupled differential equations, then (because they're uncoupled), each would represent a completely independent type of motion, without any energy transfer between the two. This would be analogous to the two pendula shown in figure 5.2.1a, but with no spring between. These two uncoupled oscillators are described by two uncoupled DEQs:

$$\left.\begin{array}{r} -\dfrac{mg}{\ell}x_1 = m\ddot{x}_1 \Leftrightarrow \ddot{x}_1 + \dfrac{g}{\ell}x_1 = 0 \\[2ex] \ddot{x}_2 + \dfrac{g}{\ell}x_2 = 0 \end{array}\right\} \text{Two uncoupled DEQs} \qquad (5.3.1)$$

In this analogy, we can set mass 1 moving, and no energy is ever transferred to mass 2. Mass 1 continues oscillating in a simple "steady-state" motion, that is, it oscillates with constant amplitude.

For the coupled pendula, can we think of a way in which the system can move in a steady state, in which each mass oscillates with constant amplitude? We will call this way of moving a "normal mode." In fact, we need to think of two normal modes, since we know that we'll need two second-order DEQs to describe the system. (After all, we started with two DEQs, equations 5.2.1a and b.)

One of these normal modes is easy to guess, as shown in figure 5.3.1. This is called the "pendulum mode"; the two masses swing in phase. Since x_1 is always equal to x_2, the spring never gets stretched or compressed, so it never exerts any force. If there is no damping, the system would continue swinging in this mode forever.

> **Concept test:** See if you can figure out the other normal mode for this system. In other words, figure out a different way in which the system can move forever in a steady state. Don't look any further until you've put an honest effort into thinking about this.

This other normal mode is called the "breathing mode," and is shown in figure 5.3.2a. This is a completely antisymmetrical motion. Therefore, if we excite this mode only, there is no way to generate the symmetrical motion that would lead to the pendulum mode. Another way to understand why this mode would continue in a steady state is that the center point of the spring never moves. Therefore, you could attach the center

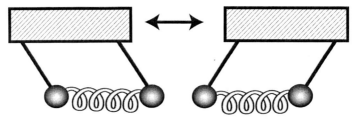

Figure 5.3.1 In the pendulum mode, the system oscillates between these two configurations. Each mass oscillates with constant amplitude.

Figure 5.3.2 a: In the breathing mode, the system oscillates between these two configurations. Each mass oscillates with constant amplitude. b: In the breathing mode, the center point of the spring doesn't move, so we could pretend that it's attached to a wall.

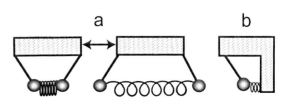

point to an immovable anchor without changing anything, turning the two pendula into two uncoupled oscillators moving in unison (though in opposite directions).

Now that we've found these two normal modes, let's find the two differential equations that describe them, and then see how they're related to the original differential equations 5.2.1a and b. We can characterize the pendulum mode by defining

$$s_p \equiv \frac{1}{\sqrt{2}} (x_1 + x_2) . \tag{5.3.2}[4]$$

When the system moves in the pendulum mode, there is an oscillation of s_p. If the system is "purely" in the pendulum mode, then $x_1 = x_2$, so $s_p = \sqrt{2}x_1 = \sqrt{2}x_2$, so this definition might seem to be of little use. However, we will soon consider more complicated motions, involving a *superposition* of the pendulum mode and the breathing mode, and then we will really need this definition. You can show in problem 5.17 that s_p always shows a simple harmonic oscillation, even when the motions of x_1 and x_2 are more complicated (because of a superposition of the two modes). Therefore, s_p is called the "normal mode coordinate" for the pendulum mode.

In the pure pendulum mode, the spring never stretches, so the motion of each pendulum is that of a simple pendulum. Therefore, we have

$$F_1 = m\ddot{x}_1 = -\frac{mg}{\ell}x_1 \Leftrightarrow \ddot{x}_1 + \frac{g}{\ell}x_1 = 0.$$

System in pendulum mode

In the pendulum mode, we have $s_p = \sqrt{2}x_1$, so that $\ddot{s}_p = \sqrt{2}\ddot{x}_1$. Therefore, we could just as well write

$$\ddot{s}_p + \frac{g}{\ell}s_p = 0. \tag{5.3.3}$$

This differential equation describes the motion of the pendulum mode. It has the form of a simple harmonic oscillator DEQ, $\ddot{x} + \frac{k}{m}x = 0$. Therefore, the angular frequency

---

4. The factor of $1/\sqrt{2}$ in this definition might seem unneeded, but eventually it will make things easier to think about.

of oscillation is given by the equivalent of $\omega = \sqrt{\dfrac{k}{m}}$:

$$\omega_{\mathrm{p}} = \sqrt{\frac{g}{\ell}} \qquad (5.3.4)$$

There is a different way of deriving the DEQ (5.3.3), which is quite revealing. Inspired by the definition of s_{p}, equation (5.3.2), we try adding together equations (5.2.1), after multiplying each by $1/\sqrt{2}$:

$$\frac{1}{\sqrt{2}}\left[\ddot{x}_1 + \frac{g}{\ell}x_1 + \frac{k}{m}\left(x_1 - x_2\right) = 0\right] \qquad (5.2.1\mathrm{a})$$

$$+ \quad \frac{1}{\sqrt{2}}\left[\ddot{x}_2 + \frac{g}{\ell}x_2 + \frac{k}{m}\left(x_2 - x_1\right) = 0\right] \qquad (5.2.1\mathrm{b})$$

$$\frac{1}{\sqrt{2}}\left[\left(\ddot{x}_1 + \ddot{x}_2\right) + \frac{g}{\ell}\left(x_1 + x_2\right) = 0\right]$$

Using the definition $s_{\mathrm{p}} = \dfrac{1}{\sqrt{2}}(x_1 + x_2)$, this becomes

$$\ddot{s}_{\mathrm{p}} + \frac{g}{\ell}s_{\mathrm{p}} = 0, \qquad (5.3.5)$$

which is just the same as equation (5.3.3). Note that to do this second derivation, we did not require that the system be in pendulum mode, so this equation must always hold, whether the system is in the pendulum mode, the breathing mode, or some more complicated motion.

Now, for the breathing mode. We can characterize the breathing mode by defining

$$s_{\mathrm{b}} \equiv \frac{1}{\sqrt{2}}\left(x_1 - x_2\right). \qquad (5.3.6)$$

When the system moves in the breathing mode, s_{b} oscillates. (If the system is "purely" in the breathing mode, then $x_2 = -x_1$, so $s_{\mathrm{b}} = \sqrt{2}x_1$.) Again, you can show in problem 5.17 that, even when the system is in a superposition of the two modes, s_{b} still moves in simple harmonic fashion. Therefore, s_{b} is called the normal mode coordinate for the breathing mode.

Since the center of the spring never moves, we could imagine replacing the right half of the apparatus by a brick wall, as shown in figure 5.3.2b; this would not affect the motion of mass 1. Therefore, the motion of the left mass is that of a simple harmonic oscillator; we just need to figure out the effective spring constant. (Of course, once we have found the motion of the left mass, we can easily find the motion of the right mass, since, in the breathing mode, $x_2 = -x_1$.) There are two forces acting on the left mass: the "pendulum force," with effective spring constant $\dfrac{mg}{\ell}$, and the force from the spring, which is effectively half its original length. Recall from section 2.3 that when the length of a spring is cut in half, the spring constant is doubled. Applying this to our analysis of the breathing mode, the effective total spring constant is $\frac{mg}{\ell} + 2k$, so that

$$F_1 = m\ddot{x}_1 = -\left(\frac{mg}{\ell} + 2k\right)x_1 \Leftrightarrow \ddot{x}_1 + \left(\frac{g}{\ell} + \frac{2k}{m}\right)x_1 = 0. \qquad (5.3.7)$$

System in breathing mode

Since (in the breathing mode), we have $s_b = \sqrt{2}x_1$, we could just as well write

$$\ddot{s}_b + \left(\frac{g}{\ell} + \frac{2k}{m}\right)s_b = 0 \qquad (5.3.8)$$

This differential equation describes the motion of the breathing mode. It has the form of a simple harmonic oscillator DEQ, $\ddot{x} + \frac{k}{m}x = 0$. Therefore, the angular frequency of oscillation is given by the equivalent of $\omega = \sqrt{\frac{k}{m}}$:

$$\omega_b = \sqrt{\frac{g}{\ell} + \frac{2k}{m}} \qquad (5.3.9)$$

We see that the breathing mode has a higher frequency than the pendulum mode, and that the difference in frequencies increases with the strength of the coupling, represented by k. We'll discuss this important feature at length in section 5.7.

As for the pendulum mode, there is a different, more mathematical way of deriving the breathing mode. Inspired by the definition of s_b, equation (5.3.6), we try subtracting equations (5.2.1), after multiplying each by $1/\sqrt{2}$:

$$\frac{1}{\sqrt{2}}\left[\ddot{x}_1 + \frac{g}{\ell}x_1 + \frac{k}{m}(x_1 - x_2) = 0\right] \qquad (5.2.1a)$$

$$- \frac{1}{\sqrt{2}}\left[\ddot{x}_2 + \frac{g}{\ell}x_2 + \frac{k}{m}(x_2 - x_1) = 0\right] \qquad (5.2.1b)$$

$$\frac{1}{\sqrt{2}}\left[(\ddot{x}_1 - \ddot{x}_2) + \frac{g}{\ell}(x_1 - x_2) + \frac{2k}{m}(x_1 - x_2) = 0\right]$$

Using the definition $s_b = \frac{1}{\sqrt{2}}(x_1 - x_2)$, this becomes

$$\ddot{s}_b + \left(\frac{g}{\ell} + \frac{2k}{m}\right)s_b = 0, \qquad (5.3.10)$$

which is just the same as equation (5.3.8). Again, we did not need to require that the system be in breathing mode, so this equation must always hold, whether the system is in the pendulum mode, the breathing mode, or some more complicated motion.

Recap: We began by describing the system of two coupled pendula in terms of the motion of each mass, as described by the pair of differential equations

DEQs describing the pendulum bobs:
$$\begin{cases} (5.2.1a): \ddot{x}_1 + \frac{g}{\ell}x_1 + \frac{k}{m}\left(x_1 - x_2\right) = 0 \\ (5.2.1b): \ddot{x}_2 + \frac{g}{\ell}x_2 + \frac{k}{m}\left(x_2 - x_1\right) = 0 \end{cases}$$

The motion of each pendulum bob can be complicated, with energy being transferred back and forth between the two bobs, and it is difficult to solve these differential equations because they are coupled.

Using physical insight, we discovered two simpler ways in which the system could move, the "pendulum mode" and the "breathing mode." When the system is in one of these modes, the behavior is a simple oscillation, with no transfer of energy from one mode to the other. We showed that the behavior of these two modes could be described by the pair of differential equations

DEQs describing the normal modes:
$$\begin{cases} (5.3.5): \ddot{s}_p + \frac{g}{\ell}s_p = 0 \\ (5.3.10): \ddot{s}_b + \left(\frac{g}{\ell} + \frac{2k}{m}\right)s_b = 0 \end{cases}$$

where

$$s_p \equiv \frac{1}{\sqrt{2}}\left(x_1 + x_2\right) \quad \text{and} \quad s_b \equiv \frac{1}{\sqrt{2}}\left(x_1 - x_2\right)$$

are the "normal mode coordinates." Since these DEQs have the form of simple harmonic oscillators, we could easily see that

$$\omega_p = \sqrt{\frac{g}{\ell}} \tag{5.3.4}$$

and

$$\omega_b = \sqrt{\frac{g}{\ell} + \frac{2k}{m}} \tag{5.3.9}$$

We showed that the second pair of DEQs could be derived from the first simply by adding or subtracting them.

Note that we don't really need to continue with steps 2 and 3 of our three-step procedure for solving the differential equations, since we've just shown that the motion of this system can be understood in terms of two simple oscillators (the pendulum mode and the breathing mode). However, we do need to get a better understanding of how these modes combine to produce the complex behavior of the pendulum bobs that you saw in the video.

5.4 Superposing normal modes

Because the pair of DEQs (5.3.5) and (5.3.10) that describes the two normal modes can be derived from the pair of DEQs (5.2.1a) and (5.2.1b) that describes the motion of the two pendulum bobs, each pair represents just as good a way of describing the *complete*

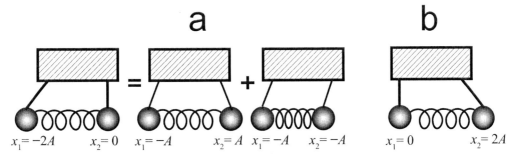

Figure 5.4.1 a: Any initial condition can be expressed as an appropriately weighted superposition of the normal modes. b: Express this initial condition as a weighted superposition of the normal modes.

behavior of the system as does the other. However, because each normal mode acts as a simple oscillator (completely independent of the other mode), it is easy to predict its behavior in time.

By choosing the phases and amplitudes of the two modes correctly, we can create any desired initial condition for the two pendulum bobs. For example, as shown in figure 5.4.1a, we can create the initial condition shown in the video clip (the one you watched at the beginning of section 5.2) by adding together equal amounts of pendulum and breathing modes.

After we let go, the pendulum and breathing modes oscillate independently. For such a mixture of modes, we can't directly observe the oscillations of each mode. Instead, we see the effect of the combination of both modes on each of the two masses:

$$s_p \equiv \frac{1}{\sqrt{2}} \left(x_1 + x_2 \right) \quad \text{and} \quad s_b \equiv \frac{1}{\sqrt{2}} \left(x_1 - x_2 \right)$$

$$\Leftrightarrow x_1 = \frac{1}{\sqrt{2}} \left(s_p + s_b \right) \quad \text{and} \quad x_2 = \frac{1}{\sqrt{2}} \left(s_p - s_b \right)$$

$$(5.4.1)$$

A remarkable insight: Thus, the "beating" behavior that we observe, say for mass 1, is due to the superposition of the pendulum mode (at angular frequency $\omega_p = \sqrt{\frac{g}{\ell}}$)

and the breathing mode (at a different angular frequency, $\omega_b = \sqrt{\frac{g}{\ell} + \frac{2k}{m}}$) ! (Recall from section 5.1 that superposing two oscillations of equal amplitude but different frequencies results in beating.)

In fact, we now understand that *any* behavior of the system can be understood in terms of a superposition of the two normal modes.

Your turn: Describe the initial condition shown in figure 5.4.1b (both pendula are initially at rest) in terms of a superposition of the two normal modes, in a way similar to figure 5.4.1a.

Example: A set of coupled pendula has $m = 0.10$ kg for both masses, but k and ℓ are unknown. At $t = 0$, the left mass is held at $x_1 = 2.0$ cm and the right mass is held at $x_2 = 0$ cm. The masses are then released. At first, the left mass oscillates with a period of about 1.1 s, and the right mass is nearly motionless. At $t = 10$ s, the two masses are oscillating with approximately equal amplitudes. At $t = 20$ s, the right mass is oscillating strongly, and the left mass is nearly motionless. At $t = 30$ s, the two masses are oscillating with approximately equal amplitude. At $t = 40$ s, the left mass is oscillating strongly, and the right mass is nearly motionless. Find approximate values for k and ℓ.

Solution: This beating behavior is caused by mixing together the two normal modes, with their two characteristic frequencies. For example, the motion of the left mass is a superposition of two different sinusoids of angular frequencies ω_b and ω_p, and the variations in the amplitude for the left pendulum are due to the beating that results from this superposition. The time between maximum amplitudes (e.g., for the left pendulum) is 40 s, so that the beat frequency is $f_{beat} = \dfrac{1}{40}$ Hz.

We have that

$$(5.1.5): f_{beat} = f_1 - f_2,$$

where in this case $f_1 = f_b = \dfrac{\omega_b}{2\pi}$ and $f_2 = f_p = \dfrac{\omega_p}{2\pi}$. So,

$$f_b - f_p = \frac{1}{40} \text{ Hz} \Rightarrow \omega_b - \omega_p = \left(\frac{\pi}{20} \text{ rad/s}\right). \qquad (5.4.2)$$

The period of the fast oscillations is given by $T_{fast} = \dfrac{2\pi}{\omega_{av}} = \dfrac{2\pi}{\dfrac{\omega_b + \omega_p}{2}} = \dfrac{4\pi}{\omega_b + \omega_p}$,

and we're told that this equals 1.1 s for this example. Therefore, $(1.1 \text{ s}) = \dfrac{4\pi}{\omega_b + \omega_p} \Leftrightarrow$

$\omega_b + \omega_p = \left(\dfrac{4\pi}{1.1} \text{ rad/s}\right)$. Subtracting this from equation (5.4.2) gives

$$\omega_b - \omega_p = \left(\frac{\pi}{20} \text{ rad/s}\right)$$

$$- \left[\omega_b + \omega_p = \left(\frac{4\pi}{1.1} \text{ rad/s}\right)\right]$$

$$\rule{6cm}{0.4pt}$$

$$-2\omega_p = \left(\frac{\pi}{20} \text{ rad/s}\right) - \left(\frac{4\pi}{1.1} \text{ rad/s}\right) \Leftrightarrow$$

$$\omega_p = 5.6 \text{ rad/s} = \sqrt{\frac{g}{\ell}}$$

$$\boxed{\Rightarrow \ell = 32 \text{ cm}}$$

Finally,

$$\omega_b - \omega_p = \frac{\pi}{20} \text{ rad/s} \Rightarrow \omega_b = 5.7 \text{ rad/s}$$

$$\omega_b = \sqrt{\frac{g}{\ell} + \frac{2k}{m}} \Rightarrow k = \frac{m}{2}\left(\omega_b^2 - \frac{g}{\ell}\right)$$

$$\boxed{\Rightarrow k = 0.088 \text{ N/m}}$$

5.5 **Normal mode analysis, and normal modes as an alternate description of reality**

The breathing mode and the pendulum mode are the two normal modes of the coupled pendulum system. We will see that we can find similar normal modes even for very complicated systems of many masses interacting through many different forces. All these normal modes share the characteristics of the breathing and pendulum modes shown above, leading us to the definition:

> **Definition of a normal mode**: A way in which the system can move in a steady state, in which all parts of the system move with the same frequency. The parts may have different (perhaps zero or negative) amplitudes.

Given any set of initial conditions, we can determine the fractions of pendulum-mode and breathing-mode which are needed to produce those initial conditions. This process is called "normal mode analysis." We can then use these normal mode amplitudes to determine the future behavior of the system as a function of time:

> **Normal mode analysis:**
>
> 1. Express the initial state of the system in terms of a superposition of the normal modes, with an appropriate amplitude and phase for each mode.
> 2. Write the time dependent expressions for each normal mode coordinate, e.g. $s_p = \text{Re}\left[A_p e^{i(\omega_p t + \varphi_p)}\right]$
> 3. If desired, form the appropriate combination of the normal mode coordinates to find the positions of the masses (x_1, x_2, etc.)

For now, we will only consider situations in which the initial velocities are zero, since this makes the math easier and shows the important ideas. However, we will treat the more general case later in this chapter. It is easiest to explain normal mode analysis with an example:

> **Core example**: A set of coupled pendula has $m = 0.10$ kg, $\ell = 0.15$ m, and $k = 5.0$ N/m. At $t = 0$, the left mass is held at $x_1 = 1.0$ cm and the right mass is held at $x_2 = 3.0$ cm. The masses are then released. What is the position of mass 1 at $t = 5.0$ s?
> **Solution**: First we make the normal modes description of the system. As we saw in chapter 1, for a single oscillator that begins at rest, the initial position is equal to the amplitude of the motion. Similarly in this case, because the initial velocities are zero, the amplitude for each normal mode is equal to the initial value of the normal mode coordinate,[5] as defined by $s_p \equiv \dfrac{1}{\sqrt{2}}(x_1 + x_2)$ and $s_b \equiv \dfrac{1}{\sqrt{2}}(x_1 - x_2)$.
>
> *continued*

---

5. You might be concerned that the two modes might have nonzero initial velocities $\dot{s}_{p0}$ and $\dot{s}_{b0}$ which, when added together, could produce zero initial velocity for both pendula.

Plugging in the initial values of x_1 and x_2, we get $s_p = \dfrac{4.0}{\sqrt{2}}$ cm $= 2\sqrt{2}$ cm and $s_b = -\dfrac{2.0}{\sqrt{2}}$ cm $= -\sqrt{2}$ cm. Each mode acts as an independent oscillator. Since the initial velocities are zero, the phase factor φ in the solution $s = \text{Re}\left[Ae^{i(\omega t + \varphi)}\right]$ is zero (as we saw in chapter 1), so we can simply use $s = A\cos \omega t$. So, we have:

$$s_p = (2\sqrt{2}\text{ cm})\cos \omega_p t \text{ and } s_b = (-\sqrt{2}\text{ cm})\cos \omega_b t.$$

From equation (5.4.1), we have $x_1 = \dfrac{1}{\sqrt{2}}(s_p + s_b)$, so

$$x_1 = (2\text{ cm})\cos \omega_p t - (1\text{ cm})\cos \omega_b t.$$

The angular frequencies of the normal modes are $\omega_p = \sqrt{\dfrac{g}{\ell}} = 8.1$ rad/s and $\omega_b = \sqrt{\dfrac{g}{\ell} + \dfrac{2k}{m}} = 13$ rad/s. Plugging these and $t = 5.0$ s into the above expression for x_1 gives $x_1(t = 5.0\text{ s}) = -1.9$ cm.

Why do we care so much about describing things in terms of normal modes? There are several reasons. First, as shown in the above "core by example" section, and as we'll see as we progress through the book, the normal modes description provides by far the easiest way of describing the behavior of a complicated system as a function of time, given the initial conditions. Secondly, our two most important senses, sight and hearing, perceive the world in a way that is closely related to normal modes. For example, when you hear a musical note being played, you don't perceive a rapid oscillation of air pressure, but rather you hear a single pitch. When you see something that emits blue light, you don't perceive a rapidly oscillating electric field; instead, you see "blue." Normal modes are also very important in understanding the spectra of chemicals; the infrared spectrum reveals the frequencies of the vibrational normal modes.

Finally, the normal mode picture presents an alternate, and a very powerful, way of looking at the world. We will see that, for systems in which the motion is one-dimensional, the number of normal modes is equal to the number of particles. For example, in the two-pendulum system there were two normal modes. In a three pendulum system, there would be three normal modes, etc. For each system, we can either choose to describe the motion of each particle, or we can instead choose to describe the motion of each normal mode. In a somewhat fantastic analogy, imagine trying to completely describe a person. One way of doing it (analogous to describing the positions of the two masses in our coupled pendula) would be to specify the positions and velocities of all the electrons and nuclei that make up the person. But, instead, we might say, "This person is a mixture of 40% Hillary Clinton, 20% Frank Sinatra, 20% Snoop Dogg, and 20% Julia Roberts." Such a description would be similar to

---

However, for the pendulum mode the velocities of the two pendula are always equal, while for the breathing mode they are always opposite. Therefore, while it is possible to add cancelling velocities for one of the two pendula, the velocities for the other would not cancel. The only way to get zero initial velocity for both pendula is to have zero initial velocity for both modes.

the normal modes way of describing the coupled pendula—the system is described in terms of its archetypical behaviors. Such a description places no limits on the system, and does not require any less information (to describe an arbitrary person in terms of personality archetypes might require 10^{25} different archetypes), but is often more revealing.

Energy of a superposition state

By equation (5.4.1), $x_1 = \dfrac{1}{\sqrt{2}}(s_p + s_b)$. So, when the system is purely in the pendulum mode, the amplitude of the left pendulum bob is $\dfrac{A_p}{\sqrt{2}}$, where A_p is the amplitude of the pendulum mode, that is, the amplitude of s_p. Therefore, the energy of the left bob is $\dfrac{1}{2}k_{\text{eff}}\left(\dfrac{A_p}{\sqrt{2}}\right)^2$, where the effective spring constant k_{eff} is entirely due to the "pendulum force," so that $k_{\text{eff}} = \dfrac{mg}{\ell}$. Since $\omega_p = \sqrt{\dfrac{g}{\ell}}$, we could also write that the energy of the left bob is $\frac{1}{4}m\omega_p^2 A_p^2$. The right bob has the same amount of energy, so that the total energy is $\frac{1}{2}m\omega_p^2 A_p^2$. For convenience, we define $E_p \equiv \frac{1}{2}m\omega_p^2\,(1\text{ m})^2$; this is the energy the pendulum mode would have for an amplitude of 1 m. Using this, the total energy of the system in the pure pendulum mode can be expressed as $\dfrac{A_p^2}{(1\text{ m})^2}E_p$.

If instead the system is in a pure breathing mode, then identical arguments show that the total energy is $\dfrac{A_b^2}{(1\text{ m})^2}E_b$, where A_b is the amplitude of motion of the left bob in this breathing mode (which of course is the same as that of the right bob), and $E_b \equiv \frac{1}{2}m\omega_b^2\,(1\text{ m})^2$ is the energy the mode would have for an amplitude of 1 m.

Now, let's consider the energy when the system is in a state in which the two modes are superposed. Since each mode can be considered as a completely independent oscillator, the total energy is simply the sum of the energies of the two modes, that is,

$$E_{\text{TOT}} = \frac{A_p^2}{(1\text{ m})^2}E_p + \frac{A_b^2}{(1\text{ m})^2}E_b \tag{5.5.1}$$

We see that this is a sort of weighted sum, with the weights given by the *squares* of the amplitudes of each mode. We'll see in the next few pages that this is closely related to the energy of a quantum mechanical system that is in a superposition of two quantum states.

Analogy between quantum superpositions and normal mode superpositions

In quantum mechanics, normal modes play an absolutely central role. For example, each electron in an atom can be in one of several "quantum states," labeled 1s, 2s, 2p, etc. Each of these states is a different normal mode for the electron. For a quantum mechanical particle, there is a simple relationship between the total energy of the particle and the angular frequency of oscillation:

$$E = \hbar\omega,$$

where $\hbar \equiv \dfrac{h}{2\pi}$. (Both $\hbar \approx 1.055 \times 10^{-34}$ J s and $h \approx 6.63 \times 10^{-34}$ J s are called "Planck's constant.") Therefore, each normal mode in quantum mechanics not only has a characteristic angular frequency of oscillation, but also a corresponding energy. This explains why the normal modes of a quantum system are called "energy eigenstates." ("Eigen" means "characteristic" in German.) When the system is in one of these energy eigenstates, the quantum mechanical wavefunction Ψ oscillates at the same frequency at all points, just as the two pendula oscillate at the same frequency when the system is in the pendulum mode (or the breathing mode). Just as in the coupled pendulum system, it is also possible to add together or superpose energy eigenstates to create more complicated states. For example, say that Ψ_a represents a state with energy E_a, and Ψ_b represents a state with energy E_b. We could create a state with a superposition of these two:

$$\Psi_{mix} = \left(A_a e^{i\varphi_a}\right)\Psi_a + \left(A_b e^{i\varphi_b}\right)\Psi_b$$

Here, we have explicitly indicated the amplitudes A_a and A_b being used in the superposition, as well as the phases φ_a and φ_b. The quantum mechanical wavefunction is inherently complex (!), so we don't take the real part. This mixed state is exactly analogous to mixing together a pendulum mode of amplitude A_p and a breathing mode of amplitude A_b, for the coupled pendulum system.

Just as the energy in the coupled pendulum mixed state "sloshes" back and forth between the two pendula, the probability distribution for an electron in such a quantum state is complicated, and the peak of probability "sloshes around" from one place to another.

However, there is one area in which this otherwise very good analogy doesn't work. If you measure the energy of the electron in this mixed quantum state, you always get E_a or E_b, and never anything in between. This indeed is a "quantum mystery," one that you will learn more about elsewhere, and one that has no analogy in the classical world. However, the probability of measuring energy E_a turns out to be A_a^2, and the probability of measuring E_b turns out to be A_b^2. Therefore, if you could prepare a million identical electrons, all initially in this same superposed state, and measure their energies, then the *average* energy of all these measurements would be

$$E_{average} = A_a^2 E_a + A_b^2 E_b$$

The average energy, defined in this way, is called the "expectation value" of the energy. (This is somewhat of a misnomer, since we don't actually *expect any* single measurement to match the expectation value. Rather, each single measurement will yield either E_a or E_b.) Comparing this with equation 5.5.1 shows that there is a close analogy between the classical energy of the coupled pendulum system and the *expectation value* of the energy for the quantum mechanical system.

5.6   Hilbert space and bra-ket notation

There is a powerful way of thinking about superposing normal modes to create more complicated behavior, or the inverse operation of analyzing complicated behavior into its simpler normal mode components. Again, for now we restrict ourselves to situations in which both pendula have zero initial velocity. We could then visualize all the possible initial conditions as points or vectors on a plane, with the initial position of mass 1, x_{10}, as the horizontal axis, and the initial position of mass 2, x_{20}, on the vertical axis. Choosing a point in this plane, that is, choosing a set of initial conditions, completely specifies the behavior of the system for all subsequent times, as shown in the example of figure 5.6.1.

This plane is the simplest example of a Hilbert space[6]; so far there is not much remarkable about it, but if we define things carefully we can generalize to much more complicated Hilbert spaces, which can be used to describe much more complicated systems, and we will gain enormous power! Mwah hah hah!!

Ahem. For example, if we had three masses in the system, we would need a three-dimensional Hilbert space to represent all the possible initial states of the system (again, sticking for now with the requirement that all the initial velocities be zero). If we had four masses, we would need a four-dimensional Hilbert space. If we had 100 masses, we would need a 100-dimensional Hilbert space. If we want to describe a continuous system, such as a rope, we could divide it up into an infinite number of infinitesimally small masses; we would need an infinite-dimensional Hilbert space to describe such a system. Choosing one point/vector in such a space would correspond to specifying one coordinate along each of the infinite number of axes which define the space. This infinite list of numbers is basically the same as a function; to completely define a function, you must define the value of the function at an infinite number of points along the x-axis. So, each different point in an infinite-dimensional Hilbert space signifies a different function.

Figure 5.6.1 If we require the initial velocities of both pendula to be zero, then any initial condition can be depicted as a vector on the plane shown in the right part of the figure. The vector can be specified in column matrix format, with x_{10} in the top line and x_{20} in the bottom line.

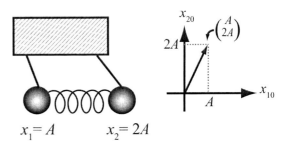

---

6. The general definition of Hilbert space is a vector space with a defined "inner product". The inner product is a rule for combining two vectors to create a new quantity; for ordinary vectors in three-dimensional space, the dot product is the inner product. Formally, a Hilbert space must be "complete" in a sense that is carefully defined by mathematicians, but this restriction is seldom important in physics.

But, for now, let's stick to our two-dimensional Hilbert space that describes our system of two coupled pendula. Each vector in this space can be represented as a column matrix, with the top entry showing the component along the x_{10} axis and the bottom entry showing the component along the x_{20} axis, as shown in the figure. This way of representing vectors may be new to you; it works well when you need to multiply a vector by a matrix, which we'll do in chapter 6.

Let's see how vector multiplication works in this notation. Here's the way you're used to taking dot products:

$$\mathbf{A} \cdot \mathbf{B} = A_x B_x + A_y B_y. \tag{5.6.1}$$

However, eventually (once we allow the initial velocities to be nonzero) we will need to deal with vectors that have complex components. For such vectors, the above definition must be modified. We want the dot product of any vector with itself to equal the square of the length of the vector. This should be real, but

$$\mathbf{A} \cdot \mathbf{A} = A_x A_x + A_y A_y$$

isn't real if A_x or A_y is complex. Therefore, we instead define an extended version of the dot product:

$$\mathbf{A} \cdot \mathbf{B} \equiv A_x^* B_x + A_y^* B_y, \tag{5.6.2}$$

where the * indicates "complex conjugate." Recall from section 1.8 that, to take the complex conjugate of any number, you simply replace any occurrence of i by $-$i. For example, if $C = a + ib = Ae^{i\varphi}$, then $C^* = a - ib = Ae^{-i\varphi}$. Thus, equation (5.6.2) is the same as equation (5.6.1) if the components are real. The complex conjugate makes it easy to find the magnitude of a complex number; if $C = a + ib = Ae^{i\varphi}$, then

$$\text{Cartesian version: } CC^* = (a + ib)(a - ib) = a^2 + b^2 = |C|^2$$

$$\text{Polar version: } CC^* = \left(Ae^{i\varphi}\right)\left(Ae^{-i\varphi}\right) = A^2$$

Using the extended definition of the dot product, equation (5.6.2), we have

$$\mathbf{A} \cdot \mathbf{A} = A_x^* A_x + A_y^* A_y = |A_x|^2 + |A_y|^2,$$

which is thus guaranteed to be real.

In the language of Hilbert space, we refer to dot products as "inner products." To take the inner product of two vectors using matrix notation, we must first take the adjoint (also called the "Hermitian transpose" or "Hermitian conjugate") of the one on the left. Taking the adjoint means: write the matrix as a row instead of a column, and take the complex conjugate of the entries. Here's how this works:

$$\mathbf{A} \equiv \begin{pmatrix} A_1 \\ A_2 \end{pmatrix} \Rightarrow \text{ adjoint of } \mathbf{A} \text{ is } \quad \mathbf{A}^\dagger \equiv \begin{pmatrix} A_1^* & A_2^* \end{pmatrix},$$

where the † superscript means "adjoint." The next step in taking the inner product of $\mathbf{A}$ and $\mathbf{B}$ is to multiply $\mathbf{A}^\dagger$ with $\mathbf{B}$ using the rules of matrix multiplication. We'll review the full rules later. For now, we simply multiply the left entry of $\mathbf{A}^\dagger$ by the top entry of

B and add this to the product of the right entry of $\mathbf{A}^{\dagger}$ and the bottom entry of **B**:

$$\mathbf{B} \equiv \begin{pmatrix} B_1 \\ B_2 \end{pmatrix} \quad \text{Inner product of } \mathbf{A} \text{ and } \mathbf{B} = \begin{pmatrix} A_1^* & A_2^* \end{pmatrix} \begin{pmatrix} B_1 \\ B_2 \end{pmatrix} = A_1^* B_1 + A_2^* B_2. \quad (5.6.3)$$

This gives the same result as the extended definition of the dot product, equation (5.6.2). (Recall that, for now, all the vectors in our Hilbert space are real, but we want to make fully general definitions.)

Bra-ket notation In 1930, Paul Dirac introduced a wonderful notation that simplfies the writing of inner products; the advantages are not immediately obvious, but this notation is almost universally used for quantum mechanics, so you may as well begin getting used to it. Instead of writing a vector in Hilbert space as **A** or $\vec{A}$, we instead write it as $|A\rangle$. **The "$|\rangle$" that surrounds the "A" plays the same role as the arrow in** $\vec{A}$ – it simply tells you that the quantity is a vector. (In fact you can see that the "$|\rangle$" sort of looks like an arrow.) A vector written with this new notation is called a "ket" – part of a little physics joke, as you'll see. Thus, we write

$$|A\rangle = \begin{pmatrix} A_1 \\ A_2 \end{pmatrix}$$

The adjoint of $|A\rangle$ is written $\langle A|$ – this is called a "bra":

$$\langle A| \equiv \begin{pmatrix} A_1^* & A_2^* \end{pmatrix}$$

Finally, the inner product of the vectors $|A\rangle$ and $|B\rangle$ is written $\langle A \mid B \rangle$:

$$|B\rangle = \begin{pmatrix} B_1 \\ B_2 \end{pmatrix}$$

$$\langle A \mid B \rangle = \begin{pmatrix} A_1^* & A_2^* \end{pmatrix} \begin{pmatrix} B_1 \\ B_2 \end{pmatrix} = A_1^* B_1 + A_2^* B_2.$$

Because this brings together the "bra" $\langle A|$ with the "ket" $|B\rangle$, it forms a "bra-ket," or bracket. Get it? Heh, heh.

Now, that you're done rolling in the aisles, let's see if you really did get it.

Concept test: If the ket $|A\rangle$ is given by $|A\rangle = \begin{pmatrix} 2 \\ -7 + 3i \end{pmatrix}$, what is the corresponding bra?

(Answer below.[7])

---

7. $\langle A| = (2 \quad -7 - 3i)$.

Self-test: If $|B\rangle = \begin{pmatrix} 4 - 2i \\ 5 \end{pmatrix}$, and $|A\rangle = \begin{pmatrix} 2 \\ -7 + 3i \end{pmatrix}$, what is the value of their inner product, $\langle A \mid B\rangle$? (Answer below.[8])

We will use the terms "vector in Hilbert space" (or simply "vector") and "ket" interchangeably.[9]

Aside: Representing kets with column matrices. Most physicists and mathematicians are comfortable writing a vector in ordinary space as $\mathbf{r} = \begin{pmatrix} x \\ y \end{pmatrix}$, where the top line of the column matrix means the position along the x-axis and the bottom line means the position along the y-axis. Perhaps $\mathbf{r}$ indicates the position of a flea on the surface of a table. If we choose to use a different coordinate system, perhaps one that is rotated by 45° clockwise relative to the original one, this does not affect the position of the flea, so that the meaning of $\mathbf{r}$ is unchanged. However, we would have to change the way $\mathbf{r}$ is *represented* by the column matrix. For example, if in the original coordinate system S we have $\mathbf{r} = \begin{pmatrix} 1 \\ 1 \end{pmatrix}$, then in the rotated coordinate system S' we would have $\mathbf{r} = \begin{pmatrix} 0 \\ \sqrt{2} \end{pmatrix}$. In order to make sense of either of these equations, we have to know what "basis" is being used, in other words whether the top line of the column matrix means the value of x (in the original coordinate system) or instead the value of x' (in the rotated coordinate system), and similarly for the bottom line. To avoid this possible confusion, we could write $\mathbf{r} \underset{S}{\rightarrow} \begin{pmatrix} 1 \\ 1 \end{pmatrix}$, where the arrow with the S under it means "in the S coordinate system, is represented by." However, this notation, although explicit, can be cumbersome. So, as long as it is clear which coordinate system is in use, we simply write $\mathbf{r} = \begin{pmatrix} 1 \\ 1 \end{pmatrix}$.

Exactly the same arguments apply for vectors in Hilbert space. In our case, when we write $|A\rangle = \begin{pmatrix} A_1 \\ A_2 \end{pmatrix}$, this is shorthand for $|A\rangle \underset{\substack{\text{positions} \\ \text{of masses}}}{\rightarrow} \begin{pmatrix} A_1 \\ A_2 \end{pmatrix}$, where the arrow means "in a coordinate system where the axes indicate the positions of the masses, is represented by." This is the coordinate system shown in the right part of figure 5.6.1. However, we could describe the same Hilbert space using a coordinate system with rotated axes. The meaning of $|A\rangle$ would be unchanged, but we would have to change the entries in the column matrix accordingly for this new basis, just as we did for the case of the flea.

continued

8.

$$\langle A \mid B\rangle = \begin{pmatrix} 2 & -7 - 3i \end{pmatrix} \begin{pmatrix} 4 - 2i \\ 5 \end{pmatrix} = 2(4 - 2i) + (-7 - 3i)5$$

$$= (8 - 35) + (-4 - 15)i = -27 - 19i.$$

9. Note that mathematicians typically write the inner product as $\langle A, B\rangle$, instead of $\langle A \mid B\rangle$, but the meaning is exactly the same.

> In this book, we will consistently use the coordinate system where each line of the column vector corresponds to the position of one of the masses, so we will simply write $|A\rangle = \begin{pmatrix} A_1 \\ A_2 \end{pmatrix}$ with this understanding. As David Griffiths writes,[10] "Technically, the 'equals' signs here [in equations such as $|A\rangle = \begin{pmatrix} A_1 \\ A_2 \end{pmatrix}$] mean 'is represented by', but I don't think any confusion will arise if we adopt the customary informal notation."

Now, back to the Hilbert space for our coupled pendulum system. Recall that the horizontal axis represents x_{10} (the initial position of the left bob), while the vertical axis represents x_{20}. Also, recall that, to start with, we are restricting ourselves to the important special case of zero initial velocities.

Some vectors in this Hilbert space represent the normal modes; two examples are shown in figure 5.6.2. Note that any vector of the form $\begin{pmatrix} a \\ a \end{pmatrix}$, that is any vector with the same entry in the top and bottom, would represent a "pure" pendulum mode. However, the one shown in the diagram (the one labeled "pendulum") is special: it has length 1. A vector of length 1 is called a "normalized" vector. The components of the vectors shown are dimensionless. That is, the length of each vector shown here is simply 1, not (1 m). This is exactly like the unit vectors $\hat{\mathbf{i}}$ and $\hat{\mathbf{j}}$ that you're used to for regular x–y space; they are vectors of length 1.

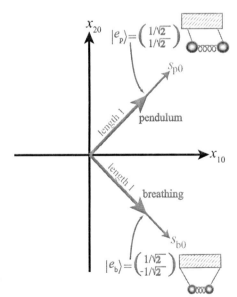

Figure 5.6.2 The eigenvectors for the symmetric coupled pendulum system.

10. *Introduction to Quantum Mechanics, 2nd Ed.*, Pearson Prentice-Hall, Upper Saddle River, NJ, 2005, p. 120.

The normalized vectors in the Hilbert space that represent the normal modes are called "eigenvectors." As you can show in problem W5.1 (on the website for this text), the eigenvectors lie along the axes that give the initial values of the pendulum mode coordinate $s_p \equiv \dfrac{1}{\sqrt{2}}(x_1 + x_2)$ and the breathing mode coordinate $s_b \equiv \dfrac{1}{\sqrt{2}}(x_1 - x_2)$. Thus, we could label these axes s_{p0} and s_{b0}, as shown in figure 5.6.2. However, it is the eigenvectors, rather than the axes, that are more important conceptually.

Note that the two eigenvectors are perpendicular or "orthogonal" to each other. Although this is obvious to the eye, we can check it by taking their inner product, which should equal zero for orthogonal vectors:

$$(1/\sqrt{2} \quad 1/\sqrt{2}) \begin{pmatrix} 1/\sqrt{2} \\ -1/\sqrt{2} \end{pmatrix} = \frac{1}{2} - \frac{1}{2} = 0 \checkmark \tag{5.6.4}$$

Let us define symbols for the two eigenvectors:

$$\left| e_p \right\rangle = \begin{pmatrix} 1/\sqrt{2} \\ 1/\sqrt{2} \end{pmatrix} = \frac{1}{\sqrt{2}} \begin{pmatrix} 1 \\ 1 \end{pmatrix} \text{ and } \left| e_b \right\rangle = \frac{1}{\sqrt{2}} \begin{pmatrix} 1 \\ -1 \end{pmatrix},$$

where the "e" stands for eigenvector. In terms of these symbols, we could write equation (5.6.4) as $\langle e_p \mid e_b \rangle = 0$.

Your turn: Verify that $\langle e_p \mid e_b \rangle = 0$, $\langle e_p \mid e_p \rangle = 1$, and $\langle e_b \mid e_b \rangle = 1$.

We could describe *any* vector in the plane by suitable combinations of the two eigenvectors. For example, the black vector shown in figure 5.6.3 has $x_{10} = (1 \text{ cm})$ and $x_{20} = (0.5 \text{ cm})$. Therefore, we can write it in terms of its components (shown by the black dashed lines) along the x_{10} and x_{20} axes: $\begin{pmatrix} 1 \text{ cm} \\ 0.5 \text{ cm} \end{pmatrix}$. However, we could also write it in terms of its components (as shown by the gray dashed lines) along the axes

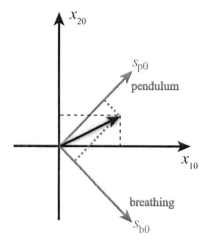

Figure 5.6.3 The black vector can be resolved into components either along the x_{10} and x_{20} axes (as shown by the black dashed lines) or along the pendulum mode and breathing mode axes (as shown by the gray dashed lines).

defined by the two eigenvectors. In other words, we could write it as

$$\begin{pmatrix} 1 \text{ cm} \\ 0.5 \text{ cm} \end{pmatrix} = \frac{(1.5 \text{ cm})}{\sqrt{2}} \begin{pmatrix} 1/\sqrt{2} \\ 1/\sqrt{2} \end{pmatrix} + \frac{(0.5 \text{ cm})}{\sqrt{2}} \begin{pmatrix} 1/\sqrt{2} \\ -1/\sqrt{2} \end{pmatrix}$$

$$= \frac{(1.5 \text{ cm})}{\sqrt{2}} |e_p\rangle + \frac{(0.5 \text{ cm})}{\sqrt{2}} |e_b\rangle$$

Your turn: Verify that the sum of vectors on the right actually adds up to the vector on the left.

This shows graphically that we can use the normal modes representation as a different and equally good way to describe the system. Either we can describe the initial condition shown in figure 5.6.3 as "the left pendulum is displaced 1 cm to the right and the right pendulum is displaced 0.5 cm to the right," or we could instead say "the pendulum mode has an amplitude of $\dfrac{(1.5 \text{ cm})}{\sqrt{2}}$ and the breathing mode has an amplitude of $\dfrac{(0.5 \text{ cm})}{\sqrt{2}}$." Note that either description requires two pieces of information (plus the information that the initial velocities are zero). The normal mode description can be seen as a way of describing the same plane of points, but with axes that are rotated by 45°.

Since we can describe *any* vector in the Hilbert space using suitable combinations of the eigenvectors, the set of two eigenvectors is called a "complete basis." We might also say that this set of vectors "spans" the space. Because the eigenvectors are normal and orthogonal, they form a "complete orthonormal basis." Often, the word "complete" is dropped, and we simply say that the set of two eigenvectors form an "orthonormal basis."

Terminology review:

Hilbert space[11]: A vector space with a defined inner product. As used in this book: a space in which each point represents a particular configuration of the system. In most applications of Hilbert space, the space has infinite dimensions, so each point represents a function. In fact, it might be helpful for you to start thinking of the vectors in our plane as functions that are evaluated only at two points (at mass 1 and at mass 2). However, if that idea confuses you now, don't start thinking that way yet.

Column matrix (also called column vector): a matrix of the form $\begin{pmatrix} A_1 \\ A_2 \end{pmatrix}$ that represents a vector: the top entry represents the component along the horizontal axis, and the bottom entry represents the component along the vertical axis.

continued

---

11. There are a number of amusing anecdotes about David Hilbert. When Hilbert spaces were just starting to be used in a variety of mathematical fields, as well as in physics, he attended a conference with the mathematician Richard Courant. Listening to a series of presentations about different uses of Hilbert space, Hilbert leaned over to Courant and asked, "Richard, exactly what *is* a Hilbert space?"

Adjoint (or Hermitian transpose or Hermitian conjugate): The adjoint of $\begin{pmatrix} A_1 \\ A_2 \end{pmatrix}$ is $(A_1^* \ A_2^*)$.

Ket: Another name for a vector in Hilbert space. For example, $|A\rangle = \begin{pmatrix} A_1 \\ A_2 \end{pmatrix}$.

Bra: The adjoint of a ket. For example, $\langle A| = (A_1^* \ A_2^*)$

Inner product: Generalized version of the dot product: $\langle A \mid B \rangle = (A_1^* \ A_2^*) \begin{pmatrix} B_1 \\ B_2 \end{pmatrix} = A_1^* B_1 + A_2^* B_2$

Normalized vector: A vector in the Hilbert space that has length 1; the length is dimensionless. (The unit vectors $\hat{\mathbf{i}}$ and $\hat{\mathbf{j}}$ are examples of normalized vectors in x–y space.) The inner product of a normalized vector with itself equals 1.

Orthogonal vectors: Vectors that are perpendicular to each other. They have an inner product of zero.

Complete basis: A set of vectors, linear combinations of which can be used to create *any* vector in the Hilbert space. For a two-dimensional Hilbert space, any two vectors that are nonparallel would form a complete basis.

Complete orthonormal basis (or simply "orthonormal basis"): A set of orthogonal normalized vectors that form a complete basis. (The unit vectors $\hat{\mathbf{i}}$ and $\hat{\mathbf{j}}$ form a complete orthonormal basis for x–y space.)

The component of an arbitrary vector $|x_0\rangle$ along the s_{p0} axis gives the amplitude of the pendulum mode needed to create the initial conditions represented by $|x_0\rangle$, and the component of $|x_0\rangle$ along the s_{b0} axis gives the amplitude of the breathing mode that is needed. To find these components, we simply take inner products (just as you can use dot products to find the components of a vector in x–y space: $D_x = \hat{\mathbf{i}} \cdot \mathbf{D}$):

$$
\begin{aligned}
\text{amplitude of pendulum mode} &= A_{\mathrm{p}} = \langle e_{\mathrm{p}} \mid x_0 \rangle \text{ where } |e_{\mathrm{p}}\rangle \equiv \frac{1}{\sqrt{2}} \begin{pmatrix} 1 \\ 1 \end{pmatrix} \\
\text{amplitude of breathing mode} &= A_{\mathrm{b}} = \langle e_{\mathrm{b}} \mid x_0 \rangle \text{ where } |e_{\mathrm{b}}\rangle \equiv \frac{1}{\sqrt{2}} \begin{pmatrix} 1 \\ -1 \end{pmatrix} \\
\Rightarrow |x_0\rangle &= \langle e_{\mathrm{p}} \mid x_0 \rangle |e_{\mathrm{p}}\rangle + \langle e_{\mathrm{b}} \mid x_0 \rangle |e_{\mathrm{b}}\rangle
\end{aligned}
$$

(5.6.5)

(zero initial velocities)

Again, this is *exactly* analogous to the process of expressing ordinary vectors in components:

$$
\begin{aligned}
\hat{\mathbf{i}} &\Leftrightarrow |e_{\mathrm{p}}\rangle \\
\hat{\mathbf{j}} &\Leftrightarrow |e_{\mathrm{b}}\rangle \\
D_x = \hat{\mathbf{i}} \cdot \mathbf{D} &\Leftrightarrow \text{amplitude of pendulum mode} = A_{\mathrm{p}} = \langle e_{\mathrm{p}} \mid x_0 \rangle \\
D_y = \hat{\mathbf{j}} \cdot \mathbf{D} &\Leftrightarrow \text{amplitude of breathing mode} = A_{\mathrm{b}} = \langle e_{\mathrm{b}} \mid x_0 \rangle \\
\mathbf{D} = (\hat{\mathbf{i}} \cdot \mathbf{D})\hat{\mathbf{i}} + (\hat{\mathbf{j}} \cdot \mathbf{D})\hat{\mathbf{j}} &\Leftrightarrow |x_0\rangle = \langle e_{\mathrm{p}} \mid x_0 \rangle |e_{\mathrm{p}}\rangle + \langle e_{\mathrm{b}} \mid x_0 \rangle |e_{\mathrm{b}}\rangle
\end{aligned}
$$

How do we use Hilbert space to help with thinking about normal mode analysis? Let's use the symbol $|x(t)\rangle$ to represent a vector with the positions of both pendula as a function of time, that is,

$$|x(t)\rangle \equiv \begin{pmatrix} x_1(t) \\ x_2(t) \end{pmatrix}$$

Consider the special case of a pure breathing mode. Then, $|x_0\rangle$ must have the form $|x_0\rangle = A_b |e_b\rangle = \dfrac{A_b}{\sqrt{2}} \begin{pmatrix} 1 \\ -1 \end{pmatrix}$, where A_b is the amplitude of the breathing mode. (Recall that $|x_0\rangle = \dfrac{A_b}{\sqrt{2}} \begin{pmatrix} 1 \\ -1 \end{pmatrix}$ means $x_{10} = A_p/\sqrt{2}$ and $x_{20} = -A_p/\sqrt{2}$.) For this simple case of a pure breathing mode, we know that both pendulum bobs just oscillate with angular frequency ω_b, so that the time dependence is $|x(t)\rangle = \dfrac{A_b}{\sqrt{2}} \cos \omega_b t \begin{pmatrix} 1 \\ -1 \end{pmatrix}$.

Similarly, if we take the special case of a pure pendulum mode, then

$$|x_0\rangle = A_p|e_p\rangle = \frac{A_p}{\sqrt{2}} \begin{pmatrix} 1 \\ 1 \end{pmatrix} \quad \text{and} \quad |x(t)\rangle = \frac{A_p}{\sqrt{2}} \cos \omega_p t \begin{pmatrix} 1 \\ 1 \end{pmatrix}.$$

Example: Let's rework the example from section 5.5 using these new ideas: A set of coupled pendula has $m = 0.10$ kg, $\ell = 0.15$ m, and $k = 5.0$ N/m. At $t = 0$, the left mass is held at $x_1 = 1.0$ cm and the right mass is held at $x_2 = 3.0$ cm. The masses are then released. What is the position of mass 1 at $t = 5.0$ s?

Solution (using Hilbert space to represent things graphically): The initial position vector in Hilbert space is shown in figure 5.6.4. The components along the eigenvectors represent the initial amplitude of each normal mode.

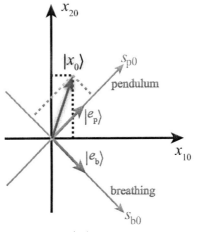

Figure 5.6.4 The initial position vector $|x_0\rangle$ is resolved into components along the axes defined by $|e_p\rangle$ and $|e_b\rangle$.

continued

Concept test (answer below[12]): We're about to find the components quantitatively, but first estimate what they are from the figure (which is drawn to scale).

To find these components quantitatively, we take the inner product of each eigenvector with the vector representing the initial condition of the system, $|x_0\rangle = \begin{pmatrix} 1.0 \text{ cm} \\ 3.0 \text{ cm} \end{pmatrix}$:

$$\text{component of pendulum mode} = \langle e_p \mid x_0 \rangle = \frac{1}{\sqrt{2}} (1 \ 1) \begin{pmatrix} 1.0 \text{ cm} \\ 3.0 \text{ cm} \end{pmatrix}$$

$$= \frac{(1 \text{ cm})}{\sqrt{2}} (1.0 + 3.0) = 2\sqrt{2} \text{ cm}$$

$$\text{component of breathing mode} = \langle e_b \mid x_0 \rangle = \frac{1}{\sqrt{2}} (1 \ {-1}) \begin{pmatrix} 1.0 \text{ cm} \\ 3.0 \text{ cm} \end{pmatrix}$$

$$= \frac{(1 \text{ cm})}{\sqrt{2}} (1.0 - 3.0) = -\sqrt{2} \text{ cm}$$

Using equation (5.6.6), we then have

$$|x(t)\rangle = \left(\sqrt{2} \text{ cm}\right) \left(2 \cos \omega_p t | e_p \rangle - \cos \omega_b t | e_b \rangle \right) = \left(\sqrt{2} \text{ cm}\right) \left[2 \cos \omega_p t \frac{1}{\sqrt{2}} \begin{pmatrix} 1 \\ 1 \end{pmatrix} \right.$$

$$\left. - \cos \omega_b t \frac{1}{\sqrt{2}} \begin{pmatrix} 1 \\ -1 \end{pmatrix} \right]$$

$$= \begin{pmatrix} 2 \cos \omega_p t - \cos \omega_b t \\ 2 \cos \omega_p t + \cos \omega_b t \end{pmatrix} \text{ cm}$$

Concept test: From the above, read off the expression for $x_1(t)$ (the position of the left mass as a function of time), and verify that it matches what we got when we worked this same example in section 5.5.

For a superposition of breathing and pendulum modes, we then have

$$|x_0\rangle = A_b |e_b\rangle + A_p |e_p\rangle = \frac{A_b}{\sqrt{2}} \begin{pmatrix} 1 \\ -1 \end{pmatrix} + \frac{A_p}{\sqrt{2}} \begin{pmatrix} 1 \\ 1 \end{pmatrix}$$

$$\text{and } |x(t)\rangle = \frac{A_b}{\sqrt{2}} \cos \omega_b t \begin{pmatrix} 1 \\ -1 \end{pmatrix} + \frac{A_p}{\sqrt{2}} \cos \omega_p t \begin{pmatrix} 1 \\ 1 \end{pmatrix}.$$

12. The component of pendulum mode is about 2.8 cm, and the component of breathing mode is about −1.4 cm.

Since $A_b = \langle e_b \mid x_0 \rangle$ and $A_p = \langle e_p \mid x_0 \rangle$, we can rewrite this as

$$\boxed{\boxed{|x(t)\rangle = \langle e_p \mid x_0 \rangle \cos \omega_p t \, |e_p\rangle + \langle e_b \mid x_0 \rangle \cos \omega_b t \, |e_b\rangle}}$$ (5.6.6)

(zero initial velocities)

This way of writing $|x(t)\rangle$ is called the "normal mode expansion."

This way of solving the problem may seem harder than the way we used in section 5.5. However, the ideas you are using, of taking inner products in Hilbert space, are *much* more powerful, and can be readily adapted to more complicated situations. Once you get used to these ideas, you will find this approach quite intuitive. (OK, I admit that that's pretty much like saying, "Once you get used to eating ketchup on your cereal, you'll be used to eating ketchup on your cereal." But Hilbert space really is a superior and clearer way of thinking about the world.)

5.7   The analogy between coupled oscillators and molecular energy levels

Our discussion of coupled oscillators is completely general, and applies to any two identical coupled oscillators. Here, again, are the expressions we found for the angular frequencies of the normal modes:

$$\omega_p = \sqrt{\frac{g}{\ell}} \quad \text{and} \quad \omega_b = \sqrt{\frac{g}{\ell} + \frac{2k}{m}}.$$

A larger k corresponds to a stronger spring connecting the two pendula, that is, to stronger coupling. From the above equations, we can draw a very important conclusion, which turns out to be true for all systems: the difference in frequency between the two normal modes (called the "frequency splitting") gets bigger as the coupling gets stronger.

$$\boxed{\text{Stronger coupling produces larger splitting of normal mode frequencies.}}$$

As discussed in section 5.5, for quantum mechanical systems, the energy of a normal mode (i.e., the energy of an "energy eigenstate") is proportional to the frequency. Therefore, we expect that when two quantum oscillators are coupled together, the resulting system will have two energy eigenstates, with a splitting in energy which increases with the strength of the coupling. This is precisely what is observed! For example, let's make a hydrogen molecule (H_2) by bringing two hydrogen atoms gradually closer together. Each hydrogen atom consists of a proton with an electron near it. The electron is bound to the nucleus by electrostatic forces, and cannot go far from the nucleus. This is analogous to the situation for one of our pendula; the pendulum bob cannot go far from the equilibrium position, because the "pendulum force" tends to push it back. The analogy is not perfect, because the potential

Two separate H-atoms:

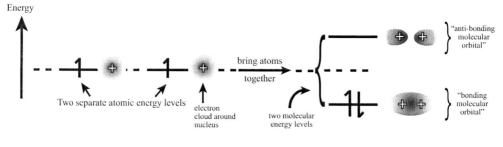

Two separate pendula:

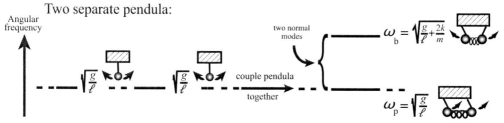

Figure 5.7.1 The analogy between molecular energy levels and coupled oscillators. Like the protons discussed in section 4.7, the electrons have spin angular momentum. Like the protons, each electron can be in a "spin up" or "spin down" state. Therefore, the electrons are shown as arrows, with a spin up electron shown as an up arrow and a spin down electron shown as a down arrow.

energy function experienced by the electron is not the same as that experienced by the pendulum, but the qualitative conclusions are the same.

The energy eigenstates of the electrons in the isolated atoms (when they are far apart) are called "atomic orbitals," and the associated energies are called "atomic energy levels." If we bring the two atoms gradually closer together, they interact more strongly. This is analogous to starting with two pendula which are completely uncoupled, then gradually increasing the strength of the coupling by increasing k. In the coupled pendula case, the system can assume one of two normal modes. In the hydrogen case, the electrons in the molecule can assume one of two molecular energy eigenstates; these are often called "molecular orbitals," and each has an associated "molecular energy level." Figure 5.7.1 shows the way this process is usually shown, with energy on the vertical axis. For stable molecules, such as H_2, one of the two molecular energy levels lies below the original atomic energy levels. If both electrons move into this one level, then the total energy of the system is lowered relative to the separate atoms, so the molecule is stable. Therefore, the associated eigenstate is called a "bonding orbital."[13]

---

13. Note that we are using a different analogy between quantum and classical systems here than at the end of section 5.5. There, we drew an analogy between a single atom and the system of two coupled pendula, with the different atomic orbitals (1s, 2s, etc.) analogous to the different normal modes of the coupled pendula. Here, instead, we think of the two pendula as analogous to two H-atoms, but we only consider a single atomic orbital for each atom. Both analogies are appropriate, but they are distinct.

5.8　Nonzero initial velocities

So far, we have concentrated on situations for which all parts of the system (i.e., both pendula) are at rest at $t = 0$. But what if they aren't? How do we determine the future behavior of the system, and the weighting of the different normal modes, if the initial velocities are nonzero?

It turns out that this question is not very important for quantum mechanics. The time derivative of the real part of the wavefunction Ψ evaluated at a particular point is analogous to the velocity of one of our pendula, and usually Ψ is chosen so that this time derivative is zero at $t = 0$ for all points. (It turns out that this corresponds to choosing Ψ to be real everywhere at $t = 0$.)

However, for mechanical systems, this is a reasonable question to ask, and it's not hard to answer. Since the two differential equations that describe the normal modes can be derived from the original coupled differential equations that describe the motion of the pendulum bobs (and vice-versa), any state of the system can be described as a superposition of the normal modes. Therefore, the most general description is

$$|x(t)\rangle = \text{Re}\left[A_{\text{p}}e^{i(\omega_{\text{p}}t + \varphi_{\text{p}})}|e_{\text{p}}\rangle + A_{\text{b}}e^{i(\omega_{\text{b}}t + \varphi_{\text{b}})}|e_{\text{b}}\rangle\right],$$

where A_{p} is the amplitude of the pendulum mode, A_{b} is the amplitude of the breathing mode, φ_{p} is the phase of the pendulum mode, and φ_{b} is the phase of the breathing mode. To write things more compactly, we can set $C_{\text{p}} \equiv A_{\text{p}}e^{i\varphi_{\text{p}}}$ and $C_{\text{b}} \equiv A_{\text{b}}e^{i\varphi_{\text{b}}}$, so that

$$|x(t)\rangle = \text{Re}\left[C_{\text{p}}e^{i\omega_{\text{p}}t}|e_{\text{p}}\rangle + C_{\text{b}}e^{i\omega_{\text{b}}t}|e_{\text{b}}\rangle\right], \tag{5.8.1}$$

where C_{p} and C_{b} are the "complex amplitudes." (The restriction of zero initial velocities which we have used up to this point is equivalent to requiring C_{p} and C_{b} to be real, as we will soon see.) Again, the above is called the "normal mode expansion." Remember that this is a vector equation: $|x(t)\rangle \equiv \begin{pmatrix} x_1(t) \\ x_2(t) \end{pmatrix}$. Our goal is to find C_{p} and C_{b}, given the initial conditions.

The velocity is obtained by taking the time derivative:

$$|\dot{x}(t)\rangle = \begin{pmatrix} \dot{x}_1(t) \\ \dot{x}_2(t) \end{pmatrix} = \text{Re}\left[i\omega_{\text{p}}C_{\text{p}}e^{i\omega_{\text{p}}t}|e_{\text{p}}\rangle + i\omega_{\text{b}}C_{\text{b}}e^{i\omega_{\text{b}}t}|e_{\text{b}}\rangle\right].$$

Plugging $t = 0$ into these expressions gives

$$|x_0\rangle \equiv \begin{pmatrix} x_{10} \\ x_{20} \end{pmatrix} = \begin{pmatrix} x_1(t=0) \\ x_2(t=0) \end{pmatrix} = \text{Re}\left[C_{\text{p}}|e_{\text{p}}\rangle + C_{\text{b}}|e_{\text{b}}\rangle\right]$$

$$\text{and } |\dot{x}_0\rangle \equiv \begin{pmatrix} \dot{x}_1(t=0) \\ \dot{x}_2(t=0) \end{pmatrix} = \text{Re}\left[i\omega_{\text{p}}C_{\text{p}}|e_{\text{p}}\rangle + i\omega_{\text{b}}C_{\text{b}}|e_{\text{b}}\rangle\right]$$

We write the complex amplitudes in Cartesian form:

$$C_{\text{p}} = \text{Re}\, C_{\text{p}} + i\, \text{Im}\, C_{\text{p}} \text{ and } C_{\text{b}} = \text{Re}\, C_{\text{b}} + i\, \text{Im}\, C_{\text{b}},$$

so that

$$\left|x_0\right\rangle = \text{Re } C_p\left|e_p\right\rangle + \text{Re } C_b\left|e_b\right\rangle \text{ and } \left|\dot{x}_0\right\rangle = -\omega_p\text{Im } C_p\left|e_p\right\rangle - \omega_b\text{Im } C_b\left|e_b\right\rangle.$$

Note that $\text{Re } C_p\left|e_p\right\rangle$ is shorthand for $\left(\text{Re } C_p\right)\left|e_p\right\rangle$. Following what we did in section 5.6, let's see what happens if we take inner products with these:

$$\left\langle e_p \mid x_0\right\rangle = \left\langle e_p\right| \left\{\text{Re } C_p\left|e_p\right\rangle + \text{Re } C_b\left|e_b\right\rangle\right\} = \left\langle e_p\right|\text{Re } C_p\left|e_p\right\rangle + \left\langle e_p\right|\text{Re } C_b\left|e_b\right\rangle$$

$$= \text{Re } C_p\underbrace{\left\langle e_p \mid e_p\right\rangle}_{1} + \text{Re } C_b\underbrace{\left\langle e_p \mid e_b\right\rangle}_{0}$$

$$\Rightarrow \left\langle e_p \mid x_0\right\rangle = \text{Re } C_p \tag{5.8.2a}$$

and

$$\left\langle e_p \mid \dot{x}_0\right\rangle = -\omega_p\text{Im}C_p. \tag{5.8.2b}$$

Similarly,

$$\left\langle e_b \mid x_0\right\rangle = \text{Re } C_b \tag{5.8.2c}$$

and

$$\left\langle e_b \mid \dot{x}_0\right\rangle = -\omega_b\text{Im } C_b. \tag{5.8.2d}$$

(As promised, you can see that if the initial velocities are zero, then C_p and C_b are real.) So, we're done; given the initial positions $\left|x_0\right\rangle$ (which, recall has x_{10} in the top row and x_{20} in the bottom row), and given the initial velocities $\left|\dot{x}_0\right\rangle$, we can determine the complex coefficients C_p and C_b in the normal mode expansion (5.8.1).

Self-test (answer below[14]): At $t = 0$, mass 1 (the pendulum bob on the left) has position 1.3 cm and velocity -1.9 cm/s, while mass 2 (the bob on the right) has position -0.3 cm and velocity -0.2 cm/s. For this system, $\omega_p = 5.0$ rad/s and $\omega_b = 7.0$ rad/s. Find the coefficients C_p and C_b that appear in the normal mode expansion of $\left|x(t)\right\rangle$.

5.9   Damped, driven coupled oscillators

Let's add damping forces to both pendula, and a drive force which is exerted only on the left pendulum (this reflects a common real-world situation in which a driving force is applied only at one point of a complicated system):

$$F_{\text{damp, 1}} = -b\dot{x}_1, F_{\text{damp, 2}} = -b\dot{x}_2, F_{\text{drive, 1}} = F_0\cos\omega_d t, F_{\text{drive, 2}} = 0$$

---

14. Answer to self-test: $C_p = (0.71 + \text{i } 0.30)$ cm , $C_b = (1.1 + \text{i } 0.17)$ cm.

Your turn: Recall that $\gamma \equiv b/m$. Show that the differential equations that describe this system are

$$\ddot{x}_1 + \gamma \dot{x}_1 + \left(\frac{k}{m} + \frac{g}{\ell} \right) x_1 - \frac{k}{m} x_2 = \frac{F_0}{m} \cos \omega_\text{d} t \qquad (5.9.1)$$

and

$$\ddot{x}_2 + \gamma \dot{x}_2 + \left(\frac{k}{m} + \frac{g}{\ell} \right) x_2 - \frac{k}{m} x_1 = 0 \qquad (5.9.2)$$

Inspired by what we did for the undamped, undriven case, we try adding these equations (after multiplying by $1/\sqrt{2}$):

$$\frac{(5.9.1)}{\sqrt{2}} - \frac{(5.9.2)}{\sqrt{2}} \Rightarrow \frac{(\ddot{x}_1 + \ddot{x}_2)}{\sqrt{2}} + \gamma \frac{(\dot{x}_1 + \dot{x}_2)}{\sqrt{2}} + \left(\frac{k}{m} + \frac{g}{\ell} \right) \frac{(x_1 + x_2)}{\sqrt{2}} - \frac{k}{m} \frac{(x_1 + x_2)}{\sqrt{2}}$$

$$= \frac{F_0}{\sqrt{2}m} \cos \omega_\text{d} t$$

$$\Rightarrow \ddot{s}_\text{p} + \gamma \dot{s}_\text{p} + \frac{g}{\ell} s_\text{p} = \frac{F_0/\sqrt{2}}{m} \cos \omega_\text{d} t, \text{ where } s_\text{p} \equiv \frac{1}{\sqrt{2}} (x_1 + x_2)$$

This is just the equation for an ordinary damped driven harmonic oscillator, with resonant frequency $\omega_0 = \sqrt{\frac{g}{\ell}} = \omega_\text{p}$. The quantity that oscillates is s_p, the normal mode coordinate for the pendulum mode. Note that the effective amplitude of the driving force is $F_0/\sqrt{2}$, reflecting the fact that the energy from the driving force is responsible for the motion of two masses instead of just one.

If instead we take the difference between equation (5.9.1) and (5.9.2) (after multiplying by $1/\sqrt{2}$), we find

$$\frac{(5.9.1)}{\sqrt{2}} + \frac{(5.9.2)}{\sqrt{2}} \Rightarrow \frac{(\ddot{x}_1 - \ddot{x}_2)}{\sqrt{2}} + \gamma \frac{(\dot{x}_1 - \dot{x}_2)}{\sqrt{2}} + \left(\frac{k}{m} + \frac{g}{\ell} \right) \frac{(x_1 - x_2)}{\sqrt{2}} - \frac{k}{m} \frac{(x_1 - x_2)}{\sqrt{2}}$$

$$= \frac{F_0}{\sqrt{2}m} \cos \omega_\text{d} t$$

$$\Rightarrow \ddot{s}_\text{b} + \gamma \dot{s}_\text{b} + \left(\frac{g}{\ell} + \frac{2k}{m} \right) s_\text{b} = \frac{F_0/\sqrt{2}}{m} \cos \omega_\text{d} t,$$

$$\text{where } s_\text{b} \equiv \frac{1}{\sqrt{2}} (x_1 - x_2),$$

Again, this is just the equation for an ordinary damped driven harmonic oscillator, with resonant frequency $\omega_0 = \sqrt{\frac{g}{\ell} + \frac{2k}{m}} = \omega_\text{b}$. The quantity that oscillates is s_b, the normal mode coordinate for the breathing mode.

The conclusion is that this system behaves like two completely independent damped, driven oscillators, one corresponding to the pendulum mode and one to

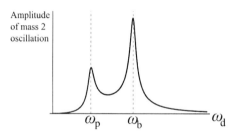

Amplitude of mass 2 oscillation

ω_p ω_b ω_d

Figure 5.9.1 For a damped driven coupled oscillator system, each part of the system shows a resonant response at the frequencies corresponding to each of the normal modes. This shows the amplitude of response of mass 2 of a two-pendulum system, with the drive force applied to mass 1. The curve for the amplitude of response for mass 1 is similar.

the breathing mode! Since $x_1 = \dfrac{1}{\sqrt{2}}\left(s_p + s_b\right)$, mass 1 shows a large amplitude of oscillation near both ω_p and ω_b. The steady-state amplitude of oscillation for mass 2 is shown in figure 5.9.1.

This behavior also occurs in more complicated coupled oscillator systems – one observes a resonant response at each normal mode frequency. This explains, for example, why the absorption spectrum for a chemical shows an absorption peak at each normal mode frequency.

Self-test (answer below)[15]: In figure 5.9.1, the peak at ω_b is higher than the one at ω_p. Explain why. *Hint: Recall that $Q = \omega_0/\gamma$.*

Concept and skill inventory for chapter 5

After reading this chapter, you should fully understand the following terms:

Beats (5.1)
Coupled differential equations (5.2)
Normal mode (5.5)
Breathing mode (5.3)
Pendulum mode (5.3)
Normal mode analysis (5.5)
Hilbert space, as applied to a system of two identical coupled oscillators (5.6)
Column matrix (5.6)
Adjoint (5.6)
ket (5.6)
Inner product (5.6)
bra-ket notation (5.6)

---

15. For light damping, the peak amplitude is about Q times the drive amplitude. Since $Q \equiv \dfrac{\omega_0}{\gamma}$, and γ is the same for both modes, the Q for the breathing mode (which has a higher characteristic frequency) is higher, so the peak is higher.

Orthogonal vectors (5.6)
Normalized vector (5.6)
Orthonormal basis (5.6)
Eigenvector (5.6)

You should know what happens when:

Two sinusoids of equal amplitude but different frequency are added together (5.1)
A system is excited into a pure normal mode (5.5)
A system is excited into a superposition of normal modes (5.4)
Two hydrogen atoms are brought close to each other (5.7)
A set of two identical coupled oscillators, with damping, is driven by a periodic force applied to one of them. (5.9)

You should understand the following connections:

Beats & Superpositions of normal modes (5.1, 5.4)
Resolving an ordinary vector into its x- and y-components & Normal mode analysis for a symmetric coupled oscillator system with zero initial velocity (5.6)
Vectors & Column matrices (5.6)
Kets & Vectors (5.6)
Bras & Row matrices (5.6)
Coupled oscillators & Molecular energy levels (5.7)
Superpositions of normal modes & Superpositions of quantum energy eigenstates (5.7)
Normal modes & Eigenvectors (5.6)
Normal mode coordinates (s_p and s_b) & the coordinates of the two oscillators (5.3)

You should be familiar with the following additional concepts:

The effect of the strength of the coupling on the separation of normal mode frequencies (5.7)

You should be able to:

Go back and forth between beat frequency and the frequencies of the underlying sinusoids (5.1)
Given the properties of a system of two identical coupled oscillators, describe the normal modes and find their frequencies. (5.3)
Compute inner products (5.6)
Given the initial positions for a system of two identical coupled oscillators, and given zero initial velocities, find the amplitudes of the normal modes. (5.6)
Similarly, for nonzero initial velocities, find the amplitudes of the normal modes. (5.8)
For either of the above, write down the positions of each oscillator as a function of time. (5.6)
For a damped, driven pair of identical coupled oscillators, calculate the resulting steady state motion (5.9)

In addition to all of the above, you should be able to combine the concepts you've learned to address new situations.

Problems

Note: Additional problems are available on the website for this text.

Instructors: Difficulty ratings for the problems, full solutions, and important additional support materials are available on the website.

5.1 **Musical scales**. The frequencies of notes in standard musical notation are defined in terms of ratios. For example, an octave is defined as a factor of two in frequency. The standard "concert A" is 440 Hz, so that one octave below concert A is 220 Hz. Each octave is divided into twelve half steps, with the same frequency ratio between any two notes separated by a half step, as shown in figure 5.P.1 on a piano keyboard. Each of the black keys on the piano has two names. For example, the black key just above concert A can be called A# (pronounced "A sharp"), meaning that it is a half-step above A, or it can be called B♭ (pronounced "B flat"), meaning that it is a half step below B.

   (a) Show that the frequency ratio between half-steps is 1.05946.[16]
   (b) The lowest note on a standard piano is the A that is four octaves below concert A. If this note is played simultaneously with the A# just above it, and both notes are played at the same volume, what beat frequency is heard?

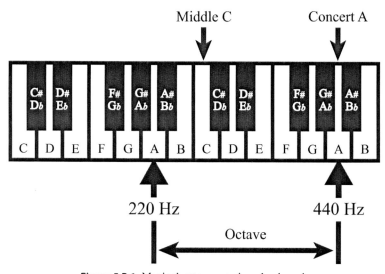

Figure 5.P.1 Musical notes on a piano keyboard.

16. This is called the "well-tempered" scale. It gives the exact desired frequency ratio for notes separated by an octave (i.e., a ratio of 2), but only gives approximations for other harmonic combinations. For example, the two notes in a "major fifth" chord should have a frequency ratio

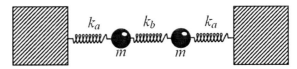

Figure 5.P.2 A pair of masses connected by springs to immovable walls.

5.2 Two tuning forks of frequencies 440 and 446 Hz are struck at the same time with the same intensity. The resulting beat pattern is recorded with a microphone (placed an equal distance from both forks) and amplified. At the output of the amplifier, the maximum *amplitude* of the rapid oscillations is 2 Volts (abbreviated 2 V).

    (a) What is the time T_{beat} between successive *amplitude* maxima?
    (b) What is the voltage of this wave as a function of time?

5.3 Two identical pendula are coupled by a spring, and are oscillating. One has a position $x(t) = (2 \text{ cm}) \cos\left[\left(\frac{3}{20}\text{rad/s}\right)t\right] \cos\left[\left(\frac{9}{2}\text{rad/s}\right)t\right]$. What are the angular frequencies of the normal modes, ω_{p} and ω_{b}?

5.4 Figure 5.P.2 shows a pair of masses, connected to immovable walls by springs. Gravity is negligible. Describe the normal modes of this system, and find their angular frequencies.

5.5 Two pendula, each with mass 0.30 kg and length 0.50 m, are coupled by a spring. One of the masses is clamped while the other is pulled aside and then released, resulting in an oscillation with $\omega = 5.00$ rad/s. Find ω_{p}, ω_{b}, and T_{beat}.

5.6 Careful observations of a pair of coupled pendula produce the plot shown in figure 5.P.3 of the *x*-position of the left pendulum bob. **(a)** Given that the time between maxima is 1.0 s (as shown), what is the *approximate* value of the pendulum length ℓ? (Assume $g = 9.80$ m/s².) **(b)** Assuming your value of ℓ from part a is exact, what is the value of k/m?

5.7 **Quantum beats.** As discussed in section 5.5, one can put a quantum system into a superposition state, made up from two fundamental states called "energy eigenstates." This is analogous to putting a coupled pendulum system into a superposition of breathing and pendulum modes. In either the quantum or the classical system, one can then observe beating phenomena, because the frequencies of the normal modes (for the classical system) or energy eigenstates (for the quantum system) are not quite the same. Just as

---

of exactly 1.5; we can get close to this using the well-tempered scale by using two notes separated by seven half steps: $1.05946^7 = 1.49831$. However, for some musicians and unusually discerning listeners, even this small departure from the ideal ratio creates a jarring effect and changes the emotional quality of the music. Therefore, instruments which do not have fixed intervals, such as violins, are sometimes played using scales with slightly different ratios between the half-steps, so that chords such as the major fifth sound more perfect. In such a scale, A# and B♭ are slightly different.

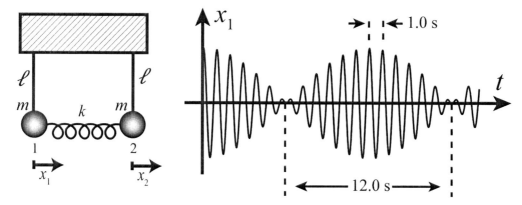

Figure 5.P.3 Left: A pair of coupled pendula. Right: The position of the left mass as a function of time.

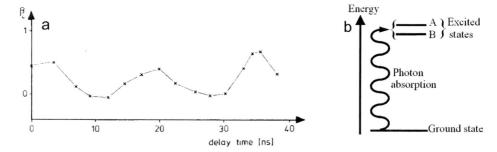

Figure 5.P.4 a: A signal proportional to the probability of ionization plotted as a function of delay time between arrival of two different photons. The oscillations shown are due to quantum beats. b: Schematic representation of the effect of the first photon, which excites an electron from the ground state into a superposition of two excited states. Part (a) reprinted from Optics Communications, **31**, G. Leuchs et al., *Quantum Beats Observed in Photoionization*, 313–316 (1979), © 1979, with Permission from Elsevier.

a guitarist can use beats to tune a guitar, an experimental physicist can use quantum beats to measure the frequency difference between two quantum states. Then, using $E = \hbar\omega$, she can deduce the energy difference between the states.

Figure 5.P.4a shows data from such an experiment. First, the outermost electron in an atom is excited from its ground state by the absorption of a photon, that is, a particle of light, as shown in figure 5.P.4b. The energy difference between the two excited states of interest (labeled A and B) is so small that the photon excites the electron into a superposition of both states. This is analogous to simultaneously exciting the breathing and pendulum modes for the coupled pendula. In that case, we observed that the system slowly evolved, first with almost all the swinging in the left pendulum bob with the right bob motionless, then with almost all the swinging in the right bob, and back again. Similarly for the electron, once it has been excited into the superposition of states A and B, it "slowly" evolves from one shape of the "electron cloud" (call it shape 1) to another (call it shape 2) and then back. In this context, "slowly" means on the order of tens of ns (nanoseconds), which is indeed slow on the timescale of many atomic events.

It turns out that, when the electron cloud is in shape 1, it is easier to remove the electron entirely from the atom than it is when the electron cloud is in shape 2. In other words, the probability for ionization of the atom (by absorption of a second photon) is higher for shape 1 than shape 2. The authors of this paper used a pair of photon pulses to exploit this fact. A photon from the first pulse excites the electron into the superposition of states A and B, and a photon from the second pulse might be able to ionize the atom, if the photon arrives when the electron cloud is in shape 1. Figure 5.P.4a shows a signal called "β_4" measured by the experimenters, which is proportional to the probability of ionization. The horizontal axis shows the time delay between the two photon pulses.

Using this data, deduce the energy difference between states A and B; express your answer in electron Volts (abbreviated eV), where $1\,eV = 1.602 \times 10^{-19}$ J. (For comparison, the energy difference between the ground state and state B in this atom is 3.68 eV. You should find that the energy difference between A and B is *much* smaller than this.) Note that the actual energy level diagram is more complex than the simplified version shown here, which explains why the graph above is only roughly sinusoidal.

5.8 State what is wrong with the following statement: "There is no difference between a normal mode and an eigenvector – the two terms are interchangeable." (As part of your response, provide a corrected version of the statement.)

5.9 State what is wrong with the following paragraph: "Consider an undamped mass on a spring. When the mass is given an initial 'kick' (by imparting some initial velocity), thereafter it always oscillates at $\omega_0 = \sqrt{k/m}$, no matter whether the initial kick is to the left or to the right, and no matter how hard the kick is. Now, consider an undamped symmetric coupled pendulum system. In the same way, if the left mass is given an initial kick (while the right mass is left initially motionless, but released immediately after the kick), the system always oscillates at one of the normal mode frequencies, either ω_p or ω_b, depending on whether the initial kick is to the right or to the left." (As part of your response, provide a corrected version of the statement.)

5.10 Two coupled identical pendula of mass 3 kg are oscillating, having started at rest with positions $x_1(0) = 0.22$ m and $x_2(0) = -0.15$ m. Mass 1 has $T_{fast} = 0.5$ s (where T_{fast} is the period of the rapid oscillations) and $T_{beat} = 10$ s.

    (a) Find the amplitudes of the breathing and pendulum modes.

    (b) Find the position of mass 1 as a function of time.

5.11 Two coupled pendula are oscillating with $|x(t)\rangle = \left(\dfrac{1}{7\sqrt{2}}\,m\right)\cos\left[(5\text{ rad/s})\,t\right]$ $\begin{pmatrix}1\\1\end{pmatrix} + \left(\dfrac{1}{3\sqrt{2}}\,m\right)\cos\left[(6.1\text{ rad/s})\,t\right]\begin{pmatrix}1\\-1\end{pmatrix}$. They started from rest. What were their initial positions?

5.12 Practice with inner products Let $|A\rangle = \begin{pmatrix} 1.2 - 3i \\ -2 \end{pmatrix}$ and $|B\rangle = \begin{pmatrix} 6.8 \\ 7.1 \end{pmatrix}$.

(a) Evaluate $\langle A \mid B \rangle$ and $\langle B \mid A \rangle$.

(b) Show that, for any two vectors $|C\rangle$ and $|D\rangle$ (which may have complex components), $\langle C \mid D \rangle = \langle D \mid C \rangle^*$. In other words, show that, to reverse the order of an inner product, you need only take the complex conjugate of the value. (The vectors are assumed to be two-dimensional, such as $|A\rangle$ and $|B\rangle$ are.)

(c) Find a vector that is orthogonal to $|B\rangle$, and verify the orthogonality using an inner product.

(d) Find a vector that points in the same direction as $|B\rangle$ but has length 1, and verify its length using an inner product.

5.13 Symmetric coupled pendula applet. Please open a browser and go to the listing for this problem on this book's web page. Click on the link labeled "Hilbert Space for Coupled Pendula," and wait for the applet to load. (It may take a couple of minutes.)

Exercises (please work through all these; written responses are needed only as indicated by **boldface**):

(a) Try to set up the pendula in a pure pendulum mode (by clicking on the Hilbert Space plot in the applet above) and then click on the 'Go!' button. Now try a few different amplitudes.

(b) Now, set up the pendula in a pure breathing mode. Again, try a few different amplitudes.

(c) **For positive breathing amplitude (downward and to the right), is the initial position of the left pendulum bob positive or negative? How about the right one?**

(d) Now set up a beat pattern so that energy is transferred back and forth between the two pendulum bobs periodically. First, make a perfect beat pattern where each bob periodically comes to a complete stop, and then set up some more complicated behavior.

(e) For the beating state that you just created, look carefully at the graphs at the bottom of the applet. **Are these graphs showing the correct behavior of the pendulum and breathing modes? Is the behavior of the system simply an addition of these two modes?**

(f) Try pressing the button labeled "Drop Lines," and **describe what this is showing.**

(g) **Describe how the colors used in the animation help to make connections between the three different pictures of the applet.**

(h) The angular frequency of the pendulum mode is $\omega_p = \sqrt{\dfrac{g}{\ell}}$. The angular frequency of the breathing mode is $\omega_b = \sqrt{\dfrac{g}{\ell} + \dfrac{2k}{m}}$. Recall what we learned about beats: $A \cos \omega_1 t + A \cos \omega_2 t = 2A \cos \omega_e t \cos \omega_{av} t$ where $\omega_{av} = \dfrac{\omega_1 + \omega_2}{2}$ and $\omega_e = \dfrac{\omega_1 - \omega_2}{2}$.

Given a linear combination of these two normal modes, what is the period and frequency of the envelope? Qualitatively, how does changing the spring constant (k) affect these results? Test this hypothesis by changing k in the animation.

(i) The normal modes are mutually orthogonal – they do not influence each other. **Why is this an illuminating way to look at a system?**

(j) **In order to predict the behavior of a system of coupled oscillators as a function of time, why is it more useful to know the amplitudes of the normal modes than the initial positions of the pendulum bobs?**

(k) **What vectors in the Hilbert space are associated with unique frequencies?**

(l) **Using the language of taking projections of vectors onto different axes, describe the process of normal mode analysis.**

5.14 For a pair of symmetric pendula (or any other symmetric coupled oscillator system) that is restricted to zero initial velocities, verify that the following peculiar-looking equation is correct:

$$|e_p\rangle\langle e_p| + |e_b\rangle\langle e_b| = 1.$$

Hint: Consider what happens when you apply each side of the equation to an arbitrary Hilbert space vector $|A\rangle$, that is, when you write $(\,|e_p\rangle\langle e_p| + |e_b\rangle\langle e_b|\,)\,|A\rangle = 1|A\rangle$. If you can explain why this equation must be true, then (because $|A\rangle$ is arbitrary), the original version of the equation must also be correct.

5.15 Each pendulum bob in a pair of symmetric coupled pendula has mass 0.25 kg, and pendulum length 2.5 m. The bobs are connected by a spring of constant 0.25 N/m. Initially, both are held immobile, with the left bob 0.25 m to the left of its equilibrium position and the right bob at its equilibrium position. At time $t = 0.25$ s, they are released. Find the position of each bob as a function of t.

5.16 A set of identical coupled pendula has $\omega_p = 5.0$ rad/s and $\omega_b = 7.0$ rad/s. The coefficients in the normal mode expansion are $C_p = (0.10\text{ m})e^{i(0.35)}$ and $C_b = (0.15\text{ m})e^{i(0.15)}$. What is the initial position and velocity of mass 2?

5.17 For the symmetric coupled pendulum system, consider an arbitrary superposition of pendulum and breathing modes, one for which the initial velocities are not necessarily zero. Show that, even in this superposition, the normal mode coordinates s_p and s_b oscillate in simple harmonic motion.

5.18 A symmetric coupled pendulum system has $m = 1.5$ kg, $\ell = 0.8$ m, and $k = 30$ N/m. At $t = 0$, the left bob is at a position of 5 cm, with a speed of -2 cm/s, while the right bob is at -2 cm with a speed of 3 cm/s.

(a) Show that $\omega_p = 3.50$ rad/s and that $\omega_b = 7.23$ rad/s.

(b) Find $|x(t)\rangle$. *Hint: after finding the real and imaginary parts of the complex amplitudes, express the complex amplitudes in polar form, then do your taking of the real part.*

5.19 A pair of symmetric coupled pendula has the following initial conditions: the left bob is at position -3.1 cm with velocity -10 cm/s, and the right bob is at position -0.2 cm with velocity 5 cm/s. Each bob has $m = 1.2$ kg and string length $\ell = 0.45$ m; the spring constant is $k = 2.5$ N/m. At $t = 5.3$ s, what is the position of each bob? Show all your work clearly.

5.20 State what is wrong with the following paragraph: "Consider a symmetric coupled pendulum system, with damping, and with a periodic drive force applied to the left pendulum bob, as discussed in section 5.9. When a drive force of angular frequency ω_d is applied, the steady-state motion $x_1(t)$ of the left bob is always the sum of an oscillation at ω_b and an oscillation at ω_p. As ω_d is varied, the relative amplitude of the responses at ω_b and ω_p changes." (As part of your response, provide a corrected version of the paragraph.)

5.21 Two lightly damped identical pendula are coupled together with a spring with spring constant 15.0 N/m. The masses of the pendula are 3.00 kg, their lengths are 1.20 m, and the damping coefficient is $b = 0.500$ kg/s. The support point S_L for the left pendulum (where the string is attached to the ceiling) can be moved left and right in a sinusoidal pattern, while the support point for the right pendulum is held motionless. **(a)** At what angular frequency should the support point S_L be moved to produce the largest steady state response in the coupled pendulum system? **(b)** Make a semi-quantitative argument showing that, if the system is driven at this angular frequency (by moving S_L), the response of the pendulum mode is negligible, bearing in mind that the system is lightly damped. *Hint: Calculate the FWHM of the power resonance curve. If the drive frequency is further away from the resonance frequency than four times the FWHM, the response will be very small indeed.* **(c)** For the angular frequency you calculated in part (a), what is the ratio of the amplitude of the right pendulum to the amplitude of motion of S_L?

5.22 (Please do problems 5.4 and 5.20 before beginning this exercise.) You are part of a team designing an experiment for the International Space Station. Part of the experiment involves a large box, labeled B in figure 5.P.5a. During the experiment, box B will float inside the space station, in an air-filled chamber. It will be tethered by a spring to a vibrating panel, labeled P, which is attached to the side of the space station. Panel P will oscillate left and right at a very low frequency (0.010 Hz); these vibrations should be transmitted via the tether to box B. However, other team members have told you that unwanted effects will cause panel P also to oscillate with the same amplitude at frequencies of 75.0 Hz. In other words, the motion of panel P will be

$$x_P = A_d \left(\cos \omega_w t + \cos \omega_u t \right),$$

where $\omega_w = 2\pi\,(0.010\text{ Hz})$ is the angular frequency of the wanted oscillations, $\omega_u = 2\pi\,(75.0\text{ Hz})$ is the angular frequency of the unwanted oscillations, and $A = 0.05$ m is the amplitude of oscillation.

Your job is to finalize the design of the tether, which must be 1.23 m long, so as to minimize the transmission of the unwanted oscillations. You are given

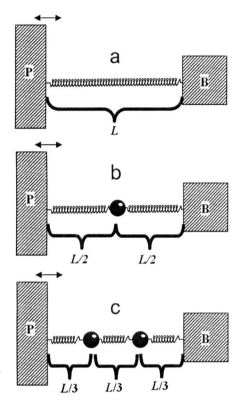

Figure 5.P.5 Three configurations for tethering box B to panel P, which is part of the International Space Station.

a spring which is 1.23 m long, and has a spring constant of 154 N/m. Box B has a mass of 54.7 kg. Tests of the air resistance of box B show that, when it is moving at 2.00 m/s, it experiences a force from the air of 1.53 N.

(a) If you simply connect box B to panel P using the spring, as shown in figure 5.P.5a, explain why the resulting steady-state behavior of box B, $x_B(t)$, is simply the steady-state behavior for $x_P = A_d \cos \omega_w t$ added to the steady state behavior for $x_P = A_d \cos \omega_u t$.

(b) If you now wait for a steady state to be achieved, what is the ratio of the amplitude of unwanted oscillations of box B (at 75.0 Hz) to the amplitude of wanted oscillations of box B (at 0.010 Hz)?

You take the results of your calculations to your team leader. "Well," she says, "that's a pretty impressive reduction of those 75 Hz oscillations. Unfortunately, this is not good enough—the undesired oscillations must be reduced in amplitude even further. Even more unfortunately, it's too late to order a different spring to use as the tether."

Then, inspiration strikes you! You could cut the spring into two pieces, and insert an extra mass in the middle, as suggested in figure 5.P.5b. This would change the resonant characteristics of the system, and so might improve rejection of the undesired oscillations. "Good idea!" says your leader, "Crunch the numbers for me."

You have a mass of 5.00 kg available; to simplify your calculation, assume it has a diameter which is negligible on the scale of the length of the spring. Air resistance tests on the 5.00 kg mass show that, when it is moving at 2.00 m/s, it experiences a force from the air of 0.33 N.

As you showed in part (b), the amplitude of oscillation of box B at 75.0 Hz is very small in any case. Therefore, to make your life simpler in determining the best configuration, you should make the assumption that the amplitude of motion of box B at the unwanted frequency (75 Hz) is essentially zero, although we will still be interested in the forces exerted on it at this frequency.

(c) Explain qualitatively why the steady-state amplitude of oscillations of box B at the *wanted* frequency of 0.010 Hz is essentially the same for configurations a and b.

(d) Compare the force exerted on box B at 75 Hz for configuration b to that for configuration a. (You should find that the force for configuration b is almost 1,000 times lower!)

(e) "That's great!" says your team leader. "But what if we take your idea even further? How about if we cut the original spring into three pieces, and add another intermediate mass?" She sketches another possibility (figure 5.P.5c). At first you think, "If one mass made the situation better, then two masses will work even better, and would reduce the 75 Hz force exerted on box B even more!" But then, you begin to realize that may not be right. Making the same assumption that box B is essentially immobile, calculate the normal mode frequencies for configuration c. Calculate the Q factor for each of the normal modes. Then, use rough approximations to show that, in fact the force transmitted at the unwanted frequency is a little larger than for configuration b.

6 Asymmetric Coupled Oscillators and the Eigenvalue Equation

Asymmetric bobs
Coupled in uneven dance—

Now I am enthralled.

—Marian McKenzie

6.1 Matrix math

What if we have a coupled oscillator system, similar to the coupled pendula of chapter 5, in which the two oscillators are not identical? For example, what if one of the masses is larger, or one of the pendulum strings is longer? This is a very important question for mechanical systems. It also has important quantum mechanical analogs, such as the formation of molecules from two different types of atoms (e.g., the CO_2 molecule).

Before we dive into the physics, it will be helpful to set up the fundamentals of matrix math. As you have seen in chapter 5, matrices provide a powerful way of describing coupled oscillators. For asymmetric systems, we will need more sophisticated matrix operations.

Matrix multiplication

It is easiest to explain matrix multiplication with an example:

$$\begin{pmatrix} A & B & C \\ D & E & F \\ G & H & I \end{pmatrix} \begin{pmatrix} j & k & l \\ m & n & o \\ p & q & r \end{pmatrix}$$

$$= \begin{pmatrix} Aj + Bm + Cp & Ak + Bn + Cq & Al + Bo + Cr \\ Dj + Em + Fp & Dk + En + Fq & Dl + Eo + Fr \\ Gj + Hm + Ip & Gk + Hn + Iq & Gl + Ho + Ir \end{pmatrix}.$$

We see that multiplying two 3×3 matrices creates a new 3×3 matrix, and that to form each entry, we multiply each of the three terms in the corresponding *row* of the left matrix with each of the three terms in the corresponding *column* of the right matrix:

$$\begin{pmatrix} A & B & C \\ D & E & F \\ G & H & I \end{pmatrix} \begin{pmatrix} j & k & l \\ m & n & o \\ p & q & r \end{pmatrix}$$

$$= \begin{pmatrix} Aj + Bm + Cp & Ak + Bn + Cq & Al + Bo + Cr \\ Dj + Em + Fp & Dk + En + Fq & Dl + Eo + Fr \\ Gj + Hm + Ip & \boxed{Gk + Hn + Iq} & Gl + Ho + Ir \end{pmatrix}.$$

This means that the width (number of columns) of the left matrix must match the height (number of rows) of the right matrix. Therefore, we can multiply a three-entry column vector by a 3×3 matrix:

$$\begin{pmatrix} A & B & C \\ D & E & F \\ G & H & I \end{pmatrix} \begin{pmatrix} j \\ k \\ l \end{pmatrix} = \begin{pmatrix} Aj + Bk + Cl \\ Dj + Ek + Fl \\ Gj + Hk + Il \end{pmatrix}.$$

We see that multiplying a vector by a matrix creates a new vector, which will ordinarily have a different length and direction. You can think of this as the matrix acting on a vector to change it into a different vector. For example, the matrix

$$\begin{pmatrix} 0 & -1 \\ 1 & 0 \end{pmatrix}$$

acts on a vector (one with two entries, of course) to rotate it by $90°$ counterclockwise, without changing its length.

Self-test: Verify that the above matrix, when applied to the vector $\begin{pmatrix} 2 \\ 3 \end{pmatrix}$, works as described.

Of course, we can also create a matrix that changes the length of any vector without rotating it: the matrix

$$\begin{pmatrix} A & 0 \\ 0 & A \end{pmatrix}$$

changes the length of any vector by a factor A:

$$\begin{pmatrix} A & 0 \\ 0 & A \end{pmatrix} \begin{pmatrix} a \\ b \end{pmatrix} = \begin{pmatrix} Aa \\ Ab \end{pmatrix} = A \begin{pmatrix} a \\ b \end{pmatrix}.$$

However, most matrices, when they act on a vector, change both the length and the direction.

The way that a matrix acts on a vector to change it into a different vector is analogous to the way a function behaves: a function acts on a number to change it into a different number. For example, the function $y(x) = x^2$ can act on the number 2 to change it into the number $2^2 = 4$. Taking this one step further, we will soon encounter "operators." An operator acts on a function to change it into a different function. For example, the operator $\dfrac{d}{dx}$ acts on the function x^3 to change it into the function $3x^2$.

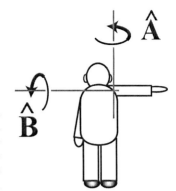

Figure 6.1.1 A view of someone (as seen from the back), with right arm extended, palm facing down. The matrix $\hat{A}$ represents a counterclockwise (as viewed from above) rotation about the vertical axis, while the matrix $\hat{B}$ represents a counterclockwise (as viewed from the left) rotation about the horizontal axis.

You should be aware that matrix multiplication doesn't commute, that is, the order of multiplication matters. For example,

$$\begin{pmatrix} A & B \\ C & D \end{pmatrix} \begin{pmatrix} e & f \\ g & h \end{pmatrix} = \begin{pmatrix} Ae + Bg & Af + Bh \\ Ce + Dg & Cf + Dh \end{pmatrix}$$

$$\text{while } \begin{pmatrix} e & f \\ g & h \end{pmatrix} \begin{pmatrix} A & B \\ C & D \end{pmatrix} = \begin{pmatrix} Ae + Cf & Be + Df \\ Ag + Ch & Bg + Dh \end{pmatrix}.$$

Since matrices can be used to represent rotations, we can see how this lack of commutation works for one particular case. Stand up now (really!). Hold your right arm up, so that it's parallel to the floor, extending to the right away from your body, with your palm facing down, as shown in figure 6.1.1. We denote matrices using capital bold-face letters that have hats. Let the matrix $\hat{A}$ represent a $90°$ rotation counterclockwise (as viewed from above) about a vertical axis passing through your right shoulder (ouch!). Let the matrix $\hat{B}$ represent a $90°$ rotation counterclockwise (as viewed from someone standing to your left) about a horizontal axis which passes through both your shoulders. Now, let the combination $\hat{A}\hat{B}$ act on your arm:

$\hat{A}\hat{B}$ (arm pointing right, palm down) $= \hat{A} [\hat{B}$ (arm pointing to right, palm down)]
$= \hat{A}$ (arm pointing right, palm pointing back)
$=$ arm pointing forward, palm pointing right

If instead you do the operations in the other order, the result is very different:

$\hat{B}\hat{A}$ (arm pointing right, palm down) $= \hat{B} [\hat{A}$ (arm pointing to right, palm down)]
$= \hat{B}$ (arm pointing forward, palm down)
$=$ arm pointing down, palm pointing back

Determinants

The study of determinants dates back to the third century BC in China, actually preceding the study of matrices. The word "determinant" was coined by Gauss in 1801, because they *determine* whether a system of equations (represented by a matrix) has a unique solution. We will use this property to find the frequencies of the normal modes.

The determinant of a 2×2 matrix is defined as follows:

$$\det \begin{pmatrix} a & b \\ c & d \end{pmatrix} \equiv ad - bc. \tag{6.1.1}$$

For larger matrices, you must use a multi-step procedure. First, you put alternating $+$ and $-$ signs above the columns. Then, multiply the top left matrix element by the $+$ sign you just wrote above it and by the determinant of the smaller matrix formed by ignoring the top row and the left column, thus forming the first term in the determinant:

$$\det \begin{pmatrix} + & - & + \\ a & b & c \\ d & e & f \\ g & h & i \end{pmatrix} = +a \det \begin{pmatrix} e & f \\ h & i \end{pmatrix} + \ldots$$

Now, move to the next entry in the top row. Multiply it by the $-$ sign you wrote above it, and by the determinant of the smaller matrix formed by ignoring the top row and the second column:

$$\det \begin{pmatrix} + & - & + \\ a & b & c \\ d & e & f \\ g & h & i \end{pmatrix} = +a \det \begin{pmatrix} e & f \\ h & i \end{pmatrix} - b \det \begin{pmatrix} d & f \\ g & i \end{pmatrix} + \ldots$$

Repeat this procedure, adding terms to the determinant, until you get to the rightmost entry of the top row; for our example, there is only one more term:

$$\det \begin{pmatrix} + & - & + \\ a & b & c \\ d & e & f \\ g & h & i \end{pmatrix} = +a \det \begin{pmatrix} e & f \\ h & i \end{pmatrix} - b \det \begin{pmatrix} d & f \\ g & i \end{pmatrix} + c \det \begin{pmatrix} d & e \\ g & h \end{pmatrix}$$

Using this procedure, you can, for example, reduce the calculation of the determinant of a 5×5 matrix to the calculation of the determinants of five 4×4 matrices. Each of these can in turn be reduced to the calculation of four 3×3 matrices. Each of these can be reduced to the calculation of three 2×2 matrices, for which you can use equation (6.1.1).

In practice, for the determinants of anything larger than a 3×3 matrix, you may prefer to use a symbolic algebra program, such as Mathematica. Instructions are given on the website for this book, under the listing for this section.

6.2   Equations of motion and the eigenvalue equation

We'll start by considering just two nonidentical oscillators, but the method we develop will be easily generalizable to a larger number. Our model system is one with two coupled pendula, where the lengths and masses might be different, as shown in figure 6.2.1. We expect that there are normal mode solutions for this system, and

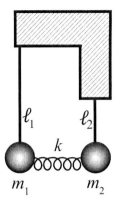

Figure 6.2.1 An asymmetric coupled pendulum system.

once we have found them, we can use the techniques of chapter 5 to perform normal mode analysis, and find the behavior of the system as a function of time, given the initial conditions. However, it is no longer obvious exactly what the normal modes are; it appears likely that there would be some type of breathing mode, and some type of pendulum mode, but perhaps the amplitudes for the two bobs would be different.

In Hilbert space terms, we are trying to find the two eigenvectors, that is, the two vectors that describe the initial positions of the bobs for the normal modes (for zero initial velocities). It will turn out that these eigenvectors are always mutually perpendicular, reflecting the fact that the different normal modes act as completely independent oscillators.[1]

To find the normal modes, we fall back on the procedure used in chapters 1–4:

Step 1: Apply $F_{\text{Total}} = m\ddot{x}$ to each object in the system to get one differential equaiton for each object.

Your turn: Show that the application of $F_{\text{Total}} = m\ddot{x}$ to the left bob gives

$$\ddot{x}_1 + \omega_A^2 x_1 - \frac{k}{m_1} x_2 = 0, \qquad (6.2.1)$$

$$\text{where } \omega_A^2 \equiv \frac{g}{\ell_1} + \frac{k}{m_1} \qquad (6.2.2)$$

Concept test: Explain why ω_A represents the angular frequency at which bob 1 would oscillate if bob 2 were held fixed at $x_2 = 0$. (Answer below.[2])

---

1. However, for systems of coupled oscillators with unequal masses, the axes of Hilbert space (i.e., the x_{10} and x_{20} axes for the case of a two-oscillator system) must be scaled according to the square root of the corresponding mass for this orthogonality to work out correctly. There is no quantum mechanical analogy for this, so we will not spend much time on it. (The analogy would be a particle whose mass depends on position.) The procedure is detailed in section 6.7.

Similarly, applying $F_{\text{Total}} = m\ddot{x}$ to the right bob gives

$$\ddot{x}_2 + \omega_{\text{B}}^2\, x_2 - \frac{k}{m_2}x_1 = 0, \qquad (6.2.3)$$

$$\text{where } \omega_{\text{B}}^2 \equiv \frac{g}{\ell_2} + \frac{k}{m_2}. \qquad (6.2.4)$$

Step 2: Guess a solution. We believe there will be normal mode solutions, that is, solutions in which all parts of the system move with the same frequency, but may have different (perhaps negative or zero) amplitudes. Also, based on our experience with the symmetric coupled pendula, we expect that the phases of all the objects in a normal mode will be the same.[3] The most general way to represent such a motion would be

$$x_1 = \text{Re } z_1 \quad z_1 \overset{?}{=} X_1 e^{i\varphi}e^{i\omega t}.$$
$$x_2 = \text{Re } z_2 \quad z_2 \overset{?}{=} X_2 e^{i\varphi}e^{i\omega t}.$$

Note that the phases and frequencies are the same for both bobs; only the amplitudes X_1 and X_2 are different. By the correct choice of when $t = 0$, we can arrange to have $\varphi = 0$, so that our guess simplifies to

$$z_1 \overset{?}{=} X_1 e^{i\omega t} \quad z_2 \overset{?}{=} X_2 e^{i\omega t}. \qquad (6.2.5)$$

With this choice of when $t = 0$, we see that the amplitudes have a simple interpretation: $X_1 = x_{10}$ and $X_2 = x_{20}$.

Step 3: Plug the guess into the DEQs, and see whether it works, and whether there are restrictions on the parameters in the guess. By analogy with the symmetric coupled pendulum system, we expect that there will be restrictions on ω and on the *relative* amplitudes, that is, on the ratio $\dfrac{X_2}{X_1}$. Plugging equation (6.2.5) into (6.2.1) gives

$$-\omega^2 X_1\, e^{i\omega t} + \omega_{\text{A}}^2 X_1\, e^{i\omega t} - \frac{k}{m_1}X_2\, e^{i\omega t} = 0$$

$$\Leftrightarrow \omega_{\text{A}}^2 X_1 - \frac{k}{m_1}X_2 = \omega^2 X_1. \qquad (6.2.6a)$$

---

2. The total force on bob 1 is the pendulum force $-\dfrac{m_1 g}{\ell_1}x_1$ plus the spring force, which depends on the relative position of the two bobs: $F_{\text{spring}} = -k\left(x_1 - x_2\right)$. Thus, if $x_2 = 0$, the total force reduces to $F_{\text{total}} = -k_{\text{eff}}x_1$, where $k_{\text{eff}} = \dfrac{m_1 g}{\ell_1} + k$. Therefore, bob 1 will oscillate with

$$\omega = \sqrt{\frac{k_{\text{eff}}}{m_1}} = \sqrt{\frac{g}{\ell_1} + \frac{k}{m_1}}.$$

3. "But wait!" you say, "In the breathing mode, aren't the bobs $180°$ out of phase?" This is indeed one correct way of describing the breathing mode of symmetric coupled pendula: positive amplitudes for both bobs, and $180°$ phase difference. However, it is completely equivalent to say that one bob has a *negative* amplitude, and that the phase factors are the same. That's the option we'll choose.

Similarly, plugging into equation (6.2.3) gives

$$\omega_B^2 X_2 - \frac{k}{m_2} X_1 = \omega^2 X_2. \tag{6.2.6b}$$

We saw in chapter 5 that matrix methods were helpful. So, let us cast equations (6.2.6) into matrix form, with the top line of the matrix equation representing (6.2.6a), and the bottom line representing (6.2.6b):

$$\begin{pmatrix} \omega_A^2 & -\dfrac{k}{m_1} \\ -\dfrac{k}{m_2} & \omega_B^2 \end{pmatrix} \begin{pmatrix} X_1 \\ X_2 \end{pmatrix} = \omega^2 \begin{pmatrix} X_1 \\ X_2 \end{pmatrix}. \tag{6.2.7}$$

Your turn: Verify that the top line of equation (6.2.7) really does represent equation (6.2.6a), and that the bottom line really does represent equation (6.2.6b).

Note that this equation looks similar in some ways to one encountered in section 6.1,

$$\begin{pmatrix} A & 0 \\ 0 & A \end{pmatrix} \begin{pmatrix} a \\ b \end{pmatrix} = A \begin{pmatrix} a \\ b \end{pmatrix}.$$

However, the important difference is that when *any* vector $\begin{pmatrix} a \\ b \end{pmatrix}$ is multiplied by the matrix $\begin{pmatrix} A & 0 \\ 0 & A \end{pmatrix}$, only the length of the vector is changed (and not its direction). In contrast, equation (6.2.7) *only works for a small number of special vectors.* For example, if we apply the matrix in equation (6.2.7) to an "arbitrary" vector such as $\begin{pmatrix} 1 \\ 0 \end{pmatrix}$, we get

$$\begin{pmatrix} \omega_A^2 & -\dfrac{k}{m_1} \\ -\dfrac{k}{m_2} & \omega_B^2 \end{pmatrix} \begin{pmatrix} 1 \\ 0 \end{pmatrix} = \begin{pmatrix} \omega_A^2 \\ -\dfrac{k}{m_2} \end{pmatrix} \neq \text{const.} \begin{pmatrix} 1 \\ 0 \end{pmatrix}.$$

Each special vector for which equation (6.2.7) *does* hold represents the initial positions X_1 and X_2 of the bobs in one of the normal modes. **Therefore, these special vectors are the eigenvectors, multiplied by an amplitude.**

Example: in the symmetric coupled pendula, we found that the breathing mode was represented by the eigenvector $|e_b\rangle = \frac{1}{\sqrt{2}} \begin{pmatrix} 1 \\ -1 \end{pmatrix}$. We could excite this mode with any overall amplitude. For example, we could set $x_{10} = 5$ cm and $x_{20} = -5$ cm. Then, (making use of the facts that, for the symmetric case $\omega_A = \omega_B$ and $m_1 = m_2 = m$) the left side of

continued

equation (6.2.7) would read

$$\begin{pmatrix} \omega_A^2 & -\dfrac{k}{m_1} \\ -\dfrac{k}{m_2} & \omega_B^2 \end{pmatrix} \begin{pmatrix} X_1 \\ X_2 \end{pmatrix} = \begin{pmatrix} \omega_A^2 & -\dfrac{k}{m} \\ -\dfrac{k}{m} & \omega_A^2 \end{pmatrix} \begin{pmatrix} 5\text{ cm} \\ -5\text{ cm} \end{pmatrix}.$$

Let's verify that equation (6.2.7) really holds for this case, bearing in mind that for this symmetric case $\omega_A^2 \equiv \dfrac{g}{\ell} + \dfrac{k}{m}$:

$$\begin{pmatrix} \omega_A^2 & -\dfrac{k}{m} \\ -\dfrac{k}{m} & \omega_A^2 \end{pmatrix} \begin{pmatrix} 5\text{ cm} \\ -5\text{ cm} \end{pmatrix} = \begin{pmatrix} \omega_A^2(5\text{ cm}) - \dfrac{k}{m}(-5\text{ cm}) \\ -\dfrac{k}{m}(5\text{ cm}) + \omega_A^2(-5\text{ cm}) \end{pmatrix} = \left(\omega_A^2 + \dfrac{k}{m}\right)\begin{pmatrix} 5\text{ cm} \\ -5\text{ cm} \end{pmatrix}$$

$$= \left[\left(\dfrac{g}{\ell} + \dfrac{k}{m}\right) + \dfrac{k}{m}\right]\begin{pmatrix} 5\text{ cm} \\ -5\text{ cm} \end{pmatrix} = \left(\dfrac{g}{\ell} + \dfrac{2k}{m}\right)\begin{pmatrix} 5\text{ cm} \\ -5\text{ cm} \end{pmatrix} = \omega_b^2 \begin{pmatrix} 5\text{ cm} \\ -5\text{ cm} \end{pmatrix}. \checkmark$$

As we had arranged, ω which appears in equation (6.2.7) is the normal mode frequency.

Each different eigenvector (or simple multiple thereof) which solves equation (6.2.7) is associated with a different squared normal mode angular frequency ω^2. Our job now becomes finding the eigenvectors and corresponding "eigenvalues" ω^2 for which equation (6.2.7) holds. This form of equation is called an "eigenvalue equation":

$$(6.2.7):\quad \underbrace{\begin{pmatrix} \omega_A^2 & -\dfrac{k}{m_1} \\ -\dfrac{k}{m_2} & \omega_B^2 \end{pmatrix}}_{\text{matrix } \hat{A}} \underbrace{\begin{pmatrix} X_1 \\ X_2 \end{pmatrix}}_{\text{eigenvector}} = \underbrace{\omega^2}_{\text{eigenvalue}} \underbrace{\begin{pmatrix} X_1 \\ X_2 \end{pmatrix}}_{\text{eigenvector}}.$$

To write this in bra-ket notation, we define $|e_u\rangle \equiv \begin{pmatrix} X_1 \\ X_2 \end{pmatrix}$, where the subscript "u" indicates unnormalized, that is, the vector $|e_u\rangle$ does not necessarily have length 1. Later, we will discuss the (simple) process for normalizing eigenvectors. With this notation, the equation becomes:

$$\hat{A}\,|e_u\rangle = \omega^2\,|e_u\rangle.$$

6.3   Procedure for solving the eigenvalue equation

Matrix eigenvalue equations such as equation (6.2.7) have been studied for a long time, and there is a well-developed recipe to find the eigenvalues and eigenvectors.

Step A: Write the characteristic equation

The "characteristic equation" is the same as the eigenvalue equation, just rearranged so that the right side is zero.

Your turn: Show that the eigenvalue equation

$$(6.2.7): \begin{pmatrix} \omega_A^2 & -\dfrac{k}{m_1} \\ -\dfrac{k}{m_2} & \omega_B^2 \end{pmatrix} \begin{pmatrix} X_1 \\ X_2 \end{pmatrix} = \omega^2 \begin{pmatrix} X_1 \\ X_2 \end{pmatrix}$$

eigenvalue equation for two coupled oscillators

is equivalent to

$$\underbrace{\begin{pmatrix} \omega_A^2 - \omega^2 & -\dfrac{k}{m_1} \\ -\dfrac{k}{m_2} & \omega_B^2 - \omega^2 \end{pmatrix}}_{\hat{B}} \underbrace{\begin{pmatrix} X_1 \\ X_2 \end{pmatrix}}_{|e_u\rangle} = 0. \qquad (6.3.1)$$

characteristic equation for two coupled oscillators

(Here, the "0" on the right side is shorthand for $\begin{pmatrix} 0 \\ 0 \end{pmatrix}$.) *Hint: Remember that each matrix equation is shorthand for two regular equations. If you can show that this pair of regular equations for (6.3.1) is equivalent to the pair for (6.2.7), then you've shown that the matrix equation (6.3.1) is equivalent to the matrix equation (6.2.7).*

Step B: Use theorem from linear algebra

There is a general theorem that states that any equation of the form of equation (6.3.1), that is, of the form $\hat{B}\,\mathbf{r} = 0$, can hold if and only if $\det \hat{B} = 0$.

Proof for 2×2 case:

$$\underbrace{\begin{pmatrix} a & b \\ c & d \end{pmatrix}}_{\hat{B}} \begin{pmatrix} e \\ f \end{pmatrix} = 0 \Rightarrow$$

$$\left. \begin{array}{l} \text{Top line}: \ ae + bf = 0 \Leftrightarrow e = -\dfrac{bf}{a} \\ \text{Bottom line}: \ ce + df = 0 \end{array} \right\} \Rightarrow c\left(-\dfrac{bf}{a}\right) + df = 0 \Leftrightarrow$$

$$-cb + ad = 0 \Leftrightarrow \det \hat{B} = 0 \checkmark$$

Applying this theorem to equation (6.3.1) gives us a quadratic equation for ω^2:

$$\hat{B}\,\mathbf{r} = 0 \Rightarrow \ \det \hat{B} = 0, \ \text{that is,}$$

$$\cdot \left(\omega_A^2 - \omega^2\right)\left(\omega_B^2 - \omega^2\right) - \dfrac{k^2}{m_1 m_2} = 0 \Leftrightarrow$$

$$\omega^4 - \omega^2\left(\omega_A^2 + \omega_B^2\right) + \left(\omega_A^2 \omega_B^2 - \dfrac{k^2}{m_1 m_2}\right) = 0.$$

Using the quadratic equation for the variable ω^2 , we obtain

$$\omega^2 = \frac{\omega_A^2 + \omega_B^2 \pm \sqrt{\left(\omega_A^2 + \omega_B^2\right)^2 - 4\left(\omega_A^2\omega_B^2 - \frac{k^2}{m_1 m_2}\right)}}{2}$$

$$= \frac{\omega_A^2 + \omega_B^2 \pm \sqrt{\left(\omega_A^2\right)^2 + 2\omega_A^2\omega_B^2 + \left(\omega_B^2\right)^2 - 4\omega_A^2\omega_B^2 + 4\frac{k^2}{m_1 m_2}}}{2}$$

$$= \frac{\omega_A^2 + \omega_B^2 \pm \sqrt{\left(\omega_A^2\right)^2 - 2\omega_A^2\omega_B^2 + \left(\omega_B^2\right)^2 + 4\frac{k^2}{m_1 m_2}}}{2}$$

$$\Rightarrow \boxed{\omega^2 = \frac{\omega_A^2 + \omega_B^2 \pm \sqrt{\left(\omega_A^2 - \omega_B^2\right)^2 + 4\frac{k^2}{m_1 m_2}}}{2}} \tag{6.3.2}$$

eigenvalues for two coupled oscillators

Note that this equation tells us both values of ω (*i.e.*, the angular frequencies of both normal modes) because of the "$\pm$."

Often, this is as far as we need to go, since often we only care about the frequencies of the normal modes. However, sometimes we also wish to know the way in which the masses are moving, which means we need to find the eigenvectors. (Again, for a pure normal mode state, the eigenvectors tell us the relative initial positions of each mass, and each mass oscillates in simple harmonic motion, with amplitude equal to its initial position.)

Step C: To find the eigenvectors, take the eigenvalues, one at a time, and substitute them into the characteristic equation. Solve for X_2 in terms of X_1.

Note that we cannot find the actual values of both X_1 and X_2, but only their relative values, since we are only making use of the underlying interactions of the system, and not of the initial conditions; each normal mode can be excited with an arbitrary amplitude, as determined by the initial conditions. Implementing step C for the general case, even for a system of only two coupled oscillators, is very messy and not very instructive. Instead, we will illustrate with a special case.

Core example: Consider the special case of coupled pendula with equal masses $m_1 = m_2 = m = 1\,\text{kg}$, but unequal lengths $\ell_1 = 4.33\,\text{m}$ and $\ell_2 = 2.30\,\text{m}$, and let $k = \sqrt{3}\,\text{N/m}$. This is shown in figure 6.3.1. Then,

$$\frac{k^2}{m^2} = 3\,\text{rad}^4/\text{s}^4,$$

and $\omega_A^2 \equiv \dfrac{g}{\ell_1} + \dfrac{k}{m_1} = (2.26 + \sqrt{3})\,\text{rad}^2/\text{s}^2 = 4\,\text{rad}^2/\text{s}^2,$

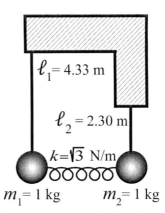

Figure 6.3.1 An example of an asymmetric coupled pendulum system with equal masses.

$$\text{while } \omega_B^2 \equiv \frac{g}{\ell_2} + \frac{k}{m_2} = (4.26 + \sqrt{3})\, \text{rad}^2/s^2 = 6\, \text{rad}^2/s^2.$$

Plugging these into equation (6.3.2) gives

$$\omega^2 = \frac{4 + 6 \pm \sqrt{(4-6)^2 + 4 \cdot 3}}{2}\, \text{rad}^2/s^2$$

$$= \frac{10 \pm \sqrt{4+12}}{2}\, \text{rad}^2/s^2 = \begin{cases} 7 \\ 3 \end{cases} \text{rad}^2/s^2 \Rightarrow \omega = \begin{cases} \sqrt{7} \\ \sqrt{3} \end{cases} \text{rad/s.}$$

We anticipate that the higher value of ω corresponds to a breathing-type mode, and the lower value to a pendulum type mode. Now, to find the eigenvectors. For the breathing mode, substitute $\omega^2 = 7\, \text{rad}^2/s^2$ into the characteristic equation (6.3.1):

$$\begin{pmatrix} \omega_A^2 - \omega^2 & -\dfrac{k}{m_1} \\ -\dfrac{k}{m_2} & \omega_B^2 - \omega^2 \end{pmatrix} \begin{pmatrix} X_1 \\ X_2 \end{pmatrix} = 0 \Rightarrow \begin{pmatrix} 4-7 & -\sqrt{3} \\ -\sqrt{3} & 6-7 \end{pmatrix} \text{rad}^2/s^2 \begin{pmatrix} X_1 \\ X_2 \end{pmatrix} = 0.$$

We now divide both sides by $1\, \text{rad}^2/s^2$, giving

$$\begin{pmatrix} 4-7 & -\sqrt{3} \\ -\sqrt{3} & 6-7 \end{pmatrix} \begin{pmatrix} X_1 \\ X_2 \end{pmatrix} = 0 \Rightarrow \text{Top line: } -3X_1 - \sqrt{3}\, X_2 = 0 \Leftrightarrow X_2 = -\sqrt{3}\, X_1$$

Note: the bottom line of the above matrix equation tells us the same thing:

$$\text{Bottom line: } -\sqrt{3}\, X_1 - X_2 = 0 \Leftrightarrow X_2 = -\sqrt{3} X_1.$$

This redundancy arises because we cannot specify the actual values of X_1 and X_2, but only their relative values, since we are only making use of the underlying interactions of the system, and not of the initial conditions. As expected for a breathing-type mode, the amplitude for right bob, X_2 is opposite in sign to that for the left bob. We also

continued

see that the amplitude for the right bob (the one with the shorter string) is larger in this mode. Expressing our result in column matrix form, we have found that the eigenvector for the breathing mode (not yet normalized) is $\left| e_{ub} \right\rangle = \begin{pmatrix} 1 \\ -\sqrt{3} \end{pmatrix}$. This is illustrated in figure 6.3.2a.

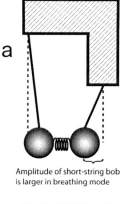

a

Amplitude of short-string bob
is larger in breathing mode

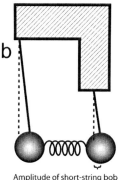

b

Amplitude of short-string bob
is smaller in pendulum mode

Figure 6.3.2 Normal modes for the system shown in figure 6.3.1. a: breathing mode. b: pendulum mode.

Your turn: Find the eigenvector for the pendulum mode by substituting $\omega^2 = 3$ rad^2/s^2 into the characteristic equation, (6.3.1), and then using the top line of the resulting matrix equation to show that $X_2 = \dfrac{1}{\sqrt{3}} X_1$. Now, show that the bottom line of the matrix equation says the same thing.

As expected for the pendulum mode, the amplitudes have the same sign for both bobs, but now the right bob has a smaller amplitude, as shown in figure 6.3.2b. Expressing our result in column matrix form, we have found that the eigenvector for the pendulum mode (not yet normalized) is $\left| e_{up} \right\rangle = \begin{pmatrix} 1 \\ 1/\sqrt{3} \end{pmatrix}$.

6.4 Systems with more than two objects

To generalize these ideas to a larger system of coupled oscillators, we use exactly the same methods.[4] Assuming there are N objects in the system, here is the procedure:

1. Write $F_{\text{TOT}} = m\ddot{x}$ for each object. This creates a system of N coupled differential equations.

2. Guess a normal mode solution, that is, guess that the position of object j is given by

$$x_j \overset{?}{=} \text{Re}\left[X_j\, e^{i\omega t}\right],$$

where ω is the angular frequency of the normal mode, and X_j is the amplitude of oscillation of object j. As before, we have chosen the time when $t \equiv 0$ so that the overall phase factor $\varphi = 0$. Therefore, the amplitude X_j of object j is equal to the initial position x_{j0} of object j.

3. Plug the guess back into the system of differential equations. All the terms will contain the $e^{i\omega t}$ factor, so this can be cancelled out, leaving a system of N coupled linear equations involving the variables X_j.

4. Write the system of equations as an eigenvalue equation, that is, a matrix equation of the form

$$\hat{\mathbf{A}}\left|e_{\text{u}}\right\rangle = \omega^2 \left|e_{\text{u}}\right\rangle,$$

where $\hat{\mathbf{A}}$ is an $N \times N$ matrix, and $\left|e_{\text{u}}\right\rangle$ is the column vector $\begin{pmatrix} X_1 \\ X_2 \\ \vdots \end{pmatrix}$.

5. Rearrange the eigenvalue equation to form the characteristic equation, that is, the equivalent equation which has the form

$$\hat{\mathbf{B}}\left|e_{\text{u}}\right\rangle = 0.$$

6. To find the eigenvalues ω^2, set

$$\det \hat{\mathbf{B}} = 0$$

This will give an Nth-order equation for ω^2, with N solutions.

---

4. The ideas described here work just as well for any system with more than two "degrees of freedom," whether there are one, two, or more objects. Here, a "degree of freedom" is a way in which the system can move that is completely independent of other degrees of freedom, at least before the coupling between degrees of freedom is added. For example, a rigid object in three-dimensional space has six degrees of freedom, since it can move in three directions, and rotate about three axes. As another example, if we had two particles, each of which was kept near a particular point in three-dimensional space by restoring forces, and these particles were coupled together by a spring, that would be a system with six degrees of freedom. In our main discussion, however, we'll stick to particles that are constrained to move in only one dimension, since this is easier to illustrate and conceptually simpler.

7. To find the eigenvectors, take each value of ω^2, one at a time, and substitute it back into the characteristic equation. Solve for X_2, X_3 , etc., in terms of X_1.

This procedure is guaranteed to work. (For systems of five or more objects, it will not usually be possible to solve for the eigenvalues analytically, but they can always be found by numerical approximation.) This is a tremendously important realization, since we have just proved that, for a system of N objects, we will always be able to find N normal modes. In principle, we could write a system of N uncoupled differential equations to represent these modes. (This is seldom necessary, but we certainly could do it.) This system of DEQs would represent the behavior of the system just as well as the system of N DEQs which describes the positions of the N objects. Therefore, we can quite generally conclude:

> **For a one-dimensional system, the number of normal modes is equal to the number of masses. Therefore, giving the amplitudes and phases of all the normal modes provides just as complete a description of the system as giving $x(t)$ for each object in the system.**

We will do a fairly generic example of a one-dimensional system with three objects, from which it will be obvious how to generalize to a one-dimensional system with any number of objects. Consider the system shown in figure 6.4.1. Masses 1 and 3 are connected by the long spring k_{13}.

1. Applying $F_{\text{TOT}} = m\ddot{x}$ to m_1 gives

$$-k_1 x_1 - \frac{m_1 g}{\ell} x_1 - k_{12}\left(x_1 - x_2\right) - k_{13}\left(x_1 - x_3\right) = m_1 \ddot{x}_1. \tag{6.4.1}$$

This is the real part of

$$-k_1 z_1 - \frac{m_1 g}{\ell} z_1 - k_{12}\left(z_1 - z_2\right) - k_{13}\left(z_1 - z_3\right) = m_1 \ddot{z}_1. \tag{6.4.2}$$

(We won't bother to apply $F_{\text{TOT}} = m\ddot{x}$ to the other masses, since the pattern will become obvious just from m_1.)

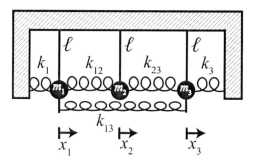

Figure 6.4.1 system of three coupled pendula with all possible couplings.

2. We guess the normal mode solution $x_j \overset{?}{=} \mathrm{Re}\left[X_j\,e^{i\omega t}\right]$, that is, $z_1 = X_1 e^{i\omega t}$, $z_2 = X_2 e^{i\omega t}$, and $z_3 = X_3 e^{i\omega t}$.

3. Plugging this guess into equation (6.4.2) gives

$$-k_1 X_1 e^{i\omega t} - \frac{m_1 g}{\ell} X_1 e^{i\omega t} - k_{12}\left(X_1 e^{i\omega t} - X_2 e^{i\omega t}\right)$$

$$-k_{13}\left(X_1 e^{i\omega t} - X_3 e^{i\omega t}\right) = -m_1 \omega^2 X_1 e^{i\omega t} \Rightarrow$$

$$\left(\frac{k_1}{m_1} + \frac{g}{\ell} + \frac{k_{12}}{m_1} + \frac{k_{13}}{m_1}\right) X_1 - \frac{k_{12}}{m_1} X_2 - \frac{k_{13}}{m_1} X_3 = \omega^2 X_1 \Rightarrow$$

$$\omega_A^2 X_1 - \frac{k_{12}}{m_1} X_2 - \frac{k_{13}}{m_1} X_3 = \omega^2 X_1, \tag{6.4.3}$$

where

$$\omega_A^2 = \frac{k_1}{m_1} + \frac{g}{\ell} + \frac{k_{12}}{m_1} + \frac{k_{13}}{m_1}$$

is the square of the angular frequency at which m_1 would oscillate if all the other masses were held fixed.

We can see that the result of applying $F_{\mathrm{TOT}} = m\ddot{x}$ to m_2 and plugging in our normal mode guess would be

$$-\frac{k_{12}}{m_2} X_1 + \omega_B^2 X_2 - \frac{k_{23}}{m_2} X_3 = \omega^2 X_2, \tag{6.4.4}$$

where ω_B^2 is the square of the angular frequency at which m_2 would oscillate if the other masses were held fixed.

Concept test (answer below[5]): What is the value of ω_B^2 in terms of ℓ, g, m_2, and the k's?

Similarly, for m_3 we would obtain

$$-\frac{k_{13}}{m_3} X_1 - \frac{k_{23}}{m_3} X_2 + \omega_C^2 X_3 = \omega^2 X_3, \tag{6.4.5}$$

where ω_C^2 is the square of the angular frequency at which m_3 would oscillate if the other masses were held fixed: $\omega_C^2 = \frac{k_{13}}{m_3} + \frac{k_{23}}{m_3} + \frac{k_3}{m_3} + \frac{g}{\ell}$.

---

5. $\omega_B^2 = \frac{k_{12}}{m_2} + \frac{g}{\ell} + \frac{k_{23}}{m_2}$

4. We can write equations (6.4.3), (6.4.4), and (6.4.5) in matrix form:

$$
\begin{pmatrix}
\omega_A^2 & -\dfrac{k_{12}}{m_1} & -\dfrac{k_{13}}{m_1} \\[2mm]
-\dfrac{k_{12}}{m_2} & \omega_B^2 & -\dfrac{k_{23}}{m_2} \\[2mm]
-\dfrac{k_{13}}{m_3} & -\dfrac{k_{23}}{m_3} & \omega_C^2
\end{pmatrix}
\begin{pmatrix} X_1 \\ X_2 \\ X_3 \end{pmatrix}
= \omega^2 \begin{pmatrix} X_1 \\ X_2 \\ X_3 \end{pmatrix}.
\tag{6.4.6}
$$

In fact, now that we've gone through this procedure, you can probably construct the equivalent of equation (6.4.6) for whatever system is given just by inspection.

From here, we could in principle go on to steps 5, 6, and 7 to find the eigenvalues and eigenvectors. However, it is much easier to do this using a symbolic algebra program (such as Mathematica), especially if there are more than three objects. Explicit instructions for how to do this, including an example, are given on the website for this text under the listing for this section.

6.5   Normal mode analysis for multi-object, asymmetrical systems

Our treatment of normal mode analysis in section 5.8 depended only on the orthonormality of the normalized eigenvectors, so we can use exactly the same methods for asymmetric systems with many objects, as long as we use the normalized eigenvectors. We will go over things again below, to make sure this is clear. However, first we need to understand the process for normalizing eigenvectors.

How to normalize eigenvectors

If we find a vector $|e_u\rangle$ which solves the eigenvalue equation $\hat{A}\,|e_u\rangle = \omega^2\,|e_u\rangle$, then the vector $D|e_u\rangle$ (where D is a constant, perhaps a complex constant) will also solve the equation, since

$$
\hat{A}\,|e_u\rangle = \omega^2\,|e_u\rangle \quad \Leftrightarrow \quad D\hat{A}\,|e_u\rangle = D\omega^2\,|e_u\rangle \quad \Leftrightarrow \quad \hat{A}\left(D\,|e_u\rangle\right) = \omega^2\left(D\,|e_u\rangle\right).
$$

Thus, there are an infinite number of vectors, all pointing along the same direction in Hilbert space but with different lengths, which solve the eigenvector equation for a particular eigenvalue ω^2. However, there is one vector that is particularly helpful: the normalized eigenvector, that is, the vector that has length 1 in this direction. (Again, just 1, not 1 m.)

Core example:

For the example from the end of section 6.3, we found that the eigenvector for

the breathing mode is of the form $|e_{ub}\rangle = \begin{pmatrix} 1 \\ -\sqrt{3} \end{pmatrix}$. To normalize it, we just divide by the length: Length $= \sqrt{1^2 + \left(-\sqrt{3}\right)^2} = 2$. So, the normalized eigenvector is

$$|e_b\rangle = \frac{|e_{ub}\rangle}{\text{Length}} = \begin{pmatrix} \frac{1}{2} \\ -\frac{\sqrt{3}}{2} \end{pmatrix}.$$

As a check, if we've correctly normalized it we should have $\langle e_b | e_b \rangle = 1$:

$$\begin{pmatrix} \frac{1}{2} & -\frac{\sqrt{3}}{2} \end{pmatrix} \begin{pmatrix} \frac{1}{2} \\ -\frac{\sqrt{3}}{2} \end{pmatrix} = \frac{1}{4} + \frac{3}{4} = 1 \checkmark$$

The above *normalization* procedure is only valid when the masses are equal; the more general procedure is treated in section 6.7.

Normal mode analysis

Here, I convince you that we only need a complete orthonormal basis in order to use the same formulas for normal mode analysis as we derived in section 5.8. Let $|e_1\rangle$ be the normalized eigenvector for mode $n = 1$, and $|e_2\rangle$ be the normalized eigenvector for mode $n = 2$, and so on. Be careful not to confuse the mode index n with the mass index j.

For example, $|e_1\rangle = \begin{pmatrix} X_1 \\ X_2 \\ \vdots \end{pmatrix}$ is a vector of length 1 specifying the initial positions of

the masses (with zero initial velocities) for normal mode $n = 1$. The subscript "1" in $|e_1\rangle$ refers to mode number 1, whereas the subscript "1" in X_1 refers to mass number 1. It is easy to tell which type of index is being used from the context: any subscript on a vector (*e.g.*, $|e_1\rangle$) or a frequency (*e.g.*, ω_1) is a mode index. We will consistently use the symbol n or m for the mode index and the symbol j for the mass index.

As detailed earlier, giving the amplitudes and phases of all the normal modes provides just as complete a description of the system as giving $x(t)$ for each object in

the system. This means that, for any state of the system, the vector $|x(t)\rangle = \begin{pmatrix} x_1(t) \\ x_2(t) \\ \vdots \end{pmatrix}$

can be written as a superposition of the normal modes, that is,

$$|x(t)\rangle = \text{Re}\left[\sum_{n=1}^{N} C_n e^{i\omega_n t} |e_n\rangle \right] \tag{6.5.1}$$

(Again, the $|e_n\rangle$'s (i.e., $|e_1\rangle$, $|e_2\rangle$, etc.) are the normalized eigenvectors representing the different normal modes.) This is the more general version of the normal mode expansion for two coupled oscillators, (5.8.1): $|x(t)\rangle = \text{Re}\left[C_p e^{i\omega_p t} |e_p\rangle + C_b e^{i\omega_b t} |e_b\rangle \right]$.

To find the complex amplitudes C_n in the normal mode expansion (6.5.1) from the initial positions and velocities of the masses, we follow the example of section 5.8.

According to equation (6.5.1), the initial positions of the objects are given by[6]

$$|x_0\rangle \equiv \begin{pmatrix} x_{10} \\ x_{20} \\ \vdots \end{pmatrix} = \mathrm{Re}\left[\sum_{n=1}^{N} C_n |e_n\rangle\right], \tag{6.5.2}$$

and the initial velocities are given by

$$|\dot{x}_0\rangle \equiv \begin{pmatrix} \dot{x}_{10} \\ \dot{x}_{20} \\ \vdots \end{pmatrix} = \frac{d}{dt}\left[|x(t)\rangle\right]_{t=0} = \frac{d}{dt}\left[\mathrm{Re}\left[\sum_{n=1}^{N} C_n e^{i\omega_n t} |e_n\rangle\right]\right]_{t=0}$$

$$= \mathrm{Re}\left[\sum_{n=1}^{N} i\omega_n C_n |e_n\rangle\right]. \tag{6.5.3}$$

Next, we write the complex amplitudes in Cartesian form:

$$C_n = \mathrm{Re}\, C_n + i\,\mathrm{Im}\, C_n.$$

This allows us to rewrite equations (6.5.2) and (6.5.3):

$$|x_0\rangle = \mathrm{Re}\left[\sum_{n=1}^{N} C_n |e_n\rangle\right] = \sum_{n=1}^{N} \mathrm{Re}\,(C_n)\, |e_n\rangle, \tag{6.5.4}$$

(since the $|e_n\rangle$'s are real) and

$$|\dot{x}_0\rangle = \mathrm{Re}\left[\sum_{n=1}^{N} i\omega_n C_n |e_n\rangle\right] = \mathrm{Re}\left[\sum_{n=1}^{N} i\omega_n\,(\mathrm{Re}\, C_n + i\,\mathrm{Im}\, C_n)\, |e_n\rangle\right]$$

$$\mathrm{Re}\left[\sum_{n=1}^{N} \omega_n\,(i\mathrm{Re}\, C_n - \mathrm{Im}\, C_n)\, |e_n\rangle\right] = -\sum_{n=1}^{N} \omega_n\,\mathrm{Im}\,(C_n)\, |e_n\rangle. \tag{6.5.5}$$

For the remainder of this section, we assume the masses of coupled oscillators are equal. This is the case of most interest, because it is most closely analogous with quantum mechanics. The general case of nonequal masses is treated in section 6.7.

Next, we need a couple of results regarding the normalized eigenvectors. The normal modes do not interact with each other, therefore the eigenvectors are mutually orthogonal, that is,

$$\langle e_m | e_n \rangle = 0 \text{ if } n \neq m$$

(We prove this more rigorously in section 6.7.) The eigenvectors $|e_n\rangle$ are normalized, that is,

$$\langle e_n | e_n \rangle = 1$$

---

6. Note that here, we use the subscript "0" to indicate $t = 0$. It does not indicate $n = 0$. (In fact, there is no mode $n = 0$, since we start numbering the modes at $n = 1$.)

We can summarize the above two equations by writing

$$\langle e_m | e_n \rangle = \delta_{mn}, \tag{6.5.6}$$

where the "Kronecker delta function" (pronounced "Crow-neck-er") is defined by

$$\boxed{\delta_{mn} \equiv \begin{cases} 1 \ \text{if} \ n = m \\ 0 \ \text{if} \ n \neq m \end{cases}} \tag{6.5.7}$$

Now, following the lead of section 5.8, we try taking the inner product of $|x_0\rangle$ with each of the normalized eigenvectors:

$$\langle e_m | x_0 \rangle = \langle e_m | \left(\sum_{n=1}^{N} \text{Re} \ (C_n) \, |e_n\rangle \right) = \sum_{n=1}^{N} \text{Re} \ (C_n) \, \langle e_m | e_n \rangle.$$

(Note that we must use $\langle e_m |$ in the left side of the inner product above, rather than $\langle e_n |$, because the subscript n is already being used in the summation on the right side of the inner product.) Because of orthonormality (as expressed in equation (6.5.6)), all the terms in the sum are zero, except the one for which $m = n$, and for that term we have $\langle e_{m=n} | e_n \rangle = 1$. Therefore,

$$\langle e_m | x_0 \rangle = \text{Re} \ (C_m) .$$

Since we have eliminated the sum over n, we could just as well write this as $\langle e_n | x_0 \rangle = \text{Re} \ (C_n)$

$$\Rightarrow \boxed{\text{Re} \ (C_n) = \langle e_n | x_0 \rangle ,} \tag{6.5.8a}$$

which is the generalized version of equation (5.8.2a): $\text{Re}\left(C_{\text{p}} \right) = \langle e_{\text{p}} | x_0 \rangle$.

Next, again following the lead of section 5.8, we try taking the inner product of $|\dot{x}_0\rangle$ with each of the normalized eigenvectors:

$$\langle e_m | \dot{x}_0 \rangle = \langle e_m | \left(-\sum_{n=1}^{N} \omega_n \text{Im} \ (C_n) \, |e_n\rangle \right) = -\sum_{n=1}^{N} \omega_n \text{Im} \ (C_n) \langle e_m | e_n \rangle = -\omega_m \text{Im} \ (C_m).$$

Again, since the sum over n has been eliminated, we could write this as $\langle e_n | \dot{x}_0 \rangle = -\omega_n \text{Im}(C_n)$

$$\Rightarrow \boxed{\text{Im} \ (C_n) = -\frac{1}{\omega_n} \langle e_n | \dot{x}_0 \rangle ,} \tag{6.5.8b}$$

which is the generalized version of equation (5.8.2b): $\text{Im}\left(C_{\text{p}} \right) = -\dfrac{1}{\omega_{\text{p}}} \langle e_{\text{p}} | \dot{x}_0 \rangle$.

Note that for the important special case of $|\dot{x}_0\rangle = 0$, $\text{Im}(C_n) = 0$, so that the amplitudes that appear in the normal mode expansion (6.5.1) reduce to $C_n = \text{Re}(C_n) = \langle e_n | x_0 \rangle$.

Stated in graphical terms, this means that, in the normal mode expansion for the special case $|\dot{x}_0\rangle = 0$, the coefficient along the Hilbert space axis corresponding to mode n is given by the inner product of the initial state of the system $|x_0\rangle$ with the normalized eigenvector for that mode, $|e_n\rangle$. Just as for the coupled pendulum, this idea of taking inner products to find the "projections" of the state of the system along the normal mode "directions" in Hilbert space (by taking these inner products) is *exactly* analogous to the process of finding the projections of an ordinary vector along the x- or y-axis (by taking dot products with $\hat{\mathbf{i}}$ or $\hat{\mathbf{j}}$).

6.6   More matrix math

In this section, we go over a few additional aspects of matrix math that are needed for section 6.7, and that are also quite important for the study of quantum mechanics.

Inverse matrices and the identity matrix

One can show that, if and only if the determinant of a matrix $\hat{\mathbf{A}}$ doesn't equal zero, then one can find the inverse matrix $\hat{\mathbf{A}}^{-1}$, such that

$$\hat{\mathbf{A}}\hat{\mathbf{A}}^{-1} = \hat{\mathbf{A}}^{-1}\hat{\mathbf{A}} = \hat{\mathbf{I}},$$

where $\hat{\mathbf{I}}$ is the identity matrix:

$$\hat{\mathbf{I}} = \begin{pmatrix} 1 & 0 & \cdots \\ 0 & 1 & 0 \\ \vdots & 0 & \ddots \end{pmatrix}.$$

Note that multiplying by $\hat{\mathbf{I}}$ has no effect on anything, just as multiplying by 1. For example, in multiplication of 2×2 matrices, we have

$$\hat{\mathbf{I}}\hat{\mathbf{A}} = \begin{pmatrix} 1 & 0 \\ 0 & 1 \end{pmatrix} \begin{pmatrix} a & b \\ c & d \end{pmatrix} = \begin{pmatrix} a & b \\ c & d \end{pmatrix}.$$

This means that the matrix $\hat{\mathbf{A}}^{-1}$ "undoes" the effect of the matrix $\hat{\mathbf{A}}$. For instance, if the matrix $\hat{\mathbf{A}}$ rotates vectors by $31°$ clockwise about the z-axis, then the matrix $\hat{\mathbf{A}}^{-1}$ rotates vectors by $31°$ counterclockwise about the z-axis.

Another feature of bra-ket notation

We've discussed how, unlike multiplication of numbers, the order matters for matrix multiplication. However, matrix multiplication *is* associative, meaning, for example, that

$$\hat{\mathbf{A}}\left(\hat{\mathbf{B}}\hat{\mathbf{C}}\right) = \left(\hat{\mathbf{A}}\hat{\mathbf{B}}\right)\hat{\mathbf{C}}.$$

We can take advantage of this in bra-ket notation. Let's define an arbitrary bra $\langle g| = \begin{pmatrix} g_1 & g_2 \end{pmatrix}$, an arbitrary ket $|x\rangle = \begin{pmatrix} x_1 \\ x_2 \end{pmatrix}$, and an arbitrary matrix $\hat{\mathbf{A}} = \begin{pmatrix} a & b \\ c & d \end{pmatrix}$. Then, we can write

$$\langle g| \hat{\mathbf{A}} |x\rangle .$$

One might think that this is ambiguous. Should we interpret this as $\left(\langle g| \, \hat{\mathbf{A}} \right) |x\rangle$, meaning first do the matrix multiplication $\langle g| \, \hat{\mathbf{A}}$, then matrix multiply the result by $|x\rangle$? Or, should we instead interpret it as $\langle g| \left(\hat{\mathbf{A}} \, |x\rangle \right)$? However, because matrix multiplication is associative, these two interpretations give the same result. (You can show this explicitly in problem 6.16.) So, when interpreting $\langle g| \, \hat{\mathbf{A}} \, |x\rangle$, we can think of $\hat{\mathbf{A}}$ "operating" to the left on $\langle g|$ or operating to right on $|x\rangle$.

Hermitian matrices

In the next section, and frequently in the study of quantum mechanics, you'll need to take the adjoint of $\hat{\mathbf{A}} \, |x\rangle$:

$$\left[\hat{\mathbf{A}} \, |x\rangle \right]^{\dagger} = ?$$

By considering a system of two masses, we'll be able to see the general pattern. Let

$$|x\rangle = \begin{pmatrix} x_1 \\ x_2 \end{pmatrix} \quad \text{and} \quad \hat{\mathbf{A}} = \begin{pmatrix} a & b \\ c & d \end{pmatrix}.$$

Then,

$$\hat{\mathbf{A}} \, |x\rangle = \begin{pmatrix} a & b \\ c & d \end{pmatrix} \begin{pmatrix} x_1 \\ x_2 \end{pmatrix} = \begin{pmatrix} ax_1 + bx_2 \\ cx_1 + dx_2 \end{pmatrix}.$$

To take the adjoint, we change it from a column matrix to a row matrix, and take the complex conjugate of the entries:

$$\left[\hat{\mathbf{A}} \, |x\rangle \right]^{\dagger} = \left(a^*x_1^* + b^*x_2^* \quad c^*x_1^* + d^*x_2^* \right).$$

Claim: We can write this as

$$\left[\hat{\mathbf{A}} \, |x\rangle \right]^{\dagger} = \langle x| \, \hat{\mathbf{A}}^{\dagger},$$

where $\hat{\mathbf{A}}^{\dagger}$ is the adjoint of $\hat{\mathbf{A}}$, that is, the matrix formed by interchanging columns (the first row becomes the first column, etc.) and then taking the complex conjugate of the entries.

Check:

$$\hat{\mathbf{A}}^{\dagger} = \begin{pmatrix} a^* & c^* \\ b^* & d^* \end{pmatrix} \Rightarrow \langle x| \, \hat{\mathbf{A}}^{\dagger} = \left(x_1^* \quad x_2^* \right) \begin{pmatrix} a^* & c^* \\ b^* & d^* \end{pmatrix}$$

$$= \left(a^* \, x_1^* + b^* \, x_2^* \quad c^* \, x_1^* + d^* \, x_2^* \right) \checkmark$$

Most of the matrices in the next section, and virtually all the matrices of interest in quantum mechanics, have the special property that they are "self adjoint," meaning that

$$\hat{\mathbf{A}}^{\dagger} = \hat{\mathbf{A}}.$$

We will use the more common term "Hermitian" for such matrices, instead of "self adjoint."

Concept test (answer below[7]): Which of the following matrices are Hermitian?

(a) $\begin{pmatrix} \omega_A^2 & -\dfrac{k}{m_1} \\ -\dfrac{k}{m_2} & \omega_B^2 \end{pmatrix}$ (b) $\begin{pmatrix} Ai & 0 \\ 0 & A \end{pmatrix}$ (c) $\begin{pmatrix} -3 & -\sqrt{3} \\ -\sqrt{3} & -1 \end{pmatrix}$ rad^2/s^2

In fact, we can see that any 2×2 Hermitian matrix can be written as

$$\hat{\mathbf{A}} = \begin{pmatrix} a & b \\ b* & c \end{pmatrix}. \tag{6.6.1}$$

Form of a 2×2 Hermitian matrix, with a and c real.

For Hermitian matrices, we have

$$\left[\hat{\mathbf{A}} \, |x\rangle\right]^{\dagger} = \langle x| \, \hat{\mathbf{A}}^{\dagger} = \langle x| \, \hat{\mathbf{A}}. \tag{6.6.2}$$

Adjoint of products of Hermitian matrices.

In the next chapter section (and in quantum mechanics), we will need to take the adjoint of $\hat{\mathbf{A}}\hat{\mathbf{B}} \, |x\rangle$, where $\hat{\mathbf{A}}$ and $\hat{\mathbf{B}}$ are both Hermitian. We will show that $\left[\hat{\mathbf{A}}\hat{\mathbf{B}} \, |x\rangle\right]^{\dagger} = \langle x| \, \hat{\mathbf{B}}\hat{\mathbf{A}}$. To begin, we define

$$|\beta\rangle \equiv \hat{\mathbf{B}} \, |x\rangle.$$

so that

$$\left[\hat{\mathbf{A}}\hat{\mathbf{B}} \, |x\rangle\right]^{\dagger} = \left[\hat{\mathbf{A}} \, |\beta\rangle\right]^{\dagger}.$$

Using equation (6.6.2), this means that

$$\left[\hat{\mathbf{A}}\hat{\mathbf{B}} \, |x\rangle\right]^{\dagger} = \langle \beta| \, \hat{\mathbf{A}}. \tag{6.6.3}$$

Since $\hat{\mathbf{B}}$ is Hermitian, we have that

$$\langle \beta| \equiv \left[\hat{\mathbf{B}} \, |x\rangle\right]^{\dagger} = \langle x| \, \hat{\mathbf{B}}^{\dagger} = \langle x| \, \hat{\mathbf{B}}.$$

Plugging this into equation (6.6.3) gives

$$\left[\hat{\mathbf{A}}\hat{\mathbf{B}} \, |x\rangle\right]^{\dagger} = \langle x| \, \hat{\mathbf{B}}\hat{\mathbf{A}}. \tag{6.6.4}$$

True if $\hat{\mathbf{A}}$ and $\hat{\mathbf{B}}$ are Hermitian.

We see that, when taking the adjoint, we must reverse the order of the two matrices. Therefore, $\left[\hat{\mathbf{A}}\hat{\mathbf{B}}\right]^{\dagger} = \hat{\mathbf{B}}\hat{\mathbf{A}}$. Since this is not necessarily equal to $\hat{\mathbf{A}}\hat{\mathbf{B}}$, we see that the product of two Hermitian matrices need not be Hermitian.

---

7. Only (c) is Hermitian. However, (a) would be Hermitian if $m_1 = m_2$.

Orthogonality of normal modes, normal mode coordinates, degeneracy, and scaling of Hilbert space for unequal masses

For the symmetric coupled pendulum system, we saw in chapter 5 that the quantity $s_p \equiv \frac{1}{\sqrt{2}}(x_1 + x_2)$, which characterizes the motion of the pendulum mode, always oscillates at ω_p, even when the system is in a superposition of pendulum and breathing modes, and that the quantity $s_b \equiv \frac{1}{\sqrt{2}}(x_1 - x_2)$, which characterizes the breathing mode, always oscillates at ω_b. The quantities s_p and s_b are called "normal mode coordinates"; each is the special linear combination of the coordinates of the masses that oscillates at the normal mode angular frequency.

Later in the chapter, we found the eigenvectors for these modes, $|e_p\rangle = \frac{1}{\sqrt{2}} \begin{pmatrix} 1 \\ 1 \end{pmatrix}$ and $|e_b\rangle = \frac{1}{\sqrt{2}} \begin{pmatrix} 1 \\ -1 \end{pmatrix}$. Therefore, defining $|x(t)\rangle \equiv \begin{pmatrix} x_1(t) \\ x_2(t) \end{pmatrix}$, we can write

$$s_p = \langle e_p | x(t)\rangle \quad \text{and} \quad s_b = \langle e_b | x(t)\rangle.$$

We will show in this section[8] that, for the case of equal masses, the analogous expressions for normal mode coordinates hold even for asymmetrical systems with many masses, that is, that the coordinate

$$s_n \equiv \langle e_n | x(t)\rangle, \tag{6.7.1}$$

where $|e_n\rangle$ is one of the eigenvectors of the system, oscillates at the associated angular frequency ω_n. (As for the symmetric coupled pendulum system discussed in chapter 5, the coordinate s_n is associated with an s_{n0} axis in Hilbert space, and the direction of this axis is defined by $|e_n\rangle$.)

When the masses are unequal, we must modify the recipe for finding normal mode coordinates. We will show that the normal mode coordinate for unequal masses is

$$\boxed{s_n \equiv \langle e_n | \hat{\mathbf{M}} | x(t)\rangle}, \tag{6.7.2}$$

Normal mode coordinate for the case of unequal masses.

where the "mass matrix" is $\hat{\mathbf{M}} = \begin{pmatrix} m_1 & 0 & \cdots \\ 0 & m_2 & 0 \\ \vdots & 0 & \ddots \end{pmatrix}$ and $|x(t)\rangle \equiv \begin{pmatrix} x_1(t) \\ x_2(t) \\ \vdots \end{pmatrix}$. Note

that $\hat{\mathbf{M}}$ is Hermitian. Also, when all the masses equal m, equation (6.7.2) becomes $x_n \equiv m\langle e_n | x(t)\rangle$; this is the same as equation (6.7.1) except for the constant m, which only affects the overall scaling of the normal mode coordinate.

Now, we must show that equation (6.7.2) is indeed the correct definition for the normal mode coordinate in the general case, meaning that s_n oscillates at frequency

---

8. Most of this presentation is based on that in *The Physics of Waves,* by Howard Georgi (Prentice-Hall, Englewood Cliffs, NJ, 1993), pp. 81–2.

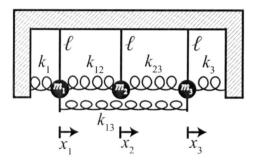

Figure 6.7.1 The same system of three coupled pendula as shown in figure 6.4.1.

ω_n, so that

$$s_n \propto \cos\left(\omega_n t + \varphi_n\right). \tag{6.7.3}$$

To show this, we need only prove that

$$\ddot{s}_n = -\omega_n^2 s_n. \tag{6.7.4}$$

We consider the three-mass system shown in figure 6.4.1, reproduced here for convenience as figure 6.7.1. (It will be clear how to extend the arguments to a system with any number of masses.) Recall that the result of applying $F_{\text{TOT}} = m\ddot{x}$ to mass 1 was (6.4.1):

$$-k_1 x_1 - \frac{m_1 g}{\ell} x_1 - k_{12}\left(x_1 - x_2\right) - k_{13}\left(x_1 - x_3\right) = m_1 \ddot{x}_1 \Rightarrow$$

$$-\left(k_{11} + k_{12} + k_{13}\right) x_1 + k_{12} x_2 + k_{13} x_3 = m_1 \ddot{x}_1,$$

where k_{11} is the total spring constant associated with restoring forces that act on m_1 but are not associated with coupling to m_2 or m_3. In the example of figure 6.4.1, $k_{11} = k_1 + m_1 g/\ell$. We could also write the above as

$$-k_A x_1 + k_{12} x_2 + k_{13} x_3 = m_1 \ddot{x}_1, \tag{6.7.5a}$$

where $k_A = m_1 \omega_A^2 = k_{11} + k_{12} + k_{13}$. Applying $F_{\text{TOT}} = m\ddot{x}$ to m_2 and to m_3 would similarly result in

$$k_{12} x_1 - k_B x_2 + k_{23} x_3 = m_2 \ddot{x}_2 \tag{6.7.5b}$$

and

$$k_{13} x_1 + k_{23} x_2 - k_C x_3 = m_3 \ddot{x}_3, \tag{6.7.5c}$$

where $k_B = m_1 \omega_B^2 = k_{12} + k_{22} + k_{23}$, and $k_{22} = m_2 g/\ell$ is the total spring constant associated with restoring forces that act on m_2, but are not associated with coupling to m_1 or m_2. (The constant k_C is defined analogously to k_A.)

Your turn: Show that we can then rewrite equations (6.7.5) as

$$-\hat{\mathbf{K}} \left|x\left(t\right)\right\rangle = \hat{\mathbf{M}} \frac{d^2}{dt^2} \left|x\left(t\right)\right\rangle, \tag{6.7.6}$$

where

$$\hat{\mathbf{K}} = \begin{pmatrix} k_A & -k_{12} & -k_{13} \\ -k_{12} & k_B & -k_{23} \\ -k_{13} & -k_{23} & k_C \end{pmatrix},$$

and is Hermitian.

Taking the inner product of both sides of equation (6.7.6) with $|e_n\rangle$ gives

$$-\langle e_n|\hat{\mathbf{K}}|x(t)\rangle = \langle e_n|\hat{\mathbf{M}}\frac{d^2}{dt^2}|x(t)\rangle.$$

Since neither $\langle e_n|$ nor $\hat{\mathbf{M}}$ has any time dependence, we can rearrange the right side, giving

$$-\langle e_n|\hat{\mathbf{K}}|x(t)\rangle = \frac{d^2}{dt^2}\langle e_n|\hat{\mathbf{M}}|x(t)\rangle,$$

Recall that our proposed s_n is given by equation (6.7.2): $s_n \equiv \langle e_n|\hat{\mathbf{M}}|x(t)\rangle$. Therefore, the above gives

$$\ddot{s}_n = -\langle e_n|\hat{\mathbf{K}}|x(t)\rangle.$$

Comparing this with what we need to prove, (6.7.4): $\ddot{s}_n = -\omega_n^2 s_n$, we see that we must show

$$\langle e_n|\hat{\mathbf{K}}|x(t)\rangle = \omega_n^2 s_n. \tag{6.7.7}$$

The eigenvalue equation for this system is (6.4.6):

$$\begin{pmatrix} \omega_A^2 & -\dfrac{k_{12}}{m_1} & -\dfrac{k_{13}}{m_1} \\ -\dfrac{k_{12}}{m_2} & \omega_B^2 & -\dfrac{k_{23}}{m_2} \\ -\dfrac{k_{13}}{m_3} & -\dfrac{k_{23}}{m_3} & \omega_C^2 \end{pmatrix} \begin{pmatrix} X_1 \\ X_2 \\ X_3 \end{pmatrix} = \omega^2 \begin{pmatrix} X_1 \\ X_2 \\ X_3 \end{pmatrix}.$$

We could rewrite this as

$$\hat{\mathbf{A}}|e_n\rangle = \omega_n^2|e_n\rangle, \tag{6.7.8}$$

where

$$\hat{\mathbf{A}} = \begin{pmatrix} \omega_A^2 & -\dfrac{k_{12}}{m_1} & -\dfrac{k_{13}}{m_1} \\ -\dfrac{k_{12}}{m_2} & \omega_B^2 & -\dfrac{k_{23}}{m_2} \\ -\dfrac{k_{13}}{m_3} & -\dfrac{k_{23}}{m_3} & \omega_C^2 \end{pmatrix}.$$

It will be helpful to separate out the part involving the masses:

$$\hat{\mathbf{A}} = \hat{\mathbf{M}}^{-1}\hat{\mathbf{K}}, \tag{6.7.9}$$

where

$$\hat{\mathbf{M}}^{-1} = \begin{pmatrix} \dfrac{1}{m_1} & 0 & 0 \\ 0 & \dfrac{1}{m_2} & 0 \\ 0 & 0 & \dfrac{1}{m_3} \end{pmatrix},$$

and is Hermitian.

Your turn: Do the matrix multiplication to verify that $\hat{\mathbf{A}} = \hat{\mathbf{M}}^{-1}\hat{\mathbf{K}}$.

Therefore, we can rewrite equation (6.7.8) as

$$\hat{\mathbf{M}}^{-1}\hat{\mathbf{K}}|e_n\rangle = \omega_n^2 |e_n\rangle.$$

The adjoint of this equation is

$$\left[\hat{\mathbf{M}}^{-1}\hat{\mathbf{K}}|e_n\rangle\right]^{\dagger} = \omega_n^2 \langle e_n|.$$

Since both $\hat{\mathbf{M}}^{-1}$ and $\hat{\mathbf{K}}$ are Hermitian, we can apply equation (6.6.4): $\left[\hat{\mathbf{A}}\hat{\mathbf{B}}|x\rangle\right]^{\dagger} = \langle x|\hat{\mathbf{B}}\hat{\mathbf{A}}$, giving

$$\langle e_n|\hat{\mathbf{K}}\hat{\mathbf{M}}^{-1} = \omega_n^2 \langle e_n|.$$

Taking the inner product with $\hat{\mathbf{M}}|x(t)\rangle$ gives

$$\langle e_n|\hat{\mathbf{K}}\hat{\mathbf{M}}^{-1}\hat{\mathbf{M}}|x(t)\rangle = \omega_n^2 \langle e_n|\hat{\mathbf{M}}|x(t)\rangle.$$

The right side of this includes $s_n \equiv \langle e_n|\hat{\mathbf{M}}|x(t)\rangle$, so we can rewrite the above as

$$\langle e_n|\hat{\mathbf{K}}\hat{\mathbf{M}}^{-1}\hat{\mathbf{M}}|x(t)\rangle = \omega_n^2 s_n$$

Note that $\hat{\mathbf{M}}\hat{\mathbf{M}}^{-1} = \hat{\mathbf{M}}^{-1}\hat{\mathbf{M}} = \hat{\mathbf{I}}$. Therefore,

$$\langle e_n|\hat{\mathbf{K}}\hat{\mathbf{M}}^{-1}\hat{\mathbf{M}}|x(t)\rangle = \langle e_n|\hat{\mathbf{K}}|x(t)\rangle \Rightarrow \langle e_n|\hat{\mathbf{K}}|x(t)\rangle = \omega_n^2 s_n,$$

which is (6.7.7), completing the proof that s_n, as defined by equation (6.7.2): $s_n \equiv \langle e_n|\hat{\mathbf{M}}|x(t)\rangle$, is indeed the normal mode coordinate.

Let us be clear about the interpretation of $s_n \equiv \langle e_n|\hat{\mathbf{M}}|x(t)\rangle$. This does *not* mean that, in order to excite normal mode number 1, we must position the masses in an initial pattern of positions proportional to $\hat{\mathbf{M}}|e_1\rangle$. Rather, as usual, we should position them in a pattern proportional to $|e_1\rangle$; when they are then released, the system will be in a pure mode 1 oscillation. However, if we instead excite the system into a superposition of modes, then (assuming unequal masses) the coordinate that displays a simple sinusoidal motion at angular frequency ω_1 is $s_1 \equiv \langle e_1|\hat{\mathbf{M}}|x(t)\rangle$. Other coordinates, such as $\langle e_1|x(t)\rangle$, do not show simple sinusoidal motion when the system is in a superposition. (They *do* show simple sinusoidal motion when the system is in a pure mode, as do all coordinates.)

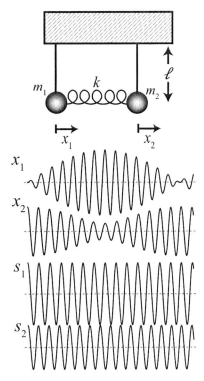

Figure 6.7.2 Top: Coupled pendula with unequal masses. For the example discussed in the text, $m_1 = 1$ kg, $m_2 = 2$ kg, $\ell = 1$ m, $k = 1$ N/m, and $g = 10$ m/s². Bottom: Position *versus*. time for each of the two masses (x_1 and x_2) and for each of the normal mode coordinates (s_1 and s_2).

As an example, consider the two-pendulum system shown in the top of figure 6.7.2, in which $m_1 = 1$ kg, $m_2 = 2$ kg, $\ell = 1$ m, $k = 1$ N/m, and $g = 10$ m/s². You can show in problem 6.17 that the (unnormalized) eigenvectors for this system are

$$|e_1\rangle = \begin{pmatrix} 1 \\ 1 \end{pmatrix} \text{ and } |e_2\rangle = \begin{pmatrix} -2 \\ 1 \end{pmatrix}.$$

Therefore, to excite mode 2 (which is a breathing-like mode), we could displace m_1 2 cm to the left of equilibrium, displace m_2 1 cm to the right of equilibrium, and release them. The system would then oscillate in mode 2, in which m_1 moves with an amplitude of 2 cm and m_2 moves with an amplitude of 1 cm, both with an angular frequency ω_2. In this pure mode, any linear combination of x_1 and x_2 oscillates at ω_2, since both x_1 and x_2 oscillate at ω_2. If instead we excite the system by holding m_1 at equilibrium, displacing m_2 to the right by 1 cm, and releasing, then both modes are excited. (Unlike the case for equal masses, this does not excite both modes equally; we'll see below how to calculate the amplitudes for the two modes.) The resulting behavior is therefore complicated, as shown in figure 6.7.2. However, s_1 shows simple harmonic oscillation at angular frequency ω_1 and s_2 shows simple harmonic oscillation at the slightly higher angular frequency ω_2, as shown in the figure.

Next, we wish to show that, for the case of equal masses, the normal modes are mutually orthogonal. We'll keep our arguments general, so that we'll easily be able to see how to adapt things for unequal masses. Again, we'll consider a system of three masses, but it will be obvious how to generalize the argument to any number of masses. We consider an arbitrary state of the system, which is formed by a superposition of the normal modes:

$$|x(t)\rangle = A_1 \cos(\omega_1 t + \varphi_1)|e_1\rangle + A_2 \cos(\omega_2 t + \varphi_2)|e_2\rangle + A_3 \cos(\omega_3 t + \varphi_3)|e_3\rangle$$

$$= \sum_p A_p \cos(\omega_p t + \varphi_p)|e_p\rangle,$$

where p is the mode index, and ranges from 1 to 3. Taking the inner product with $\hat{\mathbf{M}}|e_n\rangle$ gives

$$\langle e_n|\hat{\mathbf{M}}|x(t)\rangle = \langle e_n|\hat{\mathbf{M}}\sum_p A_p \cos(\omega_p t + \varphi_p)|e_p\rangle$$

$$= \sum_p A_p \cos(\omega_p t + \varphi_p)\langle e_n|\hat{\mathbf{M}}|e_p\rangle.$$

We know from equations (6.7.2) and (6.7.3) that $s_n = \langle e_n|\mathbf{M}|x(t)\rangle \propto \cos(\omega_n t + \varphi_n)$. Inserting this into the above gives

$$\cos(\omega_n t + \varphi_n) \propto \sum_p A_p \cos(\omega_p t + \varphi_p)\langle e_n|\hat{\mathbf{M}}|e_p\rangle.$$

Since both sides of this equation must oscillate at the same angular frequency, we must have

$$\langle e_n|\hat{\mathbf{M}}|e_p\rangle = 0 \quad \text{for} \quad \omega_n \neq \omega_p. \tag{6.7.10}$$

We cannot quite yet write $\langle e_n|\hat{\mathbf{M}}|e_p\rangle = 0$ for $n \neq p$, since it is possible that two or more of the normal modes have the same angular frequency. Such modes are called "degenerate." (The quantum mechanical analog is a situation in which two or more of the stable states have the same energy.) For example, with appropriate choices of spring constants and masses, one might be able to arrange $\omega_1 = \omega_2$, so that using equation (6.7.8): $\hat{\mathbf{A}}|e_n\rangle = \omega_n^2|e_n\rangle$ we would have

$$\hat{\mathbf{A}}|e_1\rangle = \omega_1^2|e_1\rangle \tag{6.7.11}$$

and

$$\hat{\mathbf{A}}|e_2\rangle = \omega_2^2|e_2\rangle = \omega_1^2|e_2\rangle. \tag{6.7.12}$$

Forming the combination β_1 (6.7.11) $+ \beta_2$ (6.7.12), where β_1 and β_2 are constants gives

$$\beta_1\hat{\mathbf{A}}|e_1\rangle + \beta_2\hat{\mathbf{A}}|e_2\rangle = \beta_1\omega_1^2|e_1\rangle + \beta_2\omega_1^2|e_2\rangle \Rightarrow$$

$$\hat{\mathbf{A}}[\beta_1|e_1\rangle + \beta_2|e_2\rangle] = \omega_1^2[\beta_1|e_1\rangle + \beta_2|e_2\rangle].$$

Thus, the vector $[\beta_1|e_1\rangle + \beta_2|e_2\rangle]$ is also an eigenvector with eigenvalue ω_1^2. In other words, any linear combination of degenerate eigenvectors is also an eigenvector with the same eigenvalue. In particular, the vector

$$|e_2'\rangle \equiv |e_2\rangle - \frac{\langle e_1|\hat{\mathbf{M}}|e_2\rangle}{\langle e_1|\hat{\mathbf{M}}|e_1\rangle}|e_1\rangle$$

is an eigenvector with eigenvalue ω_1^2.

> **Your turn:** Show that $\langle e_1 | \hat{\mathbf{M}} | e_2' \rangle = 0$.

Thus, we can always choose a set of eigenvectors that fulfill

$$\boxed{\langle e_n | \hat{\mathbf{M}} | e_p \rangle = 0 \text{ for } n \neq p.}$$ (6.7.13)

Equivalent of orthogonality for unequal masses

For equal masses, we have

$$\langle e_n | \hat{\mathbf{M}} | e_p \rangle = m \langle e_n | e_p \rangle \Rightarrow$$

$$\boxed{\langle e_n | e_p \rangle = 0 \text{ for } n \neq p,}$$ (6.7.14)

Orthogonality of normal modes for equal masses

which is the general orthogonality condition we set out to prove.

When the masses are not equal, we can still use all the results from section 6.5 simply by inserting the $\hat{\mathbf{M}}$ matrix inside each inner product. The normalization condition becomes

$$\boxed{\langle e_n | \hat{\mathbf{M}} | e_n \rangle = 1}$$ (6.7.15)

Equivalent of normalization for the case of unequal masses

If the eigenvectors are scaled so that the above holds, then the equivalent of (6.5.6), $\langle e_m | e_n \rangle = \delta_{mn}$ becomes

$$\boxed{\langle e_n | \hat{\mathbf{M}} | e_n \rangle = \delta_{mn}.}$$ (6.7.16)

Equivalent of orthonormality for the case of unequal masses

All the subsequent arguments of section 6.5 follow just the same way, simply inserting $\hat{\mathbf{M}}$ inside each inner product. Thus, assuming the eigenvectors have been scaled to satisfy equation (6.7.15), the coefficients in the superposition (6.5.1),

$$|x(t)\rangle = \text{Re}\left[\sum_{n=1}^{N} C_n e^{i\omega_n t} |e_n\rangle \right],$$

are given by the equivalent of (6.5.8a), $\text{Re}(C_n) = \langle e_n | x_0 \rangle$,

$$\text{Re}(C_n) = \langle e_n | \hat{\mathbf{M}} | x_0 \rangle,$$ (6.7.17a)

and by the equivalent of (6.5.8b), $\text{Im}(C_n) = -\dfrac{1}{\omega_n} \langle e_n | \dot{x}_0 \rangle$,

$$\text{Im}(C_n) = -\frac{1}{\omega_n} \langle e_n | \hat{\mathbf{M}} | \dot{x}_0 \rangle.$$ (6.7.17b)

One way of interpreting this insertion of $\hat{\mathbf{M}}$ inside each inner product is to consider Hilbert space to be rescaled. We can define

$$\hat{\mathbf{M}}^{1/2} = \begin{pmatrix} \sqrt{m_1} & 0 & 0 \\ 0 & \sqrt{m_2} & 0 \\ 0 & 0 & \sqrt{m_3} \end{pmatrix}, \text{ so that } \hat{\mathbf{M}} = \hat{\mathbf{M}}^{1/2}\,\hat{\mathbf{M}}^{1/2}.$$

Therefore, for example,

$$\langle e_n | \hat{\mathbf{M}} | x_0 \rangle = \langle e_n | \hat{\mathbf{M}}^{1/2}\,\hat{\mathbf{M}}^{1/2} | x_0 \rangle.$$

Since $\hat{\mathbf{M}}^{1/2}$ is Hermitian, we can think of the above as the inner product of the vector $\hat{\mathbf{M}}^{1/2} | x_0 \rangle$ with the vector $\hat{\mathbf{M}}^{1/2} | e_n \rangle$. Thus, we could consider each of the axes of Hilbert space, corresponding to the motion of x_1, x_2, and x_3, to be scaled by the factor $\sqrt{m_1}$, $\sqrt{m_2}$, or $\sqrt{m_3}$ respectively, so that the x_1 axis becomes the $x_1\sqrt{m_1}$ axis, and so on. In this scaled Hilbert space $|e_1\rangle$ would be perpendicular to $|e_2\rangle$.

Example: Returning to the example shown in figure 6.7.2, the two (unnormalized) eigenvectors $|e_1\rangle = \begin{pmatrix} 1 \\ 1 \end{pmatrix}$ and $|e_2\rangle = \begin{pmatrix} -2 \\ 1 \end{pmatrix}$ are not perpendicular to each other in an unscaled Hilbert space space, as shown in figure 6.7.3a. However, $\langle e_1 | \hat{\mathbf{M}} | e_2 \rangle = 0$; we can show this graphically using Hilbert space axes that are scaled by the square root of the mass, as shown in figure 6.7.3b. Also shown in figure 6.7.3b are versions of the eigenvectors after the equivalent of normalization has been applied, for example:

$$\text{"normalized" version of } |e_1\rangle = \frac{|e_1\rangle}{\sqrt{\langle e_1 | \hat{\mathbf{M}} | e_1 \rangle}}.$$

Concept and skill inventory for chapter 6

After reading this chapter, you should fully understand the following terms:

Determinant (6.1)
Eigenvalue, eigenvector, eigenvalue equation (6.2)
Characteristic equation (6.3)
Mode index (6.5)
Kronecker delta function (6.5)
Inverse of a matrix (6.6)
Identity matrix (6.6)
Hermitian matrix (6.6)
Mass matrix (6.7)
Degenerate modes (6.7)

You should know what happens when:

One takes the adjoint of a product of Hermitian matrices (6.6)

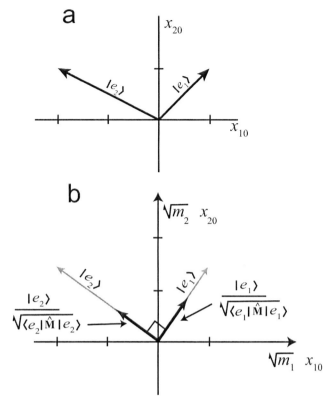

Figure 6.7.3 a: When the masses for a coupled oscillator system are unequal, the eigenvectors are not orthogonal in an unscaled Hilbert space. b: If the axes are scaled by the square root of the mass, then the eigenvectors are orthogonal. Also shown are the "normalized" versions of the eigenvectors, that is, the versions that have length 1 in the scaled Hilbert space.

You should understand the following connections:

The number of modes & the number of masses for a 1D system (6.4)

You should understand the difference between:

The eigenvalue & the angular frequency for a mode (6.2)
Mass index & mode index (6.5)

You should be familiar with the following additional concepts:

Matrix multiplication doesn't commute. (6.1)
Matrix multiplication is associative. (6.6)
Scaling of Hilbert space for unequal masses (6.7)

You should be able to:

Multiply matrices. (6.1)
Find determinants by hand for 2×2 and 3×3 matrices. (6.1)

Explain what an eigenvalue equation is. (6.2)

For a system with several objects and couplings, derive the set of DEQs that describes the system, and go from that set of DEQs to the eigenvalue equation. (6.4)

Given the eigenvalue equation, check a proposed vector to see if it's an eigenvector and (if it is) to find the associated eigenvalue. (6.4)

Given the eigenvalue equation, find the eigenvalues and eigenvectors for systems of two or three coupled oscillators. (6.3–6.4)

Normalize an eigenvector. (6.5)

Given the eigenvectors and eigenvalues for a system of coupled equal masses, and given the initial positions and velocities, express the state of the system as a superposition of normal modes. Then be able to find the position as a function of time for each mass. (6.5)

Given the eigenvectors and eigenvalues for a system of coupled unequal masses, and given the initial positions and velocities, express the state of the system as a superposition of normal modes. Then, be able to find the position as a function of time for each mass. (6.7)

In addition to all of the above, you should be able to combine the concepts you've learned to address new situations.

6.8 Problems

Note: Additional problems are available on the website for this text.

Instructor: Ratings of problem difficulty, full solutions, and important additional support materials are available on the website.

6.1 (a) Find the matrix that reflects a conventional two-dimensional vector $\mathbf{r} = \begin{pmatrix} x \\ y \end{pmatrix}$ across the y-axis, as shown in figure 6.P.1. Show your reasoning clearly. (b) Find the matrix that reflects a vector across the x-axis. (c) Find the matrix that reflects a vector across the line $y = x$.

6.2 **2D rotation matrix.** Derive the matrix that rotates a conventional two-dimensional vector counterclockwise through an angle φ. In other words, if $\mathbf{r} = \begin{pmatrix} x \\ y \end{pmatrix}$, then find the matrix $\hat{\mathbf{A}}$ such that $\hat{\mathbf{A}}\mathbf{r}$ is a vector with the same length as $\mathbf{r}$, but rotated counterclockwise by φ. Show your reasoning clearly. *Hint: start by writing* $\mathbf{r} = r \begin{pmatrix} x/r \\ y/r \end{pmatrix}$, *where r is the length of* $\mathbf{r}$.

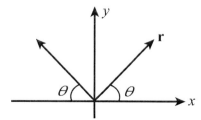

Figure 6.P.1 The vector $\mathbf{r}$ is reflected across the y-axis.

6.3 Inverse of a matrix. As we have discussed, a matrix $\hat{\mathbf{A}}$ can operate on a vector, either in conventional space or Hilbert Space, to transform it into another vector through a combination of rotation, scaling, and reflections. In many cases, one can find an inverse matrix $\hat{\mathbf{A}}^{-1}$, which undoes these changes. In other words, if $\hat{\mathbf{A}}\mathbf{r} = \mathbf{r}'$, then $\hat{\mathbf{A}}^{-1}\mathbf{r}' = \mathbf{r}$. We could also write this as $\hat{\mathbf{A}}^{-1}\hat{\mathbf{A}}\mathbf{r} = \mathbf{r}$. This means that the combination $\hat{\mathbf{A}}^{-1}\hat{\mathbf{A}}$ has no net effect, so that we could write $\hat{\mathbf{A}}^{-1}\hat{\mathbf{A}} = \hat{\mathbf{I}}$, where $\hat{\mathbf{I}}$ is the "identity matrix," a matrix with 1's on the diagonal and 0 elsewhere. For a two-dimensional space, $\hat{\mathbf{I}} = \begin{pmatrix} 1 & 0 \\ 0 & 1 \end{pmatrix}$. **(a)** It is not always possible to find an inverse of a matrix. If the operation of $\hat{\mathbf{A}}$ destroys information about the original vector $\mathbf{r}$, then it will be impossible to reconstruct $\mathbf{r}$ from the vector $\hat{\mathbf{A}}\mathbf{r} = \mathbf{r}'$. Give an example, for a two-dimensional system, of a matrix $\hat{\mathbf{A}}$ that has no inverse, and explain how your matrix destroys information. **(b)** One can show (see any text on linear algebra) that $\hat{\mathbf{A}}$ has an inverse if and only if $\det \hat{\mathbf{A}} \neq 0$. Given this, which of the following matrices does not have an inverse: **(i)** $\begin{pmatrix} 2 & 1 \\ 1 & -2 \end{pmatrix}$ **(ii)** $\begin{pmatrix} 1 & 2 \\ 1 & 2 \end{pmatrix}$ **(iii)** $\begin{pmatrix} 1 & 2 \\ -1 & 2 \end{pmatrix}$. **(c)** Show that the determinant for your matrix from part a is zero. **(d)** Verify that the inverse of $\begin{pmatrix} 1 & 2 \\ 3 & 4 \end{pmatrix}$ is $\begin{pmatrix} -2 & 1 \\ \frac{3}{2} & -\frac{1}{2} \end{pmatrix}$.

6.4 (You should complete problems 6.1 and 6.2 before doing this problem.) **(a)** Create a single matrix that rotates a conventional two-dimensional vector counterclockwise through an angle φ and then reflects it across the y-axis. Verify that your matrix has the expected effect on the vector $\begin{pmatrix} x_0 \\ 0 \end{pmatrix}$ for the case $\varphi = 45°$. **(b)** Create a single matrix that reflects a conventional two-dimensional vector across the y-axis, and then rotates it counterclockwise through an angle φ. Verify that your matrix has the expected effect on the vector $\begin{pmatrix} x_0 \\ 0 \end{pmatrix}$ for the case $\varphi = 45°$.

6.5 3D rotation matrices. (a) The matrix $\hat{\mathbf{R}}_x = \begin{pmatrix} 1 & 0 & 0 \\ 0 & \cos\theta & -\sin\theta \\ 0 & \sin\theta & \cos\theta \end{pmatrix}$ rotates a conventional three-dimensional vector $\mathbf{r}$ through an angle θ counterclockwise around the x-axis (as viewed looking from the positive x-axis toward the origin). Demonstrate that this matrix works as expected on the vector $\mathbf{r} = \begin{pmatrix} x_0 \\ y_0 \\ 0 \end{pmatrix}$ for the case $\theta = 45°$. (Recall that $\hat{\mathbf{i}} \times \hat{\mathbf{j}} = \hat{\mathbf{k}}$.)

(b) The matrix $\hat{\mathbf{R}}_y = \begin{pmatrix} \cos\theta & 0 & \sin\theta \\ 0 & 1 & 0 \\ -\sin\theta & 0 & \cos\theta \end{pmatrix}$ rotates a conventional three-dimensional vector $\mathbf{r}$ through an angle θ counterclockwise around the y-axis (as viewed looking from the positive y-axis toward the origin). Create a single matrix that rotates a vector by $90°$ counterclockwise about the y-axis and then $90°$ counterclockwise about the x-axis. Show your reasoning clearly, and then show that your matrix has the expected effect on the vector $\begin{pmatrix} 0 \\ 0 \\ z_0 \end{pmatrix}$.

(c) Create a single matrix that rotates a vector by $90°$ counterclockwise about the x-axis and then $90°$ counterclockwise about the y-axis. Show your reasoning clearly, and then show that your matrix has the expected effect on the vector $\begin{pmatrix} 0 \\ 0 \\ z_0 \end{pmatrix}$.

6.6 For a particular coupled oscillator system, of the form shown in figure 6.2.1, $m_1 = 0.200$ kg, $m_2 = 0.400$ kg, and $k = 3.00$ N/m. If m_2 is fixed and m_1 is displaced from equilibrium and then released, one observes that m_1 oscillates with a period of 1.068 s. If instead m_1 is fixed and m_2 is displaced from equilibrium and then released, one observes that m_2 oscillates with a period of 0.919 s. What are the periods for the normal modes of this system (with neither mass held fixed)?

6.7 The coupled oscillator system shown in figure 6.P.2 has $m_1 = 0.300$ kg, $m_2 = 0.500$ kg, and $k_R = 4.00$ N/m. The masses slide on a frictionless surface. If m_2 is fixed and m_1 is displaced from equilibrium and then released, one observes that m_1 oscillates with a period of 1.257 s. If instead m_1 is fixed and m_2 is displaced from equilibrium and then released, one observes that m_2 oscillates with a period of 2.221 s. What are the periods for the normal modes of this system (with neither mass held fixed)?

6.8 **(a)** Derive the set of coupled differential equations that describes the system of coupled pendula in figure 6.P.3. The position of the left mass is x_1, and is measured relative to its equilibrium position. Similarly, the position of the right mass is x_2, and is measured relative to its equilibrium position.

(b) Derive the eigenvalue equation for this system. Show all your steps explicitly.

Hint: Answer is

$$\begin{pmatrix} \omega_A^2 & -\omega_0^2 \\ -\omega_0^2 & \omega_A^2 \end{pmatrix} \begin{pmatrix} X_1 \\ X_2 \end{pmatrix} = \omega^2 \begin{pmatrix} X_1 \\ X_2 \end{pmatrix},$$

where $\omega_A^2 \equiv \dfrac{g}{\ell} + \dfrac{2k}{m}$ and $\omega_0^2 \equiv \dfrac{k}{m}$

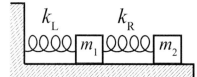

Figure 6.P.2 Two masses coupled by springs slide on a frictionless surface.

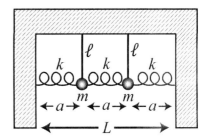

Figure 6.P.3 A set of two coupled pendula.

(c) Show that the frequencies of the normal modes for this system are given by

$$\omega^2 = \begin{cases} \dfrac{g}{\ell} + \dfrac{k}{m} \\ \dfrac{g}{\ell} + \dfrac{3k}{m} \end{cases}$$

(d) Show that the corresponding normalized eigenvectors are:

$$\text{for } \omega^2 = \frac{g}{\ell} + \frac{k}{m} : \frac{1}{\sqrt{2}} \begin{pmatrix} 1 \\ 1 \end{pmatrix} \text{ and}$$

$$\text{for } \omega^2 = \frac{g}{\ell} + \frac{3k}{m} : \frac{1}{\sqrt{2}} \begin{pmatrix} 1 \\ -1 \end{pmatrix}$$

(e) Describe the motion of each normal mode using words and pictures.
(f) Using our usual definition of the inner product for discrete systems, show that the normal modes of the above system are orthogonal to each other. Note: the math for this is very simple.

6.9 Consider the double-pendulum system shown in figure 6.P.4. As usual, we'll only consider small displacements. Recall that in this limit, the combined forces of gravity and string tension for a generic pendulum give a spring-like restoring force for lateral displacements, with effective spring constant $\dfrac{mg}{L}$.

Therefore, the restoring force for a generic pendulum is $F_{\text{pendulum}} = -\dfrac{mg}{L}x$, where x is the lateral displacement of the pendulum bob *relative to* the lateral

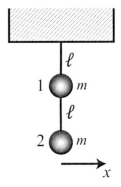

Figure 6.P.4 A double pendulum system.

position of the support point. (**a**) Derive the eigenvalue equation (in matrix form) for this system. (**b**) Now write the characteristic equation, and use it to find the normal mode frequencies. (**c**) Finally, find the eigenvectors.

6.10 The CO_2 molecule. Carbon dioxide is a linear molecule with a carbon atom at the center which is double-bonded to two oxygen atoms, as suggested in figure 6.P.5. The two springs are identical, and have spring constant $k = 3628$ N/m. We'll assume that the carbon atom is a ^{12}C isotope (the most common kind, having six protons and six neutrons), meaning that it has a mass of exactly 12 u $= 1.993 \times 10^{-26}$ kg, and that the oxygen atoms are both ^{16}O, meaning that the mass of each is 2.66×10^{-26} kg. Find the eigenvectors and normal mode frequencies for this system. For each normal mode, describe the oscillation in words as well as by giving the eigenvector. (You are encouraged to use a symbolic algebra program such as Mathematica or Maple for this problem, though it is not necessary.)

6.11 (You should do problem 6.6 before this problem.) Find the normalized eigenvectors for the system described in problem 6.6.

6.12 (You should do problem 6.7 before this problem.) Find the normalized eigenvectors for the system described in problem 6.7.

6.13 Consider the system shown in figure 6.P.6. (**a**) Find the matrix $\hat{A}$ that appears in the eigenvalue equation $\hat{A}\,|e\rangle = \omega^2\,|e\rangle$, where $|e\rangle$ is an eigenvector. (**b**) For the special case $m_1 = m_2 = 1$ kg, $m_3 = m_4 = 2$ kg, $\ell_\alpha = 1$ m, $\ell_\beta = 2$ m,

$$k_{13} = 2 \text{ N/m}, k_{24} = 4 \text{ N/m, and } g = 4 \text{ m/s}^2, \text{ show that } |e_a\rangle = \begin{pmatrix} 0.653 \\ -0.271 \\ 0.653 \\ -0.271 \end{pmatrix} \text{ is}$$

an eigenvector for this system, **without using a symbolic algebra program**.

Figure 6.P.5 Model for a CO_2 molecule.

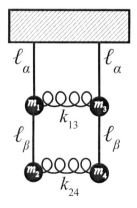

Figure 6.P.6 A system of four coupled pendula.

(c) Determine the angular frequency of oscillation ω_a for the mode described by $|e_a\rangle$. (d) Make approximate sketches of the positions of the masses for this mode at $t = 0$ and at $t = \pi/\omega_a$. (e) Show that $|e_a\rangle$ is normalized, without using a symbolic algebra program.

6.14 For the system shown in figure 6.P.6, consider the special case $m_1 = 1$ kg, $m_2 = 2$ kg, $m_3 = 3$ kg, $m_4 = 4$ kg, $\ell_\alpha = 1$ m, $\ell_\beta = 2$ m, $k_{13} = 3$ N/m, $k_{24} = 6$ N/m, and $g = 10$ m/s^2. Use a symbolic algebra program to find the normalized eigenvectors and corresponding angular frequencies. Make clear which eigenvector goes with which angular frequency. For each eigenvector, make a rough sketch of the positions of the masses at $t = 0$ and at $t = \pi/\omega$, where ω is the angular frequency for that mode.

6.15 Do not use a symbolic algebra program or calculator for this problem. The asymmetric coupled pendulum system shown in fig. 6.2.1 has $l_1 = 1.00$ m, $l_2 = 0.500$ m, $k = 1.50$ N/m, $m_1 = m_2 = 0.500$ kg. The masses are released from rest at $t = 0$, with initial positions $x_1 = -0.025$ m and $x_2 = 0.045$ m. What are the positions of the two masses as a function of time?

6.16 Let's define an arbitrary bra $\langle g| = \begin{pmatrix} g_1 & g_2 \end{pmatrix}$, an arbitrary ket $|x\rangle = \begin{pmatrix} x_1 \\ x_2 \end{pmatrix}$, and an arbitrary matrix $\hat{\mathbf{A}} = \begin{pmatrix} a & b \\ c & d \end{pmatrix}$. Show explicitly that $\langle g| \hat{\mathbf{A}} |x\rangle$ is unambiguous, that is, show that $\left(\langle g| \hat{\mathbf{A}} \right) |x\rangle$ is the same as $\langle g| \left(\hat{\mathbf{A}} |x\rangle \right)$.

6.17 Consider the two-pendulum system shown in figure 6.7.2, in which $m_1 = 1$ kg, $m_2 = 2$ kg, $\ell = 1$ m, $k = 1$ N/m, and $g = 10$ m/s^2. Verify that the (unnormalized) eigenvectors for this system are $|e_1\rangle = \begin{pmatrix} 1 \\ 1 \end{pmatrix}$ and $|e_2\rangle = \begin{pmatrix} -2 \\ 1 \end{pmatrix}$.

7 String Theory

Midnight. No waves,
no wind, the empty boat
is flooded with moonlight.
                    —Dogen (1200–1253)

Each string I touch
Sends out its love in waves
And now the boat is full.
                    —Marian McKenzie

In some of the most complicated theories of modern physics, elementary particles are represented in terms of the normal modes of extremely short (10^{-35} m) strings. Usually, these theories are hyperdimensional, that is, they use more than the normal number (four) of spacetime dimensions; several string theories have ten dimensions or more. Often, string theorists investigate a slice through this multidimensional space. A slice through three-dimensional space has two dimensions, and is called a "membrane." A slice through a five-dimensional space might have four dimensions, and would be called a "4-brane." In general, a slice with p dimensions is called, get this, a "p-brane." This proves that even the most advanced theoretical physicists have a sense of humor! So far, it has been very difficult to test string theories experimentally.

In this chapter, we will study ordinary macroscopic strings. This will lead us to an understanding of the normal modes of continuous systems, and of how the normal modes are modified when a system is not truly continuous (e.g., when it consists of atoms). Surprisingly, our study of strings will also lead us to a fundamental understanding of Fourier analysis, a completely mathematical idea in which any arbitrary function can be constructed by adding together sine waves. Finally, as in previous chapters, we will point out the deep connections between these macroscopic oscillating systems and quantum mechanics.

7.1 The beaded string

We begin by considering a massless string of length L stretched between two walls, so that it is under tension T. The string has N beads (each of mass m) spaced at even

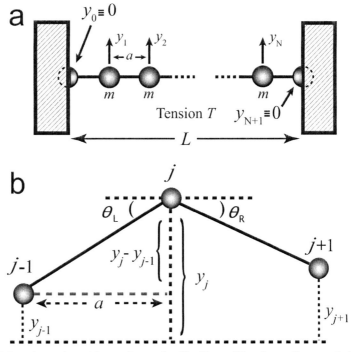

Figure 7.1.1 a: A massless string under tension T with small beads equally spaced along its length. b: Geometry needed for calculating the force on bead j.

intervals a, as shown in figure 7.1.1a. We assume gravity is unimportant. If the string is plucked, it vibrates. Each bead feels a force from the string coupling it to the bead on its left, and a second force from the string coupling it to the bead on its right. Therefore, this is a system of N coupled oscillators. As in chapter 6, we expect that such a system will display normal modes, and we will use a similar procedure to find out what they are. For mathematical convenience, we introduce two fictitious beads at the ends of the string, as shown.

We wish to find the ways in which this system can move. We follow our multi-step procedure.

Step 1: Write $F_{\text{TOT}} = m\ddot{y}$ for each mass. We will take the approximation of small displacements from equilibrium. In this limit, when the system oscillates the change in the magnitude of the string tension T is negligible. So, from figure 7.1.1b, we see that the y-component of the force acting on mass j is

$$F_{\text{TOT, y}} = -T \sin \theta_{\text{L}} - T \sin \theta_{\text{R}}, \tag{7.1.1}$$

where

$$\sin \theta_{\text{L}} = \frac{y_j - y_{j-1}}{\sqrt{a^2 + \left(y_j - y_{j-1}\right)^2}}$$

For small displacements from equilibrium, $\left(y_j - y_{j-1} \right) \ll a$, so we can write

$$\sin \theta_{\mathrm{L}} \cong \frac{y_j - y_{j-1}}{a} \quad \text{and similarly} \quad \sin \theta_{\mathrm{R}} \cong \frac{y_j - y_{j+1}}{a}$$

Substituting these into equation (7.1.1) gives

$$F_{\mathrm{TOT},\, y} = -\frac{T}{a} \left(y_j - y_{j-1} + y_j - y_{j+1} \right) = m \ddot{y}_j$$

$$\Leftrightarrow -\frac{T}{ma} \left(-y_{j-1} + 2 y_j - y_{j+1} \right) = \ddot{y}_j \tag{7.1.2}$$

Let us define

$$\boxed{\; \omega_{\mathrm{A}} \equiv \sqrt{\frac{2T}{ma}}. \;} \tag{7.1.3}$$

This is the angular frequency at which one mass would oscillate if its neighbors were held fixed, as you can see by setting $y_{j-1} = y_{j+1} = 0$ in equation (7.1.2). The simplest complex equation whose real part is the above would be

$$-\frac{\omega_{\mathrm{A}}^2}{2} \left(-z_{j-1} + 2 z_j - z_{j+1} \right) = \ddot{z}_j, \tag{7.1.4}$$

where $y_j = \mathrm{Re}\, z_j$. Since j can range from 1 to N, this represents a system of N coupled DEQs.

Step 2: Use physical and mathematical intuition to guess a solution. Based on our experience with the system of two coupled oscillators, we hope that this system may display normal modes, in which each bead moves with the same frequency. The most general possible such guess would be

$$z_j \overset{?}{=} Y_j\, e^{i\omega_n t}. \tag{7.1.5}$$

Where ω_n is the angular frequency of the normal mode, n is the mode index, and Y_j is the amplitude of bead j. (As before, we have chosen the definition of the moment when $t = 0$ so that the phase of the normal mode $\varphi = 0$.) We expect for a system of N beads that there will be N normal modes, each of which has its own characteristic frequency.

Step 3: Plug the guess into the system of DEQs. Plugging our normal modes guess (7.1.5) into (7.1.4) gives

$$-\frac{\omega_{\mathrm{A}}^2}{2} \left(-Y_{(j-1)} e^{i\omega_n t} + 2 Y_j e^{i\omega_n t} - Y_{(j+1)} e^{i\omega_n t} \right) = -\omega_n^2 Y_j e^{i\omega_n t}$$

$$\Rightarrow \frac{\omega_{\mathrm{A}}^2}{2} \left(-Y_{(j-1)} + 2 Y_j - Y_{(j+1)} \right) = \omega_n^2 Y_j \tag{7.1.6}$$

This is a system of N coupled linear equations. In principle, we could solve it using the methods of section 6.3, that is, we could write it as a matrix eigenvalue equation, rearrange it to a characteristic equation, set the determinant equal to zero to find the eigenvalues, and plug these back into the characteristic equation to find the eigenvectors.

7.2   Standing wave guess: Boundary conditions quantize the allowed frequencies

Obviously, this procedure would be tedious for a system of more than a few beads. However, we can get to the answer much more quickly by using additional physical insight to guess what the eigenvectors are, that is, to guess what the initial positions of the beads are for a normal mode. We have all played with ropes, telephone cords, and so on, and have observed "standing waves," in which the rope oscillates between the positions shown by the solid lines and the dashed lines in figure 7.2.1. Each part of the rope oscillates up and down with the same frequency, so these represent the normal modes of the system. It is reasonable to expect that the normal modes of the beaded string would be similar.

In these standing waves, each part of the string oscillates with a different amplitude. From figure 7.2.1c we see that the amplitude of oscillation for a point at position x is given by,

$$\text{amplitude} = A_n \sin\left(2\pi \frac{x}{\lambda_n}\right),$$

where λ_n is the wavelength of the standing wave and A_n is the amplitude of the standing wave.

Therefore, we will make the following guess for the positions of the beads on the beaded string:

$$\text{amplitude} = Y_j \overset{?}{=} A_n \sin\left(2\pi \frac{x_j}{\lambda_n}\right), \tag{7.2.1}$$

where x_j is the x-position of bead j. This is our "standing wave guess" for the form of the eigenvectors; each different value of j in the above equation (corresponding to each different bead) gives a different line of the column matrix representing the eigenvector.

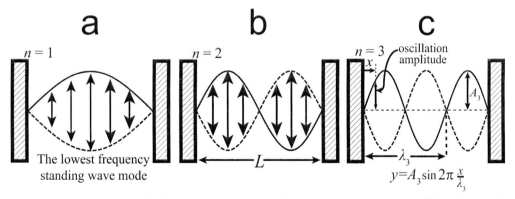

Figure 7.2.1 a: In the lowest-frequency standing wave ($n = 1$), the rope oscillates between the shape shown by the solid curve and that shown by the dashed curve. b: Similar curves for the $n = 2$ standing wave. c: The $n = 3$ standing wave.

Because the string must go to $y = 0$ at the boundaries, we must fit an integer number of half-wavelengths between the walls, that is,

$$n\frac{\lambda_n}{2} = L \Leftrightarrow$$

$$\lambda_n = \frac{2L}{n}. \tag{7.2.2}$$

Since the wavelength cannot take on a continuous range of values, but only those allowed by the above equation, we see that the imposition of boundary conditions has "quantized" the allowed values of λ. To save writing (and to follow well-established convention), we define the "wavenumber"

$$\boxed{k_n \equiv \frac{2\pi}{\lambda_n} = n\frac{\pi}{L}.} \tag{7.2.3}$$

Thus, the wavenumber is 2π divided by the periodicity in space (λ), just as the angular frequency ω is 2π divided by the periodicity in time (T). Using this, equation (7.2.1) can be written

$$Y_j \stackrel{?}{=} A_n \sin\left(k_n x_j\right) \tag{7.2.4}$$

(Again, n is the mode index and j is the mass index.)

Figure 7.2.1 shows that the $n = 2$ mode has one "node," that is, one spot in the middle where the rope doesn't move. The $n = 3$ mode has two nodes, and so on. Quite generally:

$$\boxed{\text{Number of nodes } + 1 = n.}$$

To see if our standing wave guess for the eigenvectors works, we plug equation (7.2.4) into (7.1.6). (Recall that equation (7.1.6) was obtained by combining $F_{\text{TOT}} = m\ddot{y}$ with our normal mode guess.) After canceling the common factor of A_n, this gives

$$-\frac{\omega_{\text{A}}^2}{2}\sin\left(k_n x_{j-1}\right) + \omega_{\text{A}}^2 \sin\left(k_n x_j\right) - \frac{\omega_{\text{A}}^2}{2}\sin\left(k_n x_{j+1}\right) \stackrel{?}{=} \omega_n^2 \sin\left(k_n x_j\right).$$

Now, $x_{j-1} = x_j - a$, and $x_{j+1} = x_j + a$, so this becomes

$$-\frac{\omega_{\text{A}}^2}{2}\left[\sin\left(k_n x_j - k_n a\right) + \sin\left(k_n x_j + k_n a\right)\right] + \omega_{\text{A}}^2 \sin\left(k_n x_j\right) \stackrel{?}{=} \omega_n^2 \sin\left(k_n x_j\right).$$

Next we use the standard formula $\sin(A + B) = \sin A \cos B + \cos A \sin B$, and recall that $\cos(-\theta) = \cos\theta$, while $\sin(-\theta) = -\sin\theta$, to obtain

$$-\frac{\omega_{\text{A}}^2}{2}\left[\left(\sin k_n x_j \cos k_n a - \cos k_n x_j \sin k_n a\right) + \left(\sin k_n x_j \cos k_n a + \cos k_n x_j \sin k_n a\right)\right]$$

$$+ \omega_{\text{A}}^2 \sin\left(k_n x_j\right) \stackrel{?}{=} \omega_n^2 \sin\left(k_n x_j\right)$$

$$\Leftrightarrow -\omega_{\text{A}}^2 \sin k_n x_j \cos k_n a + \omega_{\text{A}}^2 \sin\left(k_n x_j\right) \stackrel{?}{=} \omega_n^2 \sin\left(k_n x_j\right)$$

$$\Leftrightarrow -\omega_{\text{A}}^2 \cos k_n a + \omega_{\text{A}}^2 \stackrel{?}{=} \omega_n^2$$

So, our standing wave guess *does* work, but only if[1]

$$\omega_n^2 = \omega_A^2 \left(1 - \cos k_n a\right). \tag{7.2.5}$$

Using the trigonometric identity[2]

$$1 - \cos\theta = 2\sin^2\frac{\theta}{2},$$

Equation (7.2.5) becomes

$$\omega_n^2 = 2\omega_A^2 \sin^2\frac{k_n a}{2} = 2\omega_A^2 \sin^2\left(n\frac{\pi a}{2L}\right). \tag{7.2.6}$$

Recall that the imposition of the boundary conditions required that λ_n can only take on discrete values, as given by equation (7.2.2), or equivalently that the wavenumber k_n can only take on discrete values as given by equation (7.2.3). Now, we can see that the boundary conditions also force ω_n to take on discrete values. This is a very general and very important observation:

> Imposition of boundary conditions quantizes the normal mode frequencies.

Exactly the same thing happens in quantum mechanics: imposition of boundary conditions (i.e., restricting the electron to a finite region of space, such as the region near an atomic nucleus) quantizes the allowed frequencies of the wavefunction Ψ, and since $E = \hbar\omega$, this is equivalent to quantizing the allowed energies. Thus, we see that one of the most famous results of quantum mechanics, the idea that an electron in an atom can only take on certain allowed levels of energy, is *directly* the result of attributing a wave nature to the electron, and restricting this wave to a small region in space.

We can see from equation (7.2.6) that the smaller the region is (i.e., the smaller L is), the greater the interval between allowed frequencies. Similarly in quantum mechanics, the more tightly we confine the electron, the larger the interval between allowed energies. This effect is important in determining the energy levels of atoms.

---

1. This is an example of a "dispersion relation," that is, a relation between ω and k. It's likely that all the waves you have studied in previous courses, including standing waves in organ pipes or on violin strings, had linear dispersion relations, that is, relations in which ω is directly proportional to k. For example, for electromagnetic waves in vacuum, we have $c = \dfrac{\lambda}{T} = \dfrac{2\pi/k}{2\pi/\omega} = \dfrac{\omega}{k} \Leftrightarrow$ $\omega = ck$. Clearly, the dispersion relation for standing waves on a beaded string, equation (7.2.5) is nonlinear. We'll discuss dispersion relations in more depth in chapter 9, and explain why they are called "dispersion relations."

2. It is not hard to prove this from more fundamental identities:

$$\cos\theta = \cos\left(\frac{\theta}{2} + \frac{\theta}{2}\right) = \cos^2\frac{\theta}{2} - \sin^2\frac{\theta}{2} \Rightarrow 1 - \cos\theta = \underbrace{1 - \cos^2\frac{\theta}{2}}_{\sin^2\frac{\theta}{2}} + \sin^2\frac{\theta}{2} = 2\sin^2\frac{\theta}{2}$$

It also dictates the behavior of electrons that are not tightly attached to any single atom, such as electrons in metals. Such electrons can roam throughout the metal. Using electron beam lithography, experimentalists can make metal samples that are very small, down to only about 100 atoms on a side! As the sample is made smaller, they can observe the transition from having a "continuum" of energy levels (because the energy interval between levels is too small to observe) to having clearly quantized levels. Such "mesoscopic" samples (between the truly microscopic atomic scale and the macroscopic scale) display many exciting properties, and may form the basis of new types of extremely small transistors.

Let us collect all our results. Plugging the standing wave eigenvectors (7.2.4) into our "normal mode guess" equation (7.1.5): $z_j = Y_j e^{i\omega_n t}$ gives the complete description of the normal modes:

$$z_j = A_n \sin\left(k_n x_j\right) e^{i\omega_n t}.$$

The actual displacement of bead j would be given by the real part of this, so that

$$y_j = A_n \underbrace{\sin\left(k_n x_j\right)}_{\substack{\text{spatial} \\ \text{variation}}} \underbrace{\cos\left(\omega_n t\right)}_{\substack{\text{time} \\ \text{variation}}}$$

$$x_j = ja \quad \lambda_n = \frac{2L}{n} \quad n = 1,\ 2,\ 3,\ \cdots \quad k_n = n\frac{\pi}{L}$$

$$\omega_n = \sqrt{2}\,\omega_A \sin\left(n\frac{\pi a}{2L}\right) \quad \omega_A \equiv \sqrt{\frac{2T}{ma}}$$

(7.2.7)

System in pure normal mode n.

7.3 The highest possible frequency; connection to waves in a crystalline solid

We know that the number of normal modes should match the number of objects[3] in the system (for a system such as this in which the objects can only move in one dimension), that is, that the maximum value of the mode index n should be N. However, from the above, there is no apparent limit on n. Let's look more carefully. From equation (7.2.7), we have that the frequencies of the normal modes are given by

$$\omega_n = \sqrt{2}\,\omega_A \sin\left(n\frac{\pi}{2}\frac{a}{L}\right). \tag{7.3.1}$$

We see right away that the normal mode frequencies can never exceed $\sqrt{2}\,\omega_A$, so that we already have a hint that the number of normal modes might actually be limited.

---

3. More generally, the number of normal modes should match the number of degrees of freedom.

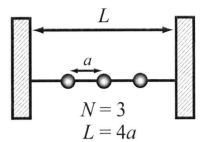

Figure 7.3.1 A string with three beads.

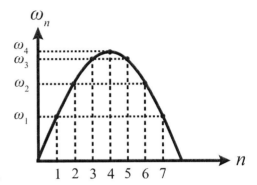

Figure 7.3.2 ω_n *versus* n for $N = 3$.

From figure 7.3.1 we see that

$$L = (N + 1)\,a. \tag{7.3.2}$$

Plugging this into equation (7.3.1) gives

$$\omega_n = \sqrt{2}\,\omega_A \sin\left(\frac{\pi}{2}\frac{n}{N+1}\right). \tag{7.3.3}$$

If we plot ω_n as a function of n, it clearly reaches a maximum at $n = N + 1$; this is shown in figure 7.3.2 for the case $N = 3$. We can already see graphically[4] that $\omega_5 = \omega_3$, or more generally that $\omega_{N+2} = \omega_N$. This means that at least the time dependence of the $n = N + 2$ "mode" is the same as that of the $n = N$ mode, suggesting that the $n = N + 2$ is not a new mode at all, but rather is really the same as the $n = N$ mode. (Similarly, we can see that the time dependence of the $n = N + 3$ "mode" is the same as that of the $n = N - 1$ mode, etc.)

---

4. This is also easy to show symbolically: from equation (7.3.3) we have $\omega_{N+2} = \sqrt{2}\,\omega_A$
$\sin\left(\frac{\pi}{2}\frac{N+2}{N+1}\right) = \sqrt{2}\,\omega_A \sin\left(\frac{\pi \cdot 2(N+1)}{2(N+1)} - \frac{\pi \cdot N}{2(N+1)}\right) = \sqrt{2}\,\omega_A \sin\left(\pi - \frac{\pi \cdot N}{2(N+1)}\right).$
Now, $\sin(\pi - x) = -\sin(-x) = \sin x$. So, $\omega_{N+2} = \sqrt{2}\,\omega_A \sin\left(\frac{\pi}{2}\frac{N}{N+1}\right) = \omega_N.$

Does the spatial dependence of the $n = N + 2$ mode bear out this hunch? From equation (7.2.7), we have that for mode $n = N + 2$

$$y_j = A_{N+2} \sin k_{N+2} x_j \cos \omega_{N+2} t.$$

Where A_{N+2} is the overall amplitude of the mode. The spatial dependence of this is

$$y_j(t=0) = A_{N+2} \sin k_{N+2} x_j = A_{N+2} \sin \left(\pi \frac{N+2}{L} x_j \right) = A_{N+2} \sin \left(2\pi \frac{2(N+1)-N}{2L} x_j \right)$$

$$= A_{N+2} \sin \left\{ 2\pi \left[\frac{2(N+1)x_j}{2(N+1)a} - \frac{N x_j}{2L} \right] \right\} = A_{N+2} \sin \left\{ 2\pi \left[\frac{2(N+1)ja}{2(N+1)a} - \frac{N x_j}{2L} \right] \right\}$$

$$= A_{N+2} \sin \left\{ 2\pi j - \pi \frac{N}{L} x_j \right\} = A_{N+2} \sin \left\{ -\pi \frac{N}{L} x_j \right\}$$

where the last step works because j is an integer. Finally, since $\sin(-x) = -\sin x$ and $k_N = \pi \frac{N}{L}$, we have

$$y_j(t = 0) = -A_{N+2} \sin \left(k_N x_j \right).$$

This is the same as the spatial dependence of the $n = N$ mode, differing only by a minus sign that can be absorbed into the amplitude.

This is shown graphically in figure 7.3.3; note that the function $A_n \sin k_n x_j$ only takes on physical reality at the positions x_j of the beads; we can draw the continuous mathematical function $A_n \sin k_n x$ as shown by the dashed lines, but this does not represent the shape of the string. Since the string is massless, it simply forms straight line segments between the beads, as shown by the solid line. The *same* bead positions

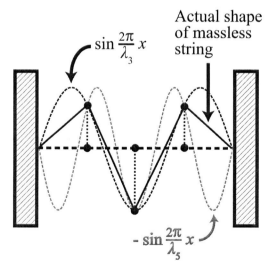

Figure 7.3.3 As shown here for the case $N = 3$, the mode $n = N + 2$ gives the same bead locations (i.e., has the same spatial dependence) as the mode $n = N$. The mathematical functions $A_3 \sin k_3 x$ (black dashed line) and $A_5 \sin k_5 x$ (gray dashed line) only take on physical reality at the positions of the beads. Because the string is massless, it follows straight lines (solid lines) between the beads.

can be characterized using the black dashed curve (with k_N) or the gray dashed curve (with k_{N+2}).

Thus, since the spatial and time dependence of the $n = N + 2$ are the same as for $n = N$, we see that the $n = N + 2$ mode is not actually a new, independent mode, but is really the same as the $n = N$ mode. We could make similar arguments to show that the spatial dependence of the $n = N + 3$ mode is the same as the $n = N - 1$ mode, and so on. So, all the modes with $n = N + 2$ or greater simply reproduce the modes with $n = N$ or smaller.

But what about the $n = N + 1$ "mode"? We see from figure 7.3.2 that the frequency for this "mode" doesn't match the frequency of any other mode, and yet we know that there should be only N modes. The solution to this dilemma comes from considering the spatial dependence. For $n = N + 1$, we have

$$k_{N+1} = (N + 1)\frac{\pi}{L} = (N + 1)\frac{\pi}{(N + 1)a} = \frac{\pi}{a}.$$

This means that $\lambda_{N+1} = \dfrac{2\pi}{k_{N+1}} = 2a$. Therefore, the spatial dependence for this mode is as shown in the top part of figure 7.3.4. Each bead is at a node, and so the beads never move. Therefore, this is not really a "mode," but rather a very complicated way of saying that the beads are allowed to remain motionless!

Here's another way of understanding why there is a highest possible mode, that is, a shortest possible wavelength. Note that, in the highest actual mode ($n = N$), the positions of the masses alternate up–down–up, and so on, as shown in figures 7.3.3 and 7.3.4 bottom. Thus, this is the "most wiggliness" that can be represented by the system, that is, the shortest wavelength.

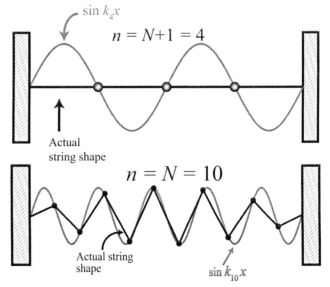

Figure 7.3.4 Top: For $n = N + 1$, each bead is at a node. Bottom: For $n = N$, there is an alternating up-down pattern, so that the string is displaying the most "wiggliness" possible.

These ideas are illustrated further in the applet for this section on the website for this text.

A crystal is a regular array of atoms. Examples include all metals (usually made of many tiny "crystallites" stuck together) and the silicon wafers used to make integrated circuits. Each atom in a crystal is in stable equilibrium, and therefore (using the arguments of chapter 1) can be modeled as a harmonic oscillator. So, one line of atoms in a crystal can be modeled as a beaded string. Since there is a well-defined highest frequency of normal mode for a beaded string, there is a well-defined highest frequency of vibration for a crystal. Because the masses are small, and the restoring forces are relatively strong, this frequency is fairly high – on the order of about 10^{13} Hz. This highest frequency can be observed experimentally – both using spectroscopic methods and using heat capacity measurements.

7.4   Normal mode analysis for the beaded string

Our analysis in section 6.5 was fully general, and applies perfectly well to the beaded string. This means that any behavior of the system can be represented as a superposition of the normal modes. Here's the way we stated that back in section 6.5:

$$(6.5.1): \quad |x(t)\rangle = \text{Re}\left[\sum_{n=1}^{N} C_n e^{i\omega_n t} |e_n\rangle\right].$$

For the beaded string, the beads move in the y-direction, so it makes more sense to write this as

$$|y(t)\rangle \equiv \begin{pmatrix} y_1(t) \\ y_2(t) \\ \vdots \end{pmatrix} = \text{Re}\left[\sum_{n=1}^{N} C_n e^{i\omega_n t} |e_n\rangle\right], \qquad (7.4.1)$$

Normal mode expansion for beaded string

where $y_1(t)$ is the position of bead 1, and $|e_n\rangle$ is the normalized eigenvector for mode n. We already know the form of the eigenvectors, though they aren't yet normalized. In section 7.2, we found that the entries in each line of the eigenvector are

$$(7.2.4): \quad Y_j = A_n \sin\left(k_n x_j\right),$$

so that the eigenvectors are given by

$$|e_n\rangle = A_n \begin{pmatrix} \sin k_n x_1 \\ \sin k_n x_2 \\ \vdots \end{pmatrix}.$$

To find the normalized eigenvectors, we set

$$\langle e_n \mid e_n \rangle = 1$$

$$\Rightarrow A_n^* \left(\sin k_n x_1 \quad \sin k_n x_2 \quad \cdots \right) A_n \begin{pmatrix} \sin k_n x_1 \\ \sin k_n x_2 \\ \vdots \end{pmatrix} = 1 \Rightarrow |A_n|^2 = \frac{1}{\displaystyle\sum_{j=1}^{N} \sin^2 k_n x_j}.$$

You can show in problem 7.6 that

$$\sum_{j=1}^{N} \sin^2 k_n x_j = \frac{N+1}{2}. \tag{7.4.2}$$

For convenience, we may as well choose A_n to be real, so that

$$A_n = \sqrt{\frac{2}{N+1}}$$

and

$$\boxed{|e_n\rangle = \sqrt{\frac{2}{N+1}} \begin{pmatrix} \sin k_n x_1 \\ \sin k_n x_2 \\ \vdots \end{pmatrix}} \tag{7.4.3}$$

Normalized eigenvectors for beaded string.

Self-test: Verify that the above is indeed correctly normalized for the case $N = 3$ and $n = 2$.

Because the analysis of section 6.5 was fully general, we can find the coefficients C_n in the normal mode expansion (7.4.1) by using the formulas

(6.5.8a): $\text{Re } C_n = \langle e_n \mid y_0 \rangle$ *and* (6.5.8b): $\text{Im } C_n = -\frac{1}{\omega_n} \langle e_n \mid \dot{y}_0 \rangle$.

(Here, we have replaced the x's by y's to fit the current situation.)

7.5 Longitudinal oscillations

As discussed in section 7.4, the beaded string is a good model for vibrations in a solid. However, the atoms in a solid can vibrate in two distinct ways. So far, we have discussed transverse vibrations, in which the motion is the y-direction, perpendicular to the direction in which the wavelength is defined (the x-direction). For a solid, this would correspond to planes of atoms sliding laterally without changing the distance between planes, as shown in the left part of figure 7.5.1. However, the planes of atoms can also vibrate along the x-direction, so that the distance between planes oscillates, as shown in the right part of figure 7.5.1. The one-dimensional model for this type of oscillation is a chain of beads connected by springs. We measure the x-position of each

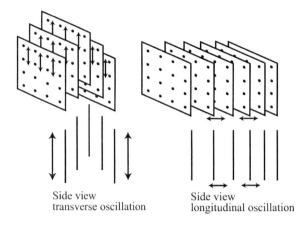

Figure 7.5.1 Transverse and longitudinal oscillations in a solid.

Side view transverse oscillation

Side view longitudinal oscillation

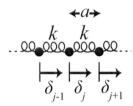

Figure 7.5.2 Quantities defined for longitudinal oscillations.

bead relative to its equilibrium position, however to avoid confusion with the variable x, we describe this displacement using the symbol δ, as shown in figure 7.5.2.

Your turn: Explain why the total force on bead j is $F_{\text{TOT}} = -k\left(\delta_j - \delta_{j-1}\right) - k\left(\delta_j - \delta_{j+1}\right)$.

Since $F_{\text{TOT}} = m\ddot{\delta}_j$, this gives

$$-\frac{k}{m}\left(-\delta_{j-1} + 2\delta_j - \delta_{j+1}\right) = \ddot{\delta}_j. \tag{7.5.1}$$

This equation is isomorphic with equation (7.1.2), the differential equation describing transverse oscillations on the beaded string:

$$-\frac{T}{ma}\left(-y_{j-1} + 2y_j - y_{j+1}\right) = \ddot{y}_j.$$

Table 7.5.1. Isomorphism between transverse and longitudinal waves

Quantity	Transverse	Longitudinal
Displacement	y_j	δ_j
Angular frequency of vibration with neighbors fixed	$\omega_A = \sqrt{\dfrac{2T}{am}}$	$\omega_A = \sqrt{\dfrac{2k}{m}}$

Thus, the solutions are exactly the same as for transverse vibrations, with the substitutions indicated in table 7.5.1 above, giving

$$\delta_j = A_n \underbrace{\sin\left(k_n x_j\right)}_{\substack{\text{spatial}\\\text{variation}}} \underbrace{\cos\left(\omega_n t\right)}_{\substack{\text{time}\\\text{variation}}}$$

$$x_j = ja \quad \lambda_n = \frac{2L}{n} \quad n = 1, 2, 3, \cdots \quad k_n = n\frac{\pi}{L}$$

$$\omega_n = \sqrt{2}\,\omega_A \sin\left(n\frac{\pi a}{2L}\right) \quad \omega_A \equiv \sqrt{\frac{2k}{m}}$$

(7.5.2)

System in pure normal mode n

Figure 7.5.3 shows the vibrations of the mode $n = 1$. At $t = 0$, all the beads are displaced to the right, resulting in a bunching near the right side. A half period later, the beads are all displaced to the left, resulting in bunching near the left side.

Let's apply this model to a thin solid rod of cross-sectional area A, which is wedged between two massive walls, with one end of the rod pushing against each wall. We treat each plane of atoms perpendicular to the axis of the rod as a bead, with mass $m = \rho A a$, where a is the spacing between atomic planes. From equation (2.3.2), the spring constant of a solid with cross-sectional area A and length a is $k = E\dfrac{A}{a}$, where

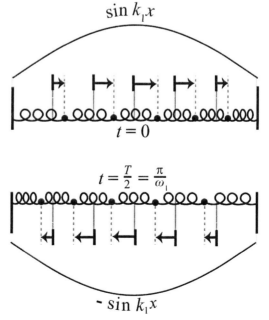

Figure 7.5.3 Displacements at $t = 0$ and half a period later for the mode $n = 1$ for longitudinal oscillations.

E is the Young's modulus. Thus, $\omega_A = \sqrt{\dfrac{2k}{m}} = \sqrt{\dfrac{2EA}{\rho A a^2}} = \dfrac{1}{a}\sqrt{\dfrac{2E}{\rho}}$. If we consider, for example, the low n modes of a rod with at least several mm of length, we have $na \ll L$, so that

$$\omega_n = \sqrt{2}\,\omega_A \sin\left(n\frac{\pi a}{2L}\right) \cong \sqrt{2}\,\omega_A n\frac{\pi a}{2L},$$

where we have used the approximation (1.6.4): $\sin\theta \cong \theta$, valid for small θ. Therefore,

$$\omega_n \cong \sqrt{2}\frac{1}{a}\sqrt{\frac{2E}{\rho}}\, n\frac{\pi a}{2L} = n\sqrt{\frac{E}{\rho}}\frac{\pi}{L}. \tag{7.5.3}$$

7.6   The continuous string

It is easy to extend the results of section 7.4 to a continuous string with finite mass. We simply imagine shrinking the distance a between the beads (by adding more beads onto the string), while also making the beads less massive, so as to keep the mass per unit length constant. Then, in the limit $a \to 0$ (and $N \to \infty$), we have a continuous, massive string.

We can readily find the frequencies of the normal modes in this limit. From equation (7.2.7), we have for the beaded string

$$\omega_n = 2\sqrt{\frac{T}{ma}}\sin\left(n\frac{\pi a}{2L}\right) = 2\sqrt{\frac{T}{ma}}\sin\left(\frac{\pi}{2}\frac{n}{N+1}\right).$$

In the limit $N \to \infty$, the argument of the sin becomes infinitesimal, so we can use

$$\sin x \xrightarrow[x\to 0]{} x.$$

Therefore, as we allow $N \to \infty$, we obtain

$$\omega_n = 2\sqrt{\frac{T}{ma}}\left(n\frac{\pi a}{2L}\right) = \sqrt{T}\sqrt{\frac{a}{m}}\, n\frac{\pi}{L}. \tag{7.6.1}$$

We define the mass per unit length:

$$\mu \equiv \frac{m}{a},$$

so that equation (7.6.1) becomes

$$\boxed{\omega_n = n\frac{\pi}{L}\sqrt{\frac{T}{\mu}} = n\omega_1,} \tag{7.6.2}$$

Normal mode frequencies for continuous string

where $\omega_1 = \dfrac{\pi}{L}\sqrt{\dfrac{T}{\mu}}$. Note that, unlike for the beaded string, these frequencies are all multiples of the fundamental frequency ω_1, and that they increase linearly with the mode index n. One way to understand this is by considering the graph of ω_n for the beaded string, as shown in figure 7.6.1. The peak of this graph is at $n = N + 1$, so

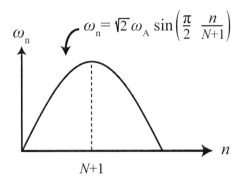

Figure 7.6.1 Angular frequency *vs.* mode index for a beaded string.

that in the limit $N \to \infty$ we never approach the peak, and we are always in the linear regime close to the origin.

As we take the limit $N \to \infty$, there is nothing fundamental changing in the physics of the situation. Therefore, the eigenvectors have the same form as for the beaded string. In section 7.2, we found that the entries in each line of the eigenvector are

$$(7.2.4): Y_j = A_n \sin\left(k_n x_j\right),$$

so that the eigenvectors are given by

$$\left|e_n\right\rangle = \begin{pmatrix} Y_1 \\ Y_2 \\ \vdots \end{pmatrix} = A_n \begin{pmatrix} \sin k_n x_1 \\ \sin k_n x_2 \\ \vdots \end{pmatrix}. \tag{7.6.3}$$

For the limit $N \to \infty$, this notation becomes awkward, since there are an infinite number of lines in the column vector. So, instead we simply write the eigenvectors as

$$y_n(x) = A_n \sin k_n x. \tag{7.6.4}$$

Note that we no longer write these explicitly as vectors; instead we simply write them as functions. Therefore, we will often call them "eigenfunctions" instead of "eigenvectors." However, it is still appropriate to think of them as vectors in an infinite-dimensional Hilbert space. (The Hilbert space has infinite dimensions, because there are an infinite number of infinitesimally small pieces making up the continuous string.)

7.7   Normal mode analysis for continuous systems

To handle continuous systems, we'll need to extend the definition of the inner product. Let's recall the general definition of the inner product for systems with a finite number of objects, such as the beaded string. For example, let's say that $\left|y_A\right\rangle$ represents one state of the beaded string (perhaps a normal mode, or perhaps a mixture of normal modes), and that $|y_B\rangle$ represents some different state. We can write

$$|y_A\rangle = \begin{pmatrix} y_{A1} \\ y_{A2} \\ \vdots \end{pmatrix} \text{ and } |y_B\rangle = \begin{pmatrix} y_{B1} \\ y_{B2} \\ \vdots \end{pmatrix},$$

where, for the state $|y_A\rangle$, y_{A1} is the position of bead 1, y_{A2} is the position of bead 2, and so on.

The inner product is defined as

$$\langle y_A \mid y_B \rangle \equiv \begin{pmatrix} y_{A1}^* & y_{A2}^* & \cdots \end{pmatrix} \begin{pmatrix} y_{B1} \\ y_{B2} \\ \vdots \end{pmatrix} = \sum_{j=1}^{N} y_{Aj}^* y_{Bj} . \tag{7.7.1}$$

For a continuous system, there would be an infinite number of objects, so this would become an infinite sum, which we can write using an integral. Therefore, we define the inner product of two continuous functions $y_A(x)$ and $y_B(x)$ to be

$$\langle y_A(x) \mid y_B(x) \rangle \equiv \int_0^L y_A^*(x) y_B(x) \, dx, \tag{7.7.2}$$

where the range $x = 0$ to $x = L$ is the size of the continuous system. For a system with a differently defined size, the limits of integration would be changed, so that the integral is still over the entire system.

Although equation (7.7.2) is the well-established convention for the definition of the inner product, it is not *quite* equivalent to the definition (7.7.1) for discrete systems, since each term in the sum which the integral represents is multiplied by the extra factor dx. It should be clear that, even with this extra factor, the eigenfunctions are still orthogonal (i.e., the inner product of two different eigenfunctions is still zero), since zero times dx is still zero.

---

5. From equation (7.6.3), we have $y_n(x) = A_n \sin k_n x$ and $y_m(x) = A_m \sin k_m x$. According to equation (7.7.2), their inner product is $\langle y_n(x) \mid y_m(x) \rangle = \int_0^L A_n^* \sin k_n x A_m \sin k_m x \, dx = A_n^* A_m \times$

$\left[\dfrac{\sin (k_n - k_m) x}{2 (k_n - k_m)} - \dfrac{\sin (k_n + k_m) x}{2 (k_n + k_m)} \right]_0^L$. Recall that $k_n = n\dfrac{\pi}{L}$, so that $k_n - k_m = (n - m)\dfrac{\pi}{L}$

and $k_n + k_m = (n+m)\dfrac{\pi}{L}$. Therefore, $\langle y_n(x) \mid y_m(x) \rangle = A_n^* A_m \left[\dfrac{\sin [(n-m)\pi]}{2 (k_n - k_m)} - \right.$

$\dfrac{\sin [(n+m)\pi]}{2 (k_n + k_m)} - 0 + 0 \left. \right]$. Since n and m are integers, this equals zero, Q.E.D.

Self-test (answer below[5]): Verify that the eigenfunctions $y_n(x)$ and $y_m(x)$ for the continuous string are indeed orthogonal if $n \neq m$. You may need this integral, given in the form you would find it in an integral table:

$$\int \sin px \, \sin qx \, dx = \frac{\sin(p-q)x}{2(p-q)} - \frac{\sin(p+q)x}{2(p+q)} \quad p \neq \pm q$$

(Although this is the common way to write this, it is somewhat ambiguous; it should be read

$$\int \sin px \, \sin qx \, dx = \frac{\sin[(p-q)x]}{2(p-q)} - \frac{\sin[(p+q)x]}{2(p+q)} \quad p \neq \pm q$$

Your turn: Show that the normalized eigenfunctions for the continuous string are given by

$$y_n(x) = \sqrt{\frac{2}{L}} \sin k_n x \tag{7.7.3}$$

Ignore the following paragraph if it confuses you, but you might find it interesting. Compare equation (7.7.3) with the normalized eigenvectors for the beaded string:

$$(7.4.3): \quad |e_n\rangle = \sqrt{\frac{2}{N+1}} \begin{pmatrix} \sin k_n x_1 \\ \sin k_n x_2 \\ \vdots \end{pmatrix}$$

Normalized eigenvectors for beaded string

By recalling that $L = (N+1)\,a$, we see that the result (7.7.3) for the continuous string is the same as that for the beaded string, except that the continuous string result is divided by $\sqrt{a}$; for the continuous string, the "spacing" between "beads" is $a = dx$, so that the normalized eigenfunctions have been divided by $\sqrt{dx}$. When we take the inner product of two such functions, we effectively divide by $\sqrt{dx}$ twice; this compensates for the "extra" factor of dx in the definition of the inner product that was discussed at the top of the previous page.

You should recall that the normalized eigenvectors for discrete systems were dimensionless. From equation (7.7.3), we can see that the normalized eigenfunctions for the continuous string have units of meters $^{-1/2}$. This might seem strange, but it is indeed correct, and necessary to make the normal mode analysis formulas work out correctly. It might make you feel better to know that the wave function for one-dimensional quantum mechanics (analogous to our one-dimensional continuous string) also has units of meters $^{-1/2}$.

As for the beaded string (and the coupled pendula), any state of the system can be written as a superposition of the normal modes, that is,

$$y(x, t) = \text{Re} \left[\sum_{n=1}^{\infty} C_n e^{i\omega_n t} y_n(x) \right]. \qquad (7.7.4)$$

Normal mode expansion for continuous string

(If the system is truly continuous, then there are an infinite number of normal modes.) Because the analysis of section 6.5 was fully general, we can find the coefficients C_n in the normal mode expansion (7.7.4) by using the formulas

$$(6.5.8a)\text{: Re } C_n = \langle y_n(x) | y_0(x) \rangle \text{ and}$$

$$(6.5.8b)\text{: Im } C_n = -\frac{1}{\omega_n} \langle y_n(x) | \dot{y}_0(x) \rangle .$$

(Here, we have replaced the x's by y's, and vectors by functions to fit the current situation.) Again, for the important special case $\dot{y}_0(x) = 0$, the process of normal mode analysis is exactly analogous to the process of taking projections of ordinary vectors onto the x- and y-axes (although now there are an infinite number of axes onto which we must take projections!).

7.8   k-Space

As for the other coupled oscillator systems that we've studied, we can fully specify the behavior of the continuous string *either* by specifying $y(x, t)$ *or* by specifying all the amplitudes C_n in the normal mode expansion. Let's consider the important special case $\dot{y}_0(x) = 0$, for which Im $C_n = 0$. Then, we can write the normal mode expansion (7.7.4) as

$$y(x, t) = \sum_{n=1}^{N} C_n \cos \omega_n t \, y_n(x),$$

where $C_n = \text{Re } C_n = \langle y_n(x) | y_0(x) \rangle$

Normal mode expansion for $\dot{y}_0(x) = 0$.

At $t = 0$, this reduces to

$$y(x, t = 0) = \sqrt{\frac{2}{L}} \sum_{n=1}^{N} C_n \sin k_n x. \qquad (7.8.1)$$

Each C_n is associated with a k_n. We could thus make a plot of C_n versus k, as shown in figure 7.8.1 for a particular $y(x, t = 0)$ which involves mixing several normal modes. The initial shape of the string can be specified *either* by plotting y versus x, *or* by plotting C_n versus k. The latter view is called "k-space." We can make a

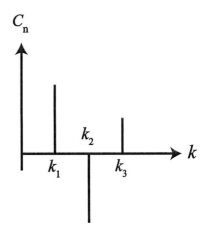

Figure 7.8.1 k-space for the continuous string.

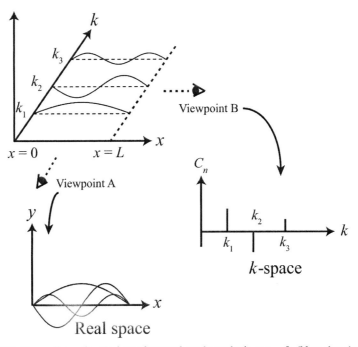

Figure 7.8.2 Three-dimensional view of normal mode analysis at $t = 0$. (Note that the picture is somewhat schematic. For example, the actual perspective from viewpoint B would not allow one to tell whether each C_n was positive or negative.)

three-dimensional plot, as shown in figure 7.8.2, which symbolizes the process of normal mode analysis at $t = 0$. Looking from viewpoint A, we can imagine adding up the sinusoids to produce $y(x, t = 0)$; this is the real space view. However, we could just as well look from viewpoint B, which gives the k-space view. This idea of looking at things in k-space turns out to be tremendously important in crystallography, solid-state physics, and many other areas.

Concept and skill inventory for chapter 7

After reading this chapter, you should fully understand the following terms:

Standing wave (7.2)
Wavenumber (7.2)
Dispersion relation (7.2)
Longitudinal waves (7.5)
k-space (7.8)
Real space (7.8)

You should understand the following connections:

Boundary conditions & allowed frequencies (7.2)
Mode index & number of nodes (7.2)
Bead spacing & maximum wiggliness (7.3)
Transverse & longitudinal oscillations (7.5)
Normal modes for beaded & continuous strings (7.6)
Hilbert space vectors & continuous functions (7.6)
Representations of a function in real space & in k-space (7.8)

You should be familiar with the following additional concepts:

Highest frequency/smallest wavelength standing wave for a beaded string (7.3)
Functions only taking on physical reality at certain positions (7.3)

You should be able to:

Find the frequency for a given normal mode of a beaded string (7.2)
Explain why there is a minimum wavelength for standing waves on a beaded string (7.3)
Analyze a given pattern of initial velocities and positions for a beaded string into a superposition of normal modes (7.4)
Find the frequency for a given normal mode of a continuous string (7.6)
Analyze a given pattern of initial velocities and positions for a continuous string into a superposition of normal modes (7.7)
Take inner products of continuous functions (7.7)

In addition to all of the above, you should be able to combine the concepts you've learned to address new situations.

Problems

Note: Additional problems are available on the website for this text.

Instructor: Ratings of problem difficulty, full solutions, and important additional support materials are available on the website.

7.1 State what is wrong with the following: "When a beaded string is in pure normal mode 2, the displacement of bead 1 is given by $A \sin k_1 x_2 \cos \omega_1 t$, where A is the amplitude of motion of the bead." (There might be more than one thing wrong.)

7.2 Orthogonality of normal modes for the beaded string

    (a) Find the normalized eigenvectors for the beaded string with $N = 3$. Let's call them $|e_1\rangle$, $|e_2\rangle$, and $|e_3\rangle$. (The subscripts here refer to the mode index n, *not* to the mass index j.)

    (b) Show that $\langle e_n \mid e_n \rangle = 1$ for all possible values of n, thus verifying that you have correctly normalized the eigenvectors.

    (c) Show that $\langle e_n \mid e_m \rangle = 0$ for all possible combinations of n and m where $n \neq m$, thus verifying that the eigenvectors are mutually orthogonal.

7.3 Beaded string applet

Go to the website for this text, and under chapter 7 open the "Beaded String Applet."

Write out answers to all **questions in boldface.**

    (a) This applet demonstrates some properties of standing waves on a beaded string. You may select the number of beads (masses) and the desired normal mode to display with the two sliders at the bottom of the applet. Additionally, you may at any time choose to slow down the animation with the "Slow" check box.

    (b) First, look at a string with 4 beads on it. Set that up using the "masses" slider. Now look at the normal modes for this configuration by varying the "mode" slider. **How many normal modes are there?** Try a few other configurations consisting of different numbers of beads. **In general, for this one dimensional case (beads can only move up and down), how does the number of normal modes relate to the number of beads?**

    (c) The main goal of this applet is to visually demonstrate the fact that there is a maximum meaningful normal mode index n for waves on a beaded string. As we have just seen, the maximum index relates to the number of beads on the string. **What happens if you try to generate normal modes with indices higher than the maximum possible mode for a situation?** Try it. What **do the blue and red waves represent? Explain why these two waveforms do not represent the actual string. What would the actual string look like?** Take a moment to try the animation with a number of different normal mode frequencies and numbers of beads. Think about what you have seen.

    (d) **Explain what the graph in the lower left represents. Why is the right part shown as a dashed line?**

    (e) **If there is a maximum normal mode index for a beaded string, what does this imply about the possible frequencies of waves that can exist on the string? Explain. What does this imply about waves in a crystal lattice (which can be thought of as a three dimensional array of beads on springs)?**

    (f) **Given the normal modes for a beaded string, could one construct a Hilbert space to represent this system? What would it look like? How many dimensions would it have?**

7.4 Driven beaded string. A string with N beads, each of mass m and spacing a, is under tension T. At the left, the string is attached to a fixed wall, as usual. However, on the right, it is attached to a wall which oscillates up and down with amplitude h and angular frequency ω_{d}. This means that the "fictitious bead" on the left has $y_0 \equiv 0$ as usual, but the fictitious bead on the right has $y_{N+1} \equiv h \cos \omega_{\mathrm{d}} t$. The differential equations of motion for this system are the same as for the undriven case (because the way the beads interact with each other is unchanged); only the boundary conditions are different.

    (a) Explain why it's reasonable to guess that a solution for this problem would be $y_j \stackrel{?}{=} Y_j \cos\left(\omega_{\mathrm{d}} t - \delta\right)$.

    (b) Given the form of the above guess, explain why we expect that Y_j need not equal y_{j0}.

    (c) Since the differential equations of motion are the same as for the undriven case, it's reasonable to guess that the amplitudes Y_j vary with position in the same way, that is, it's reasonable to guess $Y_j \stackrel{?}{=} A \sin \alpha\, x_j$, where the values of A and α are to be determined. Use $y_{N+1} \equiv h \cos \omega_{\mathrm{d}} t$ to find δ and A in terms of α, h, a, and N.

    (d) Plug our guesses $y_j \stackrel{?}{=} Y_j \cos\left(\omega_{\mathrm{d}} t - \delta\right)$ and $Y_j \stackrel{?}{=} A \sin \alpha\, x_j$ into the equation (7.1.2) (which was derived simply by applying $F = m\ddot{x}$ to mass j):

$$-\frac{T}{ma}\left(-y_{j-1} + 2y_j - y_{j+1}\right) = \ddot{y}_j$$

Verify that the guesses work, and show that $\cos\left(\alpha\, a\right) = 1 - \dfrac{\omega_{\mathrm{d}}^2}{\left(2\dfrac{T}{ma}\right)}$.

    (e) Explain why the above implies that if $\omega_{\mathrm{d}} > 2\sqrt{\dfrac{T}{ma}}$ then α is complex. *Hint: Set* $\alpha = \alpha_r + i\alpha_i$, *use* $\cos\left(\alpha\, a\right) = \mathrm{Re}\left[e^{i\alpha\, a}\right]$, *and show that for* $\omega_{\mathrm{d}} > 2\sqrt{\dfrac{T}{ma}}$, α_i *must be nonzero.*

    (f) Explain why a complex value of α means that the wave damps exponentially in space. *Hint: recall that* $\sin x = \mathrm{Im}\left[e^{ix}\right]$.

    (g) What does that phrase "damps exponentially in space" mean?

(h) Why is it reasonable that the frequency $2\sqrt{\dfrac{T}{ma}}$ should be the highest frequency which does not give exponential damping in space?

7.5 Normal mode analysis for the beaded string. (You are encouraged to use Mathematica, Maple, or other symbolic algebra program for this problem, though it is not necessary.) A beaded string has three beads, each of mass 1.00 kg, spaced at 1.00 m intervals. The string has tension 100 N. The initial positions and velocities of the beads are:

$$y_1 = -2.55033 \text{ cm} \qquad \dot{y}_1 = -35.8043 \text{ cm/s}$$

$$y_2 = 3.51028 \text{ cm} \qquad \dot{y}_2 = 1.38233 \text{ cm/s}$$

$$y_3 = -0.885746 \text{ cm} \qquad \dot{y}_3 = 15.6333 \text{ cm/s}$$

(a) Write explicit equations for $y_1(t)$, $y_2(t)$, and $y_3(t)$ that are valid for all times. Your equations should be in terms of cosines, rather than $\text{Re}\left[e^{i\omega t}\right]$. **Note: Remember that the arctan returns a result that is only defined up to an additive factor of π. Therefore, think carefully about the result your calculator or program returns to you – is it in the correct quadrant of the complex plane? If not, you need to add π to it.**

(b) Plug $t = 0$ into your expressions from part (a), and verify that they give the correct values for the initial positions.

7.6 Show that $\displaystyle\sum_{j=1}^{N} \sin^2 k_n x_j = \frac{N+1}{2}$, where $k_n = \dfrac{n\pi}{L}$, $x_j = ja$, $L = (N+1)a$, and n and N are integers. (We used this result in section 7.4.) You should probably use a symbolic algebra program for this; please see the additional instructions on the web page for this book.

7.7 Fret spacing on guitars. (Read problem 5.1 before doing this problem; you need not actually do that problem first, though.) A standard guitar has six strings. Each is held under tension, stretched between the bridge (near the middle of the main body of the guitar) and the nut (near the end of the neck), as shown in figure 7.P.1. A series of frets is positioned just below the strings. When the guitarist uses a finger just behind one of the frets to push one of the strings against the fretboard, the effective length L of the string is the distance from the bridge to the fret, as shown. The frets are spaced so that the guitarist can create notes in half-step intervals. For example, if the highest string is plucked without using any fret, then the full length of the string (from the bridge to the nut) vibrates, producing a note of E above middle C. (The main pitch produced is from the fundamental, i.e., the $n = 1$ mode.) If the guitarist pushes down behind the first fret (the one closest to the nut), and then plucks the string, it instead produces an F#. If the guitarist pushes down behind the second fret, the string produces an F#. If the length from bridge to nut is L_0, what is the equation that determines the position for fret number j, where $j = 1$ for the fret closest to the nut, and positions are

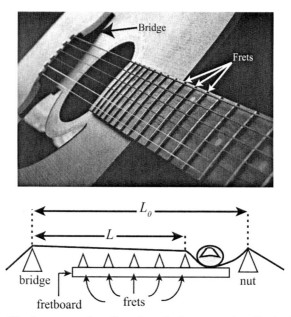

Figure 7.P.1 Top: The frets on a guitar allow the guitarist to vary the effective length of each string in a controlled way. Bottom: side view of a finger pressing a string down behind a fret. (Top image © Milinz | Dreamstime.com)

measured relative to the bridge? (Recall from problem 5.1 that the ratio of frequencies for notes that are a half-step apart is 1.05946.)

7.8 Harmonics on guitars. Rather than using the frets (see problem 7.7), guitarists sometimes use an alternate technique. Instead of using a finger to press the string against the fretboard, the finger is placed lightly on the string, at a carefully selected point. The string is plucked, and then, as quickly as possible, the lightly pressing finger is released. The string emits a note that is bell-like in tone. If the lightly pressed finger is placed at a point halfway along the length of the string, the frequency of the note sounded is twice that of the fundamental frequency of the string. Explain what is happening. (Various such "harmonics" can be produced by different placements of the lightly pressed finger. These can be used to produce notes much higher than the guitar could ordinarily make. For an example, go to the website for this text, and under this chapter find the listing for this problem. This technique is also very important for rock guitarists; you can see lots of examples by doing an internet video search using the words rock guitar harmonics.)

7.9 You can make a crude musical instrument by stretching a rubber band and plucking it. Try it. This will work best with a relatively thick band, held close to your ear. For the most reproducible results, hook your fingers through the band, rather than pinching it between your fingers. Important: you should notice that the band stretches fairly easily to a certain length, and then becomes much stiffer. We will concentrate on the range of fairly easy stretching. Start with the band stretched most of the way, and slowly let it relax as you pluck it again and again, holding it very close to your ear so you can hear clearly. Let it relax all the way to the point where it is slack. Then, stretch it out most of the way, and listen again as you let it relax. Using the ideas you've learned in this book, make a simple quantitative model for how the pitch should depend on length, and compare your model qualitatively with your observations. (You will probably find that the behavior is more complicated if you start with the band slack and stretch it out than it is if, as directed, you start with it stretched and allow it to relax. This shows that the model you have developed doesn't capture all the physics of this situation.)

7.10 A reasonably accurate model for the dispersion relation of an actual stretched piano string is

$$\omega = ak + bk^3,$$

where a and b are constants. Find an expression for all of the possible oscillatory frequencies of a piano string of length L.

7.11 A string of mass M and length L is attached to walls at either end, and is under tension T. The string is held at rest in the following shape: $y = B$ for $\frac{L}{2} < x < \frac{3L}{4}$, and $y = 0$ for the rest of the string. The string is released at $t = 0$. **(a)** Find $y(x, t)$. You may express your answer as an infinite series, so long as you have defined all the symbols in your series. **(b)** Using words, diagrams, and equations, describe what you did in part a by making analogies with a system of conventional two-dimensional vectors in ordinary $x - y$ space.

7.12 (You may wish to complete problem 7.11 first.) A string of mass M and length L is attached to walls at either end, and is under tension T. At $t = 0$, the string has the following shape: $y = B$ for $\frac{L}{2} < x < \frac{3L}{4}$, and $y = 0$ for the rest of the string. At $t = 0$ the velocity distribution is $\dot{y} = E$ for $\frac{L}{4} < x < \frac{L}{2}$, and $\dot{y} = 0$ for the rest of the string. E has units of m/s. Given these initial conditions, find $y(x, t)$. (Note that this velocity pattern is not the same as the initial position pattern – one is nonzero on the right side of the string, and the other is nonzero on the left side of the string.) You may express your answer as an infinite series, so long as you have defined all the symbols in your series.

7.13 A continuous string with mass per unit length μ is stretched with tension T between two walls, one at $x = 0$ and the other at $x = L$. At $t = 0$, its shape is given by $y(x, t = 0) = \frac{L}{100} \frac{x^2}{L^2}$, and it is at rest. What is the complex coefficient for the $n = 2$ mode in the normal mode expansion for these initial conditions? You may need the following integral (given in the form you would find it in an integral table):

$$\int x^2 \sin ax \, dx = \frac{2x}{a^2} \sin ax + \left(\frac{2}{a^3} - \frac{x^2}{a} \right) \cos ax.$$

7.14 A string of mass M and length L is attached to walls at either end, and is under tension T. Prior to $t = 0$, the string is held fixed with the following shape:

$$y_0(x) = \begin{cases} \dfrac{4}{L^2}Gx^2 & 0 \le x \le \dfrac{L}{2} \\[2mm] \dfrac{4}{L^2}G(x-L)^2 & \dfrac{L}{2} \le x \le L \end{cases}, \text{ where } G \text{ is a constant.}$$

(a) Sketch this shape.

(b) At $t = 0$, the string is released. For the normal mode expansion, briefly explain why it is unnecessary to calculate the imaginary parts of the expansion coefficients.

(c) Calculate the coefficients in the normal mode expansion. It might be helpful to note the solutions to the following integrals:

$$\int x^2 \sin Ax\, dx = \frac{(2-A^2x^2)\cos(Ax)}{A^3} + \frac{2x\sin(Ax)}{A^2}$$

$$\int (x-B)^2 \sin(Ax)\, dx = \frac{(2-A^2(B-x)^2)\cos(Ax)+2A(x-B)\sin(Ax)}{A^3}.$$

(d) You should have found that coefficients for the even n terms in the expansion are all zero. Explain why, based on symmetry.

(e) Write an expression for $y(x, t)$. You may express your answer as an infinite series, so long as you have defined all the symbols in your series.

7.15 At $t = 0$, the shape of a string under tension T which has mass/length μ is a complicated shape, as shown in figure 7.P.2, and the string is motionless at this instant. (The vertical scale is greatly exaggerated; displacement from equilibrium is actually small, so that our usual approximations work well.) The distance between the walls is L. Assume damping is negligible. It is

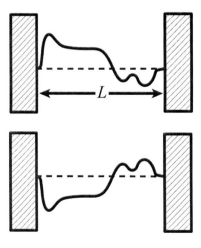

Figure 7.P.2 Top: Initial shape of string. Bottom: inverse of the original shape.

observed that at $t = \tau$, the string returns to exactly the same shape. The shape also recurs at $t = 2\tau, 3\tau$, and so on.

(a) Using ideas of normal mode analysis, explain this surprising result, and also find the value of τ in terms of the other parameters above.

(b) Explain why the inverse of the original shape, that is, the one shown in the lower part of the figure, is never observed.

7.16 The particle in a box. In section 1.11, we discussed the quantum mechanical wavefunction $\Psi(x, t)$. As you will learn in a later course on quantum mechanics, it is governed by Schrödinger's equation:

$$-\frac{\hbar^2}{2m}\frac{\partial^2 \Psi}{\partial x^2} + U(x)\,\Psi = i\hbar\frac{\partial \Psi}{\partial t},$$

where $\hbar$ is Plank's constant, m is the mass of the particle (usually an electron), $U(x)$ is the potential energy the particle would have at position x, and $\Psi = \Psi(x, t)$ is the "wave function" which describes the particle. Note that this is a linear differential equation, so that we can superpose solutions for it just as we did for the string stretched between two walls. The notation $\frac{\partial \Psi}{\partial t}$ means "partial derivative of Ψ with respect to x." This simply means "take the derivative of Ψ with respect to t, treating x as a constant." Similarly, the notation $\frac{\partial^2 \Psi}{\partial x^2}$ means "second partial derivative of Ψ with respect to x", which means "take the second derivative of Ψ with respect to x, treating t as a constant." You should not get too worried about this notation; it is needed because y is a function of both x and t, but there is really nothing mysterious about it.

You can see right away, from the presence of the "i" in the above equation that Ψ is intrinsically complex. **Therefore, in what follows there is no need to think about taking real parts of anything.**

Consider a particle in the potential $U = \begin{cases} 0 & 0 < x < L \\ \infty & x < 0 \text{ or } x > L \end{cases}$. If the particle has a finite total energy then it cannot exist outside the region $x = 0$ to $x = L$, since outside this region its potential energy would exceed its total energy by an infinite amount. Therefore, the boundary conditions are that Ψ must go to zero at $x = 0$ and at $x = L$. **This is very analogous to what happens to y for a continuous string stretched between two walls.**

(a) We can guess that the normal modes of this "particle in a box" system are given by

$$\Psi_n = \begin{cases} C_n \psi_n(x)\, e^{-i\omega_n t} & \text{inside the well} \\ 0 & \text{outside the well} \end{cases},$$

where the functions $\psi_n(x) = A \sin k_n x$ are the normalized eigen-functions, and A is a constant that you'll determine in part (c) of this problem. Verify that this guess is correct by plugging it into the Schrodinger equation and showing that it works. **As part of doing this checking, you should find the dispersion relation that**

is required to make the guess work. (Note that this guess has the factor $e^{-i\omega_n t}$, rather than the factor $e^{i\omega_n t}$ that you might have expected from our discussion of standing waves on a rope. For the rope waves, since we take the real part anyway, it wouldn't have made a difference to include the minus sign in the exponential, although we followed well-established convention by not including it. As you can see from this exercise, the minus sign really is needed in the quantum mechanical version.)

(b) What is the condition on k_n in order for the boundary conditions to be satisfied?

(c) Find the value of A that correctly normalizes the eigenfunctions for this situation. You may do this mathematically, or by referring to results we have previously obtained.

(d) Make an argument for why the eigenfunction for the lowest-frequency normal mode is orthogonal to the eigenfunction for the second lowest-frequency normal mode. Note that a response such as, "All eigenfunctions are orthogonal." is not acceptable; you must demonstrate that the eigenfunctions are orthogonal by mathematical or logical argument.

(e) Now consider this particular initial condition: At $t = 0$, $\Psi = \begin{cases} B & \dfrac{L}{4} < x < \dfrac{L}{2} \\ 0 & \text{elsewhere} \end{cases}$. Explain why this means that all the coefficients C_n that appear in the normal mode expansion of Ψ must be real. *Hint: This is a consequence of the fact that Ψ is real everywhere at $t = 0$. Also, recall that there is no Re[...] in the normal mode expansion for this situation, as indicated by the boldface text in the introduction to this problem.*

(f) Given the initial conditions of part (e), find $\Psi(x, t)$. You may express your answer in terms of an infinite series, so long as all quantities in the series are explicitly defined, including evaluation of all integrals. *Hint: As you showed in part e, all the coefficients C_n are real. Recall that, for a string stretched between two walls we get that all real coefficents if all the initial velocities are zero. Therefore, having a quantum mechanical wavefunction Ψ that is initially real everywhere is fully analogous to a stretched string with zero initial velocity everywhere.*

(g) Now, consider a different initial condition, which is formed by superposing equal amplitudes of the second-lowest-frequency normal mode and the third-lowest-frequency normal mode. The resulting superposition shows a "sloshing" back and forth of $\Psi(x, t)$. What is the period of this sloshing, in terms of $\hbar$, m, and L? Explain.

7.17 k-space picture for plucked guitar string. The tonal quality of a note sounded by a guitar string depends on where along its length the string is plucked. (For this problem, assume the string is played at full length, without using any frets.) The graphs in figure 7.P.3 show the k-space picture for two

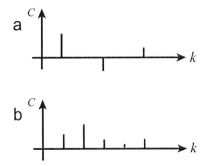

Figure 7.P.3 Two possible k-space graphs.

different plucking positions, with amplitude on the vertical axis and k on the horizontal axis. The scales are the same for both graphs. For one of the graphs, the string was plucked at the halfway point, that is, a finger halfway along the length of the string pulled it slightly away from equilibrium and then released it. For the other picture, the string was plucked one quarter of the way along its length. Which graph is which? Explain your reasoning thoroughly, including an explanation for why some of the peaks are missing from one of the graphs.

8 Fourier Analysis

You need to move past Fourier transfers, and start thinking quantum mechanics.
— Maggie Madsen, signal analyst character, the *Transformers* movie (2007).

8.1 Introduction

Although the above quote is a bit mangled (presumably the writer meant "Fourier transforms"), it accurately conveys that the *first* thing any scientist does with a complicated data set is to subject it to Fourier analysis, that is, the scientist finds how the data can be expressed as a sum of sinusoids. As with normal mode analysis, this gives a powerful and drastically different view of the data, one that is often very revealing. Fourier analysis is absolutely omnipresent in modern technology. The jpg image compression algorithm is based on Fourier analysis. The performance of fiber optic cables is evaluated using Fourier analysis. Diffraction methods used to determine the structure of proteins are based on Fourier analysis.[1]

As an example of how illuminating this method can be, consider the function $y(x)$ shown in the left part of figure 8.1.1. It appears quite irregular, and would be difficult to describe in any simple way. However, this is just the sum of the three sinusoids shown in the middle part of the figure. The right part shows a graph of the amplitude of the sinusoids (in other words the factor A in $A \cos(kx + \varphi)$) as a function of wavenumber $k = \dfrac{2\pi}{\lambda}$), and a graph of the phase φ of the sinusoids. This is just as complete a description as the graph of $y(x)$, yet it is much more revealing. In this chapter, we will

---

1. Joseph Fourier (1768–1830) lived a varied and tempestuous life. He was a strong supporter of the French revolution, but later became aghast at the excesses of the Terror, and tried to withdraw from the committee. This almost led to his beheading. He was sent to Egypt by Napoleon, along with 164 other scholars, to "civilize" the country. Fourier spent three years there cataloging antiquities and other discoveries. This exposure to warm climates may be responsible for his habit of keeping his rooms uncomfortably warm, while wearing a heavy coat. He made important contributions to the study of heat propagation, and it was in this connection that he developed the idea of summing sinusoids to represent other functions. However, this notion met with a great deal of resistance from the leading French mathematicians of the time, including Laplace, Legendre, and Poisson.

246

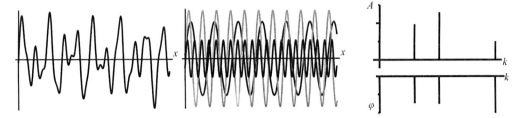

Figure 8.1.1 A seemingly complicated $y(x)$ (on the left) is actually just the sum of the three sinusoids shown in the center. The amplitude and phase of each sinusoid $A\cos(kx + \varphi)$ are shown on the right. (The amplitude is defined to be positive. For the three sinusoids here, all happen to have a negative phase.)

describe how to go from the plot as a function of x to the plot as a function of k. For a function of time, we can use the same procedure to go from the plot as a function of t to the plot as a function of $\omega = \dfrac{2\pi}{T}$.

We begin by considering functions $y(x)$ that are periodic (but complicated within the period), and then go on in section 8.5 to consider how similar procedures can be applied to functions that are not periodic.

8.2 The Fourier Expansion

In chapter 7, we saw how any behavior of a continuous string can be expressed as a superposition of the normal modes. As an important special case, we explored how any initial shape of the string with $\dot{y}_0(x) = 0$ can be expressed as a superposition of normal modes with real amplitudes. Of course, the string goes to zero at $x = 0$ and at $x = L$, as do all the functions for the modes.

Let's look at this from a purely mathematical viewpoint. The string can take on *any* shape between $x = 0$ and $x = L$. Therefore, we could think of the string shape as a mathematical function $y(x)$ which is only defined from $x = 0$ to L. Thus, over this range of x, any function that goes to zero at $x = 0$ and L can be expressed as a superposition (*i.e.*, a weighted sum) of the sinusoids that go to zero at $x = 0$ and L. Equivalently, we could say that the set of functions $\sqrt{\dfrac{2}{L}}\sin k_n x$, where $k_n = n\dfrac{\pi}{L}$, forms a complete basis for describing functions that go to zero at $x = 0$ and $x = L$, but this description is only valid over the range $x = 0$ to $x = L$.

Since these basis functions are periodic, it is reasonable to ask whether we can extend this idea; instead of only describing functions in the range $x = 0$ to L, perhaps we can describe any *periodic* function over the entire range $x = -\infty$ to $x = \infty$. For example, let's try to describe the periodic function shown as a solid gray line in figure 8.2.1, $y(x) = A\left|\sin\left(\dfrac{\pi}{L}x\right)\right|$. From $x = 0$ to L this is perfectly described by a single one of our basis functions: $y(x) = \left(A\sqrt{\dfrac{L}{2}}\right)\left(\sqrt{\dfrac{2}{L}}\sin k_1 x\right)$, where $k_1 = \dfrac{\pi}{L}$.

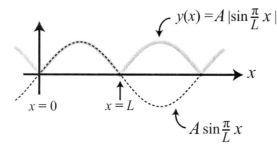

$$y(x) = A \left| \sin \frac{\pi}{L} x \right|$$

$x = 0$ $x = L$

$A \sin \frac{\pi}{L} x$

Figure 8.2.1 Solid gray line: A periodic function $y(x)$. Dashed line: An effort to represent $y(x)$ using a sinusoid works only over the range $x = 0$ to $x = L$.

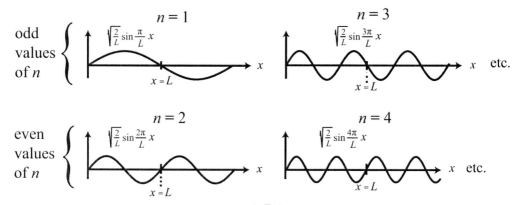

Figure 8.2.2 Functions of the form $\sin\left(n\frac{\pi}{L}x\right)$ with odd n (top graphs) don't have periodicity L, while those with even n (bottom graphs) do.

However, outside the range 0 to L, this description doesn't work at all, as shown in the figure. The problem is obvious: the basis function $\sqrt{\frac{2}{L}} \sin k_1 x$ doesn't match the periodicity of $y(x)$. In fact, of the set of basis functions $\sqrt{\frac{2}{L}} \sin k_n x$, all those with odd n don't have periodicity L, as shown in the top graphs of figure 8.2.2. However, all those with even n *do* have periodicity L, as shown in the bottom graphs.

So, it is tempting simply to discard all the basis functions with odd n. That would leave the functions $\sqrt{\frac{2}{L}} \sin\left(n\frac{\pi}{L}x\right)$, with even n. We could write this set more simply by writing them as $\sqrt{\frac{2}{L}} \sin\left(n\frac{2\pi}{L}x\right)$, where now n can take on *any* value from 1 to ∞. We can simplify the notation even further by redefining the wavenumbers for Fourier analysis:

$$k_n \equiv n\frac{2\pi}{\lambda} \text{ for Fourier analysis,} \tag{8.2.1}$$

where λ is the periodicity of the function being analyzed. (In the example above, $\lambda = L$.) Each of the wavenumbers in equation (8.2.1) corresponds to fitting an integer number n of wavelengths between 0 and λ. Compare this to our expression from normal mode analysis, $k_n = n\dfrac{\pi}{L}$, which corresponds to fitting a half-integer number of wavelengths between 0 and L. Using the Fourier analysis definition of k_n, we can write the remaining basis functions (the ones we didn't throw out) as $\sin k_n x$. (Note that we have omitted the $\sqrt{\dfrac{2}{\lambda}}$ prefactor; this is the well-established convention for Fourier analysis. Because we are not including this in the basis functions, they are no longer normalized; we'll see soon that it is easy to compensate for this.)

However, since we started with a basis complete enough to describe functions that go to zero at $x = 0$ and L, over the range $x = 0-L$, it's clear that the basis will no longer be complete once we discard half its members. To restore completeness, we need to add back in an equal number of basis functions, but this time we'll make sure they have periodicity λ. What additional functions could we add to the basis? Hmm... if only there were some function other than $\sin\left(n\dfrac{2\pi}{\lambda}x\right) = \sin k_n x$ that had periodicity λ. If only... Aha! $\cos k_n x$! In fact, it can be shown (though we will not do so here) that the set of functions $\sin k_n x$ combined with the set of functions $\cos k_n x$ forms a complete basis for describing functions $y(x)$ with periodicity λ, so long as $y(x)$ has an average value of zero. To describe functions with nonzero average value, we must add one more basis function to our set: a constant.

So, it is reasonable to expect that any function $y(x)$ with periodicity λ can be expressed as a weighted superposition of the functions $\sin k_n x$, $\cos k_n x$, and a constant. This statement expressed mathematically reads

$$y(x) = \frac{a_o}{2} + \sum_{n=1}^{\infty}\left[a_n \cos k_n x + b_n \sin k_n x\right]. \tag{8.2.2}$$

Fourier series expansion

In the above, the a_n's and b_n's are the coefficients in the expansion; in section 8.4, we will find how to determine their values, and show that they are real.

Self-test (answer below[2]): What is the basis function that is the constant part of the Fourier expansion?

We have not rigorously shown that the set $\sin k_n x$, $\cos k_n x$, and a constant forms a complete basis for all functions with periodicity λ. However, this is plausible based on our experience with the normal mode expansion (the completeness of which we showed with full rigor), and can be shown rigorously.

---

2. Each coefficient a_n or b_n multiplies the corresponding basis function. We see that a_0 multiplies $y(x) = \frac{1}{2}$, so the constant basis function in the Fourier expansion is $\frac{1}{2}$.

8.3   Expansions using nonnormalized orthogonal basis functions

As mentioned earlier, by a well-established convention, the normalization factor $\sqrt{\frac{2}{\lambda}}$ is not included in the basis functions for Fourier analysis; for example, we use $\sin k_n x$ as basis functions, rather than $\sqrt{\frac{2}{\lambda}} \sin k_n x$. It's pretty obvious how to compensate in this particular case: we simply absorb the $\sqrt{\frac{2}{\lambda}}$ into the coefficients a_n and b_n. However, compensating for nonnormalized basis functions in the more general case is easy.

Say we have a set of functions $y_n(x)$ that form a complete orthogonal basis. The "complete" part of this means that we can write

$$y(x) = \sum_n C_n y_n(x), \tag{8.3.1}$$

where $y(x)$ is any function that can be described by this basis, and the C_n's are the expansion coefficients (which may be complex). The "orthogonal" part means $\langle y_m(x) \mid y_n(x) \rangle = 0$, if $m \neq n$.

If the basis functions $y_n(x)$ are normalized, that is, $\langle y_n(x) \mid y_n(x) \rangle = 1$, then it's easy to find the C_n's:

$$\langle y_m(x) \mid y(x) \rangle = \langle y_m(x) \mid \left(\sum_n C_n \mid y_n(x) \rangle \right) = \sum_n C_n \langle y_m(x) \mid y_n(x) \rangle = C_m,$$

where in the last step we used $\langle y_m(x) \mid y_n(x) \rangle = \delta_{mn}$. Therefore,

$$C_m = \langle y_m(x) \mid y(x) \rangle.$$

At this point, we can replace the index m by n, giving

$$C_n = \langle y_n(x) \mid y(x) \rangle. \tag{8.3.2}$$

(Note that this is just what we got before for the case of the normal mode expansion with $\dot{y}_0(x) = 0$.)

However, what if the basis functions aren't normalized? In other words, what if

$$\langle y_n(x) \mid y_n(x) \rangle = F_n, \quad \text{where } F_n \neq 1?$$

Claim: The generalized version of equation (8.3.2) is

$$\boxed{C_n = \frac{\langle y_n(x) \mid y(x) \rangle}{\langle y_n(x) \mid y_n(x) \rangle}.} \tag{8.3.3}$$

Obviously, this reduces to equation (8.3.2) when $y_n(x)$ *is* normalized.
Proof:

$$\langle y_m(x) \mid y(x) \rangle = \langle y_m(x) \mid \left(\sum_n C_n \mid y_n(x) \rangle \right) = \sum_n C_n \langle y_m(x) \mid y_n(x) \rangle$$

$$= C_m \langle y_m(x) \mid y_m(x) \rangle,$$

where in the last step we used orthogonality, that is, $\langle y_m(x) \mid y_n(x) \rangle = 0$ if $m \neq n$.

$$\Rightarrow C_m = \frac{\langle y_m(x) \mid y(x) \rangle}{\langle y_m(x) \mid y_m(x) \rangle}, \quad \text{Q.E.D.}$$

Here's one way to see intuitively why equation (8.3.3) works. If $y_n(x)$ is not normalized, then it has a "length" in Hilbert space that is not equal to 1. One factor of this length appears in the normal mode expansion (8.3.1), and another factor appears in $\langle y_n(x) \mid y(x) \rangle$, so we must divide by (length)^2, that is, by $\langle y_n(x) \mid y_n(x) \rangle$, to compensate.

8.4 Finding the coefficients in the Fourier series expansion

Here, again is the Fourier series expansion from section 8.2:

$$(8.2.2): y(x) = \frac{a_0}{2} + \sum_{n=1}^{\infty} \left[a_n \cos k_n x + b_n \sin k_n x \right].$$

As we can see, the basis functions are $1/2$, the set of $\cos k_n x$, and the set of $\sin k_n x$. To find the expansion coefficients a_n and b_n, we use equation (8.3.3):

$$a_0 = \frac{\langle \frac{1}{2} \mid y(x) \rangle}{\langle \frac{1}{2} \mid \frac{1}{2} \rangle}, \quad \langle \tfrac{1}{2} \mid \tfrac{1}{2} \rangle = \int_0^\lambda \left(\tfrac{1}{2} \right)^2 dx = \frac{\lambda}{4} \quad \Rightarrow a_0 = \frac{4}{\lambda} \langle \tfrac{1}{2} \mid y(x) \rangle = \frac{2}{\lambda} \langle 1 \mid y(x) \rangle$$

$$a_n = \frac{\langle \cos k_n x \mid y(x) \rangle}{\langle \cos k_n x \mid \cos k_n x \rangle}, \quad \langle \cos k_n x \mid \cos k_n x \rangle = \int_0^\lambda \cos^2 k_n x \, dx = \frac{\lambda}{2},$$

where in the last step we made use of the fact that the average value of $\cos^2$ (or $\sin^2$) over one wavelength is ½. So,

$$a_n = \frac{2}{\lambda} \langle \cos k_n x \mid y(x) \rangle.$$

Since $k_n = n\frac{2\pi}{\lambda}$, we have $k_0 = 0$, so that $\cos k_0 x = 1$. Therefore, the above works for a_0 as well as the other a_n's. Plugging in the definition of the inner product from section 7.7, we get

$$a_n = \frac{2}{\lambda} \int_0^\lambda \cos k_n x \, y(x) \, dx.$$

We can find the coefficients of the sines in the Fourier expansion in the same way:

$$b_n = \frac{\langle \sin k_n x \mid y(x) \rangle}{\langle \sin k_n x \mid \sin k_n x \rangle}, \quad \langle \sin k_n x \mid \sin k_n x \rangle = \int_0^\lambda \sin^2 k_n x \, dx = \frac{\lambda}{2},$$

$$\Rightarrow b_n = \frac{2}{\lambda} \langle \sin k_n x \mid y(x) \rangle = \frac{2}{\lambda} \int_0^\lambda \sin k_n x \, y(x) \, dx.$$

Let's collect these results:

$$y(x) = \frac{a_0}{2} + \sum_{n=1}^{\infty} \left[a_n \cos k_n x + b_n \sin k_n x \right].$$

$$a_n = \frac{2}{\lambda} \int_0^{\lambda} \cos k_n x \, y(x) \, dx.$$

$$b_n = \frac{2}{\lambda} \int_0^{\lambda} \sin k_n x \, y(x) \, dx.$$

$$k_n = n \frac{2\pi}{\lambda}.$$

(8.4.1)

Fourier analysis for a function of x that has periodicity λ

Example: What is the Fourier series expansion for the square wave, shown as the left graph in figure 8.4.1?

Solution: We can represent the square wave $y(x)$ using the Fourier series expansion shown in the top line of equation (8.4.1). Let's begin by calculating the a_n's:

$$a_n = \frac{2}{\lambda} \int_0^{\lambda} \cos k_n x \, y(x) \, dx = \frac{2}{\lambda} \left[\int_0^{\lambda/2} \cos k_n x \ (1) \ dx + \int_{\lambda/2}^{\lambda} \cos k_n x \ (-1) \ dx \right]$$

$$= \frac{2}{\lambda} \left[\frac{1}{k_n} \sin k_n x \Big|_0^{\lambda/2} - \frac{1}{k_n} \sin k_n x \Big|_{\lambda/2}^{\lambda} \right]$$

$$= \frac{2}{\lambda k_n} \left[\sin \left(n \frac{2\pi}{\lambda} \frac{\lambda}{2} \right) - \sin 0 - \sin \left(n \frac{2\pi}{\lambda} \lambda \right) + \sin \left(n \frac{2\pi}{\lambda} \frac{\lambda}{2} \right) \right] = 0.$$

Thus, *all* of the cosine coefficients in the Fourier expansion are zero. We could have anticipated this based on symmetry: the square wave function is antisymmetric about the point $x = \lambda/2$, whereas the cosine functions are all symmetrical about this point.

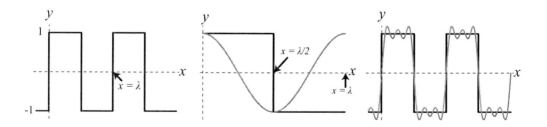

Figure 8.4.1 Left: Square wave. Middle: The square wave is antisymmetrical about $x = \lambda/2$, whereas the cosine is symmetrical. Right: Gray trace shows the sum of the first three non-zero terms in the Fourier series expansion of the square wave.

This is shown for the case $n = 1$ in the middle part of the figure. However, the sine functions do have the required symmetry, so we can anticipate that the b_n's will be nonzero:

$$b_n = \frac{2}{\lambda} \int_0^\lambda \sin k_n x \, y(x) \, dx = \frac{2}{\lambda} \left[\int_0^{\lambda/2} \sin k_n x \, (1) \, dx + \int_{\lambda/2}^\lambda \sin k_n x \, (-1) \, dx \right]$$

$$= \frac{2}{\lambda} \left[-\frac{1}{k_n} \cos k_n x \Big|_0^{\lambda/2} + \frac{1}{k_n} \cos k_n x \Big|_{\lambda/2}^\lambda \right]$$

$$= \frac{2}{\lambda k_n} \left[-\cos \left(n \frac{2\pi}{\lambda} \frac{\lambda}{2} \right) + \cos 0 + \cos \left(n \frac{2\pi}{\lambda} \lambda \right) - \cos \left(n \frac{2\pi}{\lambda} \frac{\lambda}{2} \right) \right]$$

$$= \frac{2}{\lambda n \frac{2\pi}{\lambda}} \left[-\cos n\pi + 1 + \cos n2\pi - \cos n\pi \right] = \frac{2}{n\pi} \left[1 - \cos n\pi \right].$$

If n is odd, then $b_n = \dfrac{4}{n\pi}$, while if n is even then $b_n = 0$. Plugging these results into the top line of equation (8.4.1), we obtain

$$y(x) = \frac{4}{\pi} \left(\sin \frac{2\pi}{\lambda} + \frac{1}{3} \sin \frac{3 \cdot 2\pi}{\lambda} + \cdots \right) \tag{8.4.2}$$

Fourier expansion for a square wave.

As you can see, each succeeding term is smaller (because of the factor $1/n$). The right graph in figure 8.4.1 shows the sum of the first three terms. You can perhaps see how the series begins to approximate the square wave. The approximation becomes better as more terms are added.

Self-test (answer below[3]): Use symmetry arguments to explain why all the sines with even n have zero coefficients.

You can show in problem 8.2 that an alternate version to equation (8.4.1) for the Fourier series expansion is

$$y(x) = \frac{a_0}{2} + \sum_{n=1}^\infty A_n \cos \left(k_n x + \varphi_n \right), \tag{8.4.3}$$

where $A_n = \sqrt{a_n^2 + b_n^2}$ and $\varphi_n = \tan^{-1} \left(-\dfrac{b_n}{a_n} \right)$. This is the version used for figure (8.1.1).

We can also use Fourier analysis for a function of t. In fact, this is much more common than using it for functions of x. Everything works in exactly the same way. The periodicity in x is replaced by the periodicity in t: $\lambda \to T$. The wave number is

---

3. Although these terms are antisymmetrical about $x = \lambda/2$, they are also antisymmetrical about $x = \lambda/4$, whereas the square wave is symmetrical about this point.

replaced by the angular frequency: $k_n = n\frac{2\pi}{\lambda} \rightarrow \omega_n = n\frac{2\pi}{T}$. With these substitutions, equation (8.4.1) becomes

$$y(t) = \frac{a_0}{2} + \sum_{n=1}^{\infty} \left[a_n \cos \omega_n t + b_n \sin \omega_n t \right].$$

$$a_n = \frac{2}{T} \int_0^{\lambda} \cos \omega_n t \; y(t) \, dt.$$

$$b_n = \frac{2}{T} \int_0^{\lambda} \sin \omega_n t \; y(t) \, dt.$$

$$\omega_n = n\frac{2\pi}{T}.$$

(8.4.4)

Fourier analysis for a function of t that has periodicity T

The version of equation (8.4.3) for functions of time is

$$y(t) = \frac{a_0}{2} + \sum_{n=1}^{\infty} A_n \cos \left(\omega_n t + \varphi_n \right). \tag{8.4.5}$$

8.5   Fourier Transforms and the meaning of negative frequency

Complex exponential version of Fourier analysis

In problem 8.3, you will show that the Fourier expansion can be written in terms of complex exponentials, rather than sines and cosines. You will show that we can write

$$y(x) = \sum_{n=-\infty}^{\infty} C_n e^{ik_n x}, \text{ where } C_n = \frac{1}{\lambda} \int_0^{\lambda} y(x) e^{-ik_n x} dx. \tag{8.5.1}$$

You will also show that, although the C_n's are complex, there is always cancellation of the imaginary part of the term $n = m$ with the term $n = -m$, so that there is no need to take the real part of the summation to get $y(x)$. One can show that the C_n's in this expansion are related to the a_n's and b_n's in the sine/cosine expansion:

$$C_n = \frac{1}{2} \left(a_n - ib_n \right) \text{ and } C_{-n} = \frac{1}{2} \left(a_n + ib_n \right). \tag{8.5.2}$$

This means for example, that for a case where $C_{-n} = C_n$, we have $b_n = 0$, that is, a pure cosine term, while if $C_{-n} = -C_n$, then we have $a_n = 0$, that is, a pure sine term. From equation (8.5.2), we can see that we must always have $\left| C_{-n} \right| = \left| C_n \right|$.

Fourier transforms

If we take the limit $\lambda \to \infty$, then the function $y(x)$ is no longer truly periodic. However, we can still express it as an infinite sum of sinusoids. Since $k_n = n\dfrac{2\pi}{\lambda}$, as we increase λ the k_n's get closer together. In the limit $\lambda \to \infty$, the spacing between the k_n's becomes infinitesimal. Now, instead of associating each C_n with a k_n, it is more helpful to think of a continuous function $C(k)$. By convention, this is instead called $Y(k)$. In problem 8.4, you can show that in the limit $\lambda \to \infty$, equation (8.5.1) becomes

$$y(x) = \frac{1}{\sqrt{2\pi}} \int_{-\infty}^{\infty} Y(k)e^{ikx}dk, \quad \text{where } Y(k) = \frac{1}{\sqrt{2\pi}} \int_{-\infty}^{\infty} y(x)e^{-ikx}dx \qquad (8.5.3)$$

Fourier Transforms for a function of x

$Y(k)$, which represents the Fourier amplitude as a function of k, is called the "Fourier Transform" of $y(x)$. At first, this might appear intimidating, but remember that an integral is just an infinite sum, so that we are still expressing $y(x)$ as a sum of the sinusoids e^{ikx}, each weighted by the factor $Y(k)$.

It is easy to adapt this for functions of time: $\omega = \dfrac{2\pi}{T}$ plays the role of $k = \dfrac{2\pi}{\lambda}$, so that we can express any function of time $y(t)$ as a weighted sum of the sinusoids $e^{i\omega t}$:

$$y(t) = \frac{1}{\sqrt{2\pi}} \int_{-\infty}^{\infty} Y(\omega)e^{i\omega t}d\omega, \quad \text{where } Y(\omega) = \frac{1}{\sqrt{2\pi}} \int_{-\infty}^{\infty} y(t)e^{-i\omega t}dt. \qquad (8.5.4)$$

Fourier Transforms for a function of t

Example: Fourier transform of a Gaussian. The Gaussian function is quite important in several different contexts. It is defined as

$$y(t) = Ae^{-(t-t_0)^2/2\sigma^2}, \qquad (8.5.5)$$

a bell-shaped curved centered on t_0, as shown in figure 8.5.1a. For simplicity, let's examine the case $t_0 = 0$, so that

$$y(t) = Ae^{-t^2/2\sigma^2}. \qquad (8.5.6)$$

Your turn: Verify that $y(t)$ falls to $A/2$ at $t = \pm\sigma\sqrt{-2\ln\frac{1}{2}} = \pm1.18\,\sigma$, so that the Full Width at Half Maximum (FWHM) is $2.35\,\sigma$.

Using equation (8.5.4), we see that the Fourier transform of equation (8.5.6) is

$$Y(\omega) = \frac{1}{\sqrt{2\pi}} \int_{-\infty}^{\infty} Ae^{-t^2/2\sigma^2}e^{-i\omega t}dt. \qquad (8.5.7)$$

continued

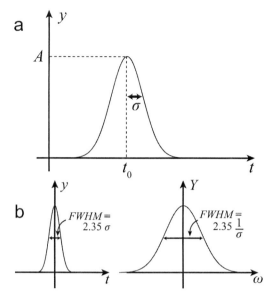

Figure 8.5.1 a: The Gaussian. b: A Gaussian centered on $t = 0$ and its Fourier transform. The width (in time) of the $y(t)$ Gaussian is inversely proportional to the width (in ω) of the $Y(\omega)$ Gaussian. Therefore a broad range of frequencies is needed to synthesize a narrow pulse.

This is a standard definite integral, which you can look up in a table (or evaluate with a symbolic calculus program):

$$\int_{-\infty}^{\infty} e^{-(ax^2+bx+c)}\,dx = \sqrt{\frac{\pi}{a}}\,e^{(b^2-4ac)/4a}. \tag{8.5.8}$$

In our case, $x \to t$, $a \to \dfrac{1}{2\sigma^2}$, $b \to i\omega$, and $c \to 0$, so

$$Y(\omega) = \frac{A}{\sqrt{2\pi}}\sqrt{2\pi\sigma^2}\,e^{-\omega^2/(2/\sigma^2)} = A\sigma e^{-\omega^2/(2/\sigma^2)}.$$

We see that this has the same form as equation (8.5.6), that is, that the Fourier transform of a Gaussian function of time $y(t)$ is a Gaussian function of angular frequency, $Y(\omega)$. The FWHM of $y(t)$ is 2.35 σ, so the FWHM of $Y(\omega)$ is $2.35\dfrac{1}{\sigma}$, as shown in figure 8.5.1b. Thus, we arrive at a quite important conclusion: to Fourier synthesize a very narrow Gaussian (one with a very small FWHM), we need a very broad range of frequencies (*i.e.*, a Gaussian $Y(\omega)$ with a very large FWHM).[4] This conclusion holds in general: to Fourier synthesize a function $y(t)$ with variations on time scales as short as Δt, we must use Fourier components covering a range of angular frequencies that is of order $\dfrac{1}{\Delta t}$.

---

4. As you can show in problem 8.18, this leads directly to the Heisenberg uncertainty principle.

Meaning of the negative frequencies in the Fourier transform

The integral in equation (8.5.4) which represents the Fourier synthesis of $y(t)$, $y(t) = \frac{1}{\sqrt{2\pi}} \int_{-\infty}^{\infty} Y(\omega)e^{i\omega t}d\omega$, includes both positive and negative angular frequencies. Let's consider a particular positive angular frequency ω_0. We can see that for each such a frequency in the integral, there is a corresponding negative frequency $-\omega_0$. Since the integral is an infinite sum, the contributions at these frequencies are added together as part of the integral: $Y(\omega_0)e^{i\omega_0 t} + Y(-\omega_0)e^{-i\omega_0 t}$. This sum must be real, since $y(t)$ is real, and this imposes a restriction on the relationship between $Y(\omega_0)$ and $Y(-\omega_0)$, as we explore below.

The function $e^{i\omega_0 t}$ represents a vector of length 1 rotating counterclockwise in the complex plane, while for $e^{-i\omega_0 t}$ the vector rotates clockwise, as shown in the top part of figure 8.5.2. If we add them together, then the imaginary part of $e^{-i\omega_0 t}$ cancels the imaginary part of $e^{i\omega_0 t}$, leaving only a real function: $e^{i\omega_0 t} + e^{-i\omega_0 t} = 2\cos\omega_0 t$. Of course, we could multiply both by any real number, and still get cancellation of the imaginary parts: $Ae^{i\omega_0 t} + Ae^{-i\omega_0 t} = 2A\cos\omega_0 t$. Finally, we can add a positive phase shift to the $e^{i\omega_0 t}$ and a negative phase shift of equal magnitude to the $e^{-i\omega_0 t}$, and still get cancellation of the imaginary parts:

$$Ae^{i(\omega_0 t+\varphi)} + Ae^{-i(\omega_0 t+\varphi)} = 2A\cos(\omega_0 t + \varphi), \tag{8.5.9}$$

as shown in the bottom part of figure 8.5.2. This is the only way to get such cancellation.

Applying these ideas to the sum $Y(\omega_0)e^{i\omega_0 t} + Y(-\omega_0)e^{-i\omega_0 t}$ which is part of the Fourier integral, we see that if we express $Y(\omega_0)$ in the form $Y(\omega_0) = Ae^{i\varphi}$, then we

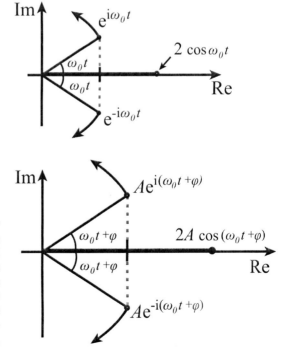

Figure 8.5.2 Top: The imaginary parts of two counter-rotating complex plane vectors can cancel only if the angular frequencies are equal and the magnitudes are equal. Bottom: The phase factors φ must be of equal magnitude and opposite sign for the imaginary parts to cancel.

must have $Y(-\omega_0) = Ae^{-i\varphi}$. Thus, using equation (8.5.9),

$$Y(\omega_0)\, e^{i\omega_0 t} + Y(-\omega_0)\, e^{-i\omega_0 t} = 2A \cos\left(\omega_0 t + \varphi\right). \qquad (8.5.10)$$

To summarize:

> $Y(\omega_0)$ is always equal in magnitude to $Y(-\omega_0)$. The complex phase difference between them determines the phase of the real oscillation $\cos\left(\omega_0 t + \varphi\right)$.

8.6   The Discrete Fourier Transform (DFT)

In most applications of Fourier analysis, one does not actually deal with continuous functions, but rather with discrete samples of a continuous function taken at regular intervals. The most common applications are for functions of time. For example, perhaps one wishes to perform Fourier analysis on an audio waveform. (Some of the reasons for doing so are detailed in section 8.7.) In this case, a microphone converts the sound waves into a time-varying voltage. The voltage is then sampled at regular time intervals Δ, and these samples are recorded on a computer; this process is shown schematically in figure 8.6.1. (We use the symbol Δ, rather than for example Δt, just for simplicity.) As long as Δ is smaller than the timescale on which the continuous function changes significantly, then the resulting discretely sampled waveform is a good approximation of the original.[5]

The samples occur at the times

$$t_j = j\Delta, \qquad (8.6.1)$$

where

$$j = 0,\ 1,\ \ldots,\ N-1, \qquad (8.6.2)$$

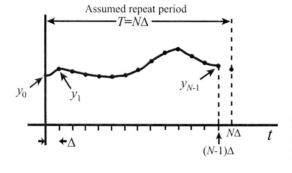

Figure 8.6.1 Sampling a continuous waveform, in preparation for the DFT.

---

5. For example, if we are sampling a sinusoidal waveform $A \cos \omega t = A \cos \dfrac{2\pi t}{T_w}$, where T_w is the period of the waveform, then we must have $\Delta \ll T_w$ in order for the graph of the sampled signal *versus* time to look approximately the same as the graph of the original signal *versus* time.

and N is the number of samples. The result is a set of N data points: $y(t_0)$, $y(t_1), \ldots, y(t_{N-1})$. (In the example of the audio waveform, y would be the voltage.) How can we represent this dataset in angular frequency space (the equivalent of k-space for functions of t), using ideas of Fourier analysis? In other words, how can we find the "Fourier spectrum" of the dataset?

The algorithm used in such a situation is called the Discrete Fourier Transform (DFT). It assumes that the function has a periodicity $T = N\Delta$. However, most often the continuous function that is being sampled (such as an audio waveform) does not actually have this periodicity, since N and Δ are ordinarily determined by the hardware and software used for data acquisition, rather than being determined by the system being examined.[6] This assumption that the DFT algorithm makes about the periodicity can cause serious inaccuracies in the Fourier transform that it calculates, particularly because the data point at the end of the set is usually at a different y-value than the data point at the beginning. Thus, the assumed repeating function has a sharp step at the beginning of each period. This would lead to spurious high-frequency peaks in the Fourier spectrum. To avoid this, the input function is usually multiplied by a "windowing function," as shown in figure 8.6.2, forcing the start and end points to zero. Although this windowing procedure does introduce some distortions into the Fourier spectrum, the benefits of eliminating the sharp step at the beginning of the period outweigh these drawbacks. You can explore the windowing operation more in problem 8.16.

Because the algorithm assumes a periodicity of $T = N\Delta$ (equal to the measurement time), the lowest angular frequency in the Fourier spectrum is

$$\omega_1 = \frac{2\pi}{N\Delta}, \tag{8.6.2}$$

and the higher angular frequencies are integer multiples of this:

$$\boxed{\omega_n = n\omega_1 = n\frac{2\pi}{N\Delta}, \text{ where } T = N\Delta \text{ is the total measurement time}} \tag{8.6.3}$$

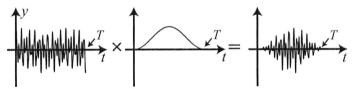

Figure 8.6.2 Because the DFT assumes the function has periodicity T, problems occur if (as is usually the case) the beginning and ending values of the sampled waveform (left) are not equal; this causes a sharp step in the assumed waveform. To avoid this problem, the sampled input waveform (left) is multiplied by a windowing function (center) to produce a waveform (right) that goes to zero at the beginning and end of the time interval T. This example shows the "Hanning window", one of the most common windowing functions; it is simply an offset cosine.

---

6. For example, if one is recording a vocalist singing a note at 440.3 Hz, corresponding to a period of 2.271 ms, it is unlikely that the equipment used to make the recording will record for a time corresponding to an exact multiple of 2.271 ms.

The resolution along the angular frequency axis of the Fourier spectrum equals the spacing between the ω_n's, so that:

> The frequency resolution of the DFT is inversely proportional to the measurement time T.

Recall in our discussion of the beaded string that there was a "maximum wiggliness" that could be represented by the system, in which the beads form an up–down–up–down pattern. In exactly the same way, there is a maximum wiggliness that can be represented by the set of N data points, corresponding to a minimum period of 2Δ, since we need at least two data points per period to represent the up–down pattern. Therefore, the maximum angular frequency in the Fourier spectrum is

$$\boxed{\omega_{max} = \frac{2\pi}{2\Delta}.}$$ (8.6.4)

By comparison with equation (8.6.3), we see that this corresponds to a maximum value of n:

$$n_{max} = \frac{N}{2}.$$

(We assume that N is even; this is almost always the case for reasons of computational efficiency, as we'll discuss briefly near the end of the section.)

Following the top line of equation (8.4.4): $y(t) = \frac{a_o}{2} + \sum_{n=1}^{\infty} [a_n \cos \omega_n t + b_n \sin \omega_n t]$, we will represent $y(t)$ as a sum of a constant term, a series of cosines, and a series of sines. However, for the reasons discussed earlier, the sum only extends to $n = \frac{N}{2}$, instead of to $n \to \infty$. Furthermore, instead of using $\frac{a_0}{2}$, $\cos \omega_n t$, and $\sin \omega_n t$ as the basis functions, it is conventional for DFT to divide them by N. Recall that, since time is measured in the discrete increments Δ, we write $y(t_j)$, where $t_j = j\Delta$ and $j = 0, 1, \ldots, N - 1$. Putting these ideas together, we write the Fourier expansion as

$$y(t_j) = \frac{1}{N} \left\{ \frac{a_o}{2} + \sum_{n=1}^{N/2} [a_n \cos \omega_n t_j + b_n \sin \omega_n t_j] \right\}.$$

However, when $n = N/2$, we have $\sin \omega_n t_j = \sin \left(n \frac{2\pi}{2\Delta} j\Delta \right) = \sin (nj\pi) = 0$, since n and j are integers. (This is the same idea as the $n = N + 1$ "mode" of the beaded string, in which all the beads are at nodes of the standing wave.) Therefore, we should omit the sin term for $n = N/2$, leaving us with

$$y(t_j) = \frac{1}{N} \left\{ \frac{a_0}{2} + \sum_{n=1}^{N/2-1} [a_n \cos \omega_n t_j + b_n \sin \omega_n t_j] + a_{N/2} \cos \omega_{N/2} t_j \right\}.$$ (8.6.5)

In the above, we write $\frac{N}{2} - 1$ as $N/2 - 1$ to save space. Note that there are N independent coefficients in the above equation: a_0, $a_{N/2}$, and a total of $2 \cdot \left(\frac{N}{2} - 1 \right) = N - 2$ a_n's and b_n's in the summation. Given that we started with N independent values of y, it is logical that we should have N independent coefficients in the Fourier expansion.

It is customary to write the DFT using complex exponentials instead of sines and cosines. The expansions for cosine and sine in terms of complex exponentials are:

$$\cos \omega_n t = \frac{e^{i\omega_n t} + e^{-i\omega_n t}}{2} \quad \text{and} \quad \sin \omega_n t = \frac{e^{i\omega_n t} - e^{-i\omega_n t}}{2i} = -i\frac{e^{i\omega_n t} - e^{-i\omega_n t}}{2}.$$

Therefore, we can write equation (8.6.5) as

$$y(t_j) = \frac{1}{N}\left\{\frac{a_0}{2} + \sum_{n=1}^{N/2-1}\left[a_n\frac{e^{i\omega_n t_j} + e^{-i\omega_n t_j}}{2} - ib_n\frac{e^{i\omega_n t_j} - e^{-i\omega_n t_j}}{2}\right] + a_{N/2}\cos\omega_{N/2}t_j\right\}.$$

(Because the $\cos\omega_{N/2}\,t$ term has no accompanying sin term, we choose not to express it as a complex exponential.)

Your turn: Show that, by defining

$$C_n \equiv \tfrac{1}{2}(a_n - ib_n) \text{ and } C_{-n} \equiv \tfrac{1}{2}(a_n + ib_n),\qquad (8.6.6)$$

we can rewrite the above equation as

$$y(t_j) = \frac{1}{N}\left\{\frac{a_0}{2} + \sum_{n=1}^{N/2-1}\left[C_n e^{i\omega_n t_j} + C_{-n}e^{-i\omega_n t_j}\right] + a_{N/2}\cos\omega_{N/2}t_j\right\}.$$

Because of equation (8.6.3): $\omega_n = n\dfrac{2\pi}{N\Delta}$, we have that $-\omega_n = \omega_{-n}$. Using this, we can compress the above equation to

$$y(t_j) = \frac{1}{N}\left\{\sum_{n=-(N/2-1)}^{N/2-1} C_n e^{i\omega_n t_j} + a_{N/2}\cos\omega_{N/2}t_j\right\}, \qquad (8.6.7)$$

where we have defined $b_0 \equiv 0$, so that the entry in the sum for $n = 0$ is $C_0 e^{i\omega_0 t_j} = \dfrac{a_0}{2}e^0 = \dfrac{a_0}{2}$.

For all values of n other than $N/2$, the basis functions in this expansion are $\left|y_n(t_j)\right\rangle \equiv \dfrac{e^{i\omega_n t_j}}{N}$. To find the coefficients C_n, we can use the version of equation (8.3.3): $C_n = \dfrac{\langle y_n(x)\,|\,y(x)\rangle}{\langle y_n(x)\,|\,y_n(x)\rangle}$ that would be appropriate for a function of time, that is,

$$C_n = \frac{\langle y_n(t_j)\,|\,y(t_j)\rangle}{\langle y_n(t_j)\,|\,y_n(t_j)\rangle}.$$

Because we have a discrete set of data $y(t_j)$, the inner products are taken in the same way as we would for a beaded string. For example,

$$\langle y_n(t_j)\,|\,y_n(t_j)\rangle = \sum_{j=0}^{N-1}\left(\frac{e^{i\omega_n t_j}}{N}\right)^*\left(\frac{e^{i\omega_n t_j}}{N}\right) = \sum_{j=0}^{N-1}\left(\frac{e^{-i\omega_n t_j}}{N}\right)\left(\frac{e^{i\omega_n t_j}}{N}\right) = \sum_{j=0}^{N-1}\frac{1}{N^2} = \frac{1}{N},$$

where the last step follows because there are N identical terms in the sum. Therefore, the coefficients C_n are given by

$$C_n = N\langle y_n(t_j) \mid y(t_j)\rangle = N \sum_{j=0}^{N-1} \frac{e^{-i\omega_n t_j}}{N} y(t_j)$$

$$\Rightarrow C_n = \sum_{j=0}^{N-1} e^{-i\omega_n t_j} y(t_j). \tag{8.6.8}$$

Similarly, to find $a_{N/2}$, we use

$$a_{N/2} = \frac{\langle y_{N/2}(t_j) \mid y(t_j)\rangle}{\langle y_{N/2}(t_j) \mid y_{N/2}(t_j)\rangle}, \tag{8.6.9}$$

where $|y_{N/2}(t_j)\rangle = \dfrac{\cos \omega_{N/2} t_j}{N}$. The denominator is

$$\langle y_{N/2}(t_j) \mid y_{N/2}(t_j)\rangle = \sum_{j=0}^{N-1} \frac{\cos^2 \omega_{N/2} t_j}{N^2}.$$

Recall from equations (8.6.3) and (8.6.1) that $\omega_n = n\dfrac{2\pi}{N\Delta}$ and $t_j = j\Delta$, so that $\omega_{N/2} t_j = \pi j$. Plugging this in gives

$$\langle y_{N/2}(t_j) \mid y_{N/2}(t_j)\rangle = \sum_{j=0}^{N-1} \frac{\cos^2 \pi j}{N^2} = \sum_{j=0}^{N-1} \frac{(\pm 1)^2}{N^2} = \frac{1}{N}.$$

Therefore, equation (8.6.9) becomes

$$a_{N/2} = N\langle y_{N/2}(t_j) \mid y(t_j)\rangle = N \sum_{j=0}^{N-1} \frac{\cos \omega_{N/2} t_j}{N} y(t_j) \Rightarrow$$

$$a_{N/2} = N \sum_{j=0}^{N-1} \frac{\cos \pi j}{N} y(t_j). \tag{8.6.10}$$

Since

$$e^{-i\omega_{N/2} t_j} = e^{-i\pi j} = \cos(-\pi j) + i \sin(-\pi j) = \cos \pi j,$$

we can write equation (8.6.10) as

$$a_{N/2} = \sum_{j=0}^{N-1} e^{-i\omega_{N/2} t_j} y(t_j).$$

Since this has the same format as equation (8.6.8), we can write

$$a_{N/2} \cos \omega_{N/2} t_j = C_{N/2} \cos \omega_{N/2} t_j = C_{N/2} \cos \pi j. \tag{8.6.11}$$

Since

$$e^{i\omega_{N/2} t_j} = e^{i\pi j} = \cos \pi j + i \sin \pi j = \cos \pi j,$$

we can write equation (8.6.11) as

$$a_{N/2} \cos \omega_{N/2} \, t_j = C_{N/2} e^{i\omega_{N/2} t_j}.$$

Now, we insert this into equation (8.6.7) to give

$$y(t_j) = \frac{1}{N} \left\{ \sum_{n=-(N/2-1)}^{N/2-1} C_n e^{i\omega_n t_j} + C_{N/2} e^{i\omega_{N/2} t_j} \right\}. \tag{8.6.12}$$

In the above, n ranges from $-\left(\dfrac{N}{2} - 1\right)$ to $\left(\dfrac{N}{2} - 1\right)$. However, it is conventional for DFT to change this range. Recall from equations (8.6.3) and (8.6.1) that $\omega_n = n\dfrac{2\pi}{N\Delta}$ and $t_j = j\Delta$, so $\omega_n t_j = \dfrac{nj}{N} 2\pi$. Therefore,

$$e^{i\omega_{n+N} t_j} = e^{i\frac{(n+N)j}{N} 2\pi} = e^{i\frac{Nj}{N} 2\pi} e^{i\frac{nj}{N} 2\pi} = e^{ij2\pi} e^{i\frac{nj}{N} 2\pi} = e^{i\frac{nj}{N} 2\pi} = e^{i\omega_n t_j}.$$

This means that we can add N to the index n for any of the terms of the sum in equation (8.6.12) without changing anything. We will do this for all the negative values of n, so that, for example,

$$n = -\left(\frac{N}{2} - 1\right) \xrightarrow{+N} \frac{N}{2} + 1 \quad \text{and} \quad n = -1 \xrightarrow{+N} N - 1. \tag{8.6.13}$$

Using this, equation (8.6.12) becomes

$$y(t_j) = \frac{1}{N} \left\{ \sum_{n=0}^{N/2-1} C_n e^{i\omega_n t_j} + \sum_{n=N/2+1}^{N-1} C_n e^{i\omega_n t_j} + C_{N/2} e^{i\omega_{N/2} t_j} \right\}$$

$$\boxed{\Rightarrow y(t_j) = \sum_{n=0}^{N-1} C_n e^{i\omega_n t_j},} \tag{8.6.14}$$

where equation (8.6.8):

$$\boxed{C_n = \sum_{j=0}^{N-1} y(t_j) \, e^{-i\omega_n t_j}.}$$

Equation (8.6.8) for the coefficients is called the Discrete Fourier Transform (DFT), while equation (8.6.14), which describes the Fourier synthesis of $y(t_j)$, is called the Inverse Discrete Fourier Transform (IDFT).

In equation (8.6.14), the term for $n = 0$ is the constant term, and the terms for $n = 1$ to $n = \dfrac{N}{2} - 1$ correspond to the positive frequencies ω_1 to $\omega_{N/2-1}$. However, because of the transformation (8.6.13), the terms from $n = \dfrac{N}{2} + 1$ to $n = N - 1$ correspond to the negative frequencies $\omega_{-(N/2-1)}$ to ω_{-1}. Finally, the term for $n = N/2$ is derived from a sum of positive and negative frequency terms. Although this is the conventional sequence for the DFT, it is not very intuitive, because the highest frequency components are those near $n = N/2$, while the components near $n = N - 1$ correspond to low

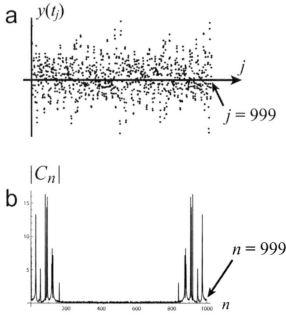

Figure 8.6.3 a: A function y sampled at 1000 times t_j. b: The magnitude of the DFT for this function. Note the symmetry about $n = N/2$. If one is only interested in the magnitude of the Fourier spectrum, then the points from $n = \dfrac{N}{2} + 1$ to $n = N - 1$ can be ignored.

frequencies (as do the components near $n = 0$). As with the Fourier Transform (see section 8.5), there is no need to take the real part of anything in any of the above discussions. There is therefore the same condition that the magnitude of the C_n for a positive frequency must equal the magnitude of the C_n for the corresponding negative frequency, so that these terms can combine to cancel the imaginary parts. Because of the transformation (8.6.13), this means that

$$\left| C_n \right| = \left| C_{N-n} \right|. \tag{8.6.15}$$

Figure 8.6.3a shows a sequence of $N = 1,000$ data points as a function of time, in which there is no obvious pattern. Figure 8.6.3b shows the magnitude of the DFT of this data (*i.e.*, $\left| C_n \right|$ as a function of n) , with the conventional sequence of n values, that is, the sequence of equation (8.6.14). The pattern is symmetrical about the point $n = N/2 = 500$. In fact, if one is only interested in the magnitudes $\left| C_n \right|$, which is often the case, then the points from $n = \dfrac{N}{2} + 1$ to $n = N - 1$ can be ignored, and one can focus just on the values for $n = 0$ to $n = \dfrac{N}{2}$.

It would appear from equations (8.6.14) and (8.6.8) that computing the DFT would take a number of steps proportional to N^2, since there are N different coefficients C_n, and to compute each of these one must do the sum (8.6.8) which has N entries. However, if $N = 2^m$, where m is an integer, then one can use a computer algorithm called the "Fast Fourier Transform" (FFT) to compute the DFT, and this takes a number of steps that is only proportional to $N \log N$, a tremendous improvement. You may have heard of research on quantum computers, in which data is represented by quantum superpositions. Such computers have not yet been fully realized, but one can show that they could be used to compute DFTs with a number of steps proportional only to $(\log N)^2$, a tremendous further improvement.

Because the DFT is of enormous scientific and commercial importance, a great deal has been written about it. To learn more, you might begin with the classic text "Numerical Recipes",[7] although you should beware that it follows the unusual convention of using $\dfrac{e^{-i\omega_n t_j}}{N}$ as the basis functions for the Fourier synthesis of y, rather than using $\dfrac{e^{i\omega_n t_j}}{N}$ as do most authors and as we have done above. The text by Julius O. Smith[8] is also very approachable, and includes reviews of all the necessary math.

8.7 Some applications of Fourier Analysis

Fourier analysis is so widespread in science and engineering that there are applications in virtually every subfield. In this section, we explore two of them briefly.

Sonograms and whale calls. For scientists studying birds and animals, it can be very helpful to have a visual representation of the characteristic calls made by a particular species. A microphone converts the sound into a time-varying voltage. The voltage is sampled at regular time intervals Δ for a measurement period T. The DFT for this dataset is calculated, and then the process is immediately repeated.

> **Self-test (answer below[9]):** (This is a hard self-test the first time you encounter it. So, don't spend too long on it before looking at the answer.) A scientist wishes to create a DFT of the sound of a whale call. She wants to make a plot of $|C_n|$ versus angular frequency, with the angular frequency ranging from 0 to $2\pi \cdot (1{,}000\,\text{Hz})$, and with a resolution in angular frequency of $2\pi \cdot (10\,\text{Hz})$. What values of the sampling interval Δ and the measurement period T should she use?

A "sonogram" is a series of DFT plots at successive times, with $|C_n|$ indicated by a gray scale, frequency $f = \omega/2\pi$ on the vertical scale, and time on the horizontal scale. (The time steps on the horizontal axis are equal to T, the time needed to collect enough data for a DFT.) Figure 8.7.1a shows the sonogram for the call of a right whale, showing that the call has three simultaneous frequency components, at about 250, 550,

---

7. *Numerical Recipes: The Art of Scientific Computing, 3rd Ed.*, by W.H. Press, S. A. Teukolsky, W. T. Vetterling, and B. P. Flannery, Cambridge University Press, Cambridge, 2007.

8. *Mathematics of the Discrete Fourier Transform with Audio Applications, 2nd Ed.*, by Julius O. Smith III, BookSurge Publishing, 2007.

9. From equation (8.6.4), we have $\omega_{max} = \dfrac{2\pi}{2\Delta}$, so we need $\dfrac{1}{2\Delta} = 1{,}000\,\text{Hz} \Leftrightarrow \Delta = 0.5\,\text{ms}$. Equation (8.6.3) states $\omega_n = n\dfrac{2\pi}{N\Delta}$, so the angular frequency resolution is $\dfrac{2\pi}{N\Delta} = \dfrac{2\pi}{T}$. Setting this equal to $2\pi \cdot (10\,\text{Hz})$ gives $T = 0.1\,\text{s}$.

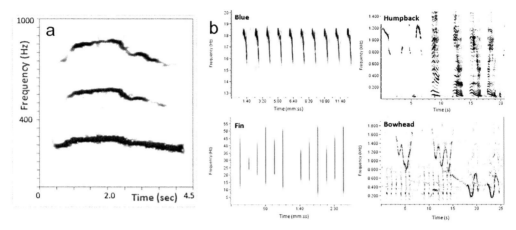

Figure 8.7.1 a: Sonogram of a Right whale. b: Sonograms for four different species of whale. Note the different time and frequency scales. Darker shades of gray darker gray indicate larger $|C_n|$. Images courtesy of and © Prof. Christopher W. Clark, Cornell University.

and 850 Hz. This would be almost impossible to tell from the graph of voltage *versus* time from the microphone. Figure 8.7.1b compares the calls from four different species of whales.

There are fewer than 400 right whales remaining in the world. Although these whales were heavily hunted in the past, collisions with ships account for many current fatalities. Scientists are working to protect these whales using underwater microphones on buoys to detect their calls, then warning ships away from possible collisions. It is essential to distinguish the calls of the right whales from other sounds in the ocean, to avoid false positives. The current generation of detection buoys uses relatively simple techniques in an effort to select out the most significant signals, but the results leave much to be desired. Scientists are hopeful that a future generation of buoys, with computerized analysis of sonograms, will lead to more accurate detection. You can learn more about this effort, including viewing a live map of the most recent right whale detections, and listening to the calls, by visiting the links listed under this section on the web page for this text.

JPEG image compression. According to the old saying, "A picture is worth a thousand words." In fact, it takes 11 kB(kilobytes) to store 1,000 words (at least in the format used by my word processor), but it takes 9 MB (megabytes) to store a reasonably high resolution (three megapixel) image if it is not compressed first. Although storage has become inexpensive, the transmission of data is still a bottleneck in many circumstances, so it is important to reduce the amount of data needed to represent photographs and other images. After compression by the jpeg algorithm, the same photo can be reduced to about 1 MB in size with no loss of quality that is visible to the eye. In fact, it can be compressed to about 100 kB with only a little loss of quality, unless one examines a magnified version. So, a more accurate version of the saying might be, "A picture is worth at least 10,000 words."

There are several image compression algorithms in wide use. The jpeg ("joint photographic experts group") method works especially well on photos, but can introduce significant artifacts into schematic diagrams and other figures with sharp edges and high contrast. In the jpeg method, the image is divided into squares of 8×8 pixels. The colors are represented by three numbers for each pixel, with one number indicating the overall brightness and the other two indicating hue. Because these numbers are functions of x and y (rather than of time), the Fourier transform is in terms of wavenumbers (rather than angular frequencies). Each of the three sets of color numbers for each 8×8 square is run through a version of the DFT called the "Discrete Cosine Transform" (DCT). This uses only cosines as basis functions, rather than both sines and cosines. However, the wavenumber spacing between the cosines of the DCT is half that of the wavenumber spacing for DFT, so that the total number of basis functions is the same. (Recall that, for normal mode analysis of a beaded string we use only sines, with $k_n = n\dfrac{\pi}{L}$, whereas for Fourier analysis of a function of x we use both sines and cosines, but with twice the wavenumber spacing: $k_n = n\dfrac{2\pi}{\lambda}$.) In addition to the difference in basis functions, the DCT used for jpeg is a two-dimensional transform; this is essentially a product of transforms in the x- and y-directions.

The advantage for jpeg of expressing the information in the 8×8 square as a sum of cosines is that the eye is less sensitive to the high wavenumber (*i.e.*, short wavelength) components, and usually these components are smaller than the low wavelength components. Therefore, after using the DCT to compute the coefficients in the sum of cosines, the higher wavenumber coefficients can be represented with very low accuracy, or even set to zero, with little perceptible change in the image after the inverse transformation (to go back from the cosines to the real-space colors of the pixels) has been applied. This is how the jpeg algorithm achieves much of its compression, although significant compression is also obtained by other steps.

Concept and skill inventory for chapter 8

After reading this chapter, you should fully understand the following terms:

Basis function (8.2)
Complete basis (8.2)
Orthogonal functions (8.3)
Fourier series expansion (8.4)
Fourier transform (8.5)
Negative frequency (8.5)
Discrete Fourier Transform (DFT) (8.6)
Sonogram (8.7)

You should understand the following connections:

Fourier analysis & normal mode analysis (8.2)
The frequency resolution of the DFT & the measurement time (8.6)
The maximum angular frequency of the DFT & the sampling interval (8.6)

You should understand the difference between:

Fourier series expansions & Fourier transforms (8.5)

You should be familiar with the following additional concepts:

The Fourier transform of a Gaussian is a Gaussian; the FWHMs are inversely related (8.5)

You should be able to:

Express any periodic function as a weighted sum of sines and cosines (8.4)

Use symmetry arguments to deduce which Fourier expansion coefficients are zero (8.4)

Given a complete set of orthogonal basis functions, express any function as a weighted sum of the basis functions, even if they aren't normalized (8.3)

Find the Fourier transform of a function that isn't periodic (8.5)

Calculate the DFT by hand for very small datasets (*e.g.*, up to four points) (8.6)

Explain the meaning of the DFT terms from $n = \dfrac{N}{2} + 1$ to $n = N - 1$ (8.6)

In addition to all of the above, you should be able to combine the concepts you've learned to address new situations.

Problems

Note: Additional problems are available on the website for this text.

Instructors: Difficulty ratings for the problems, full solutions, and important additional support materials are available on the website.

8.1 Show that all the terms in the Fourier expansion (including sines, cosines, and the constant term) of $y(x)$ are orthogonal to each other. There are five combinations you must test: sines *versus* sines, sines *versus* cosines, sines *versus* constant, cosines *versus* cosines, and cosines *versus* constant. You may need the following integrals:

$$\int \sin px \sin qx \, dx = \frac{\sin (p - q)x}{2(p - q)} - \frac{\sin (p + q)x}{2(p + q)} \quad p \neq \pm q$$

$$\int \cos px \cos qx \, dx = \frac{\sin (p - q)x}{2(p - q)} + \frac{\sin (p + q)x}{2(p + q)} \quad p \neq \pm q$$

$$\int \sin px \cos qx \, dx = -\frac{\cos (p - q)x}{2(p - q)} - \frac{\cos (p + q)x}{2(p + q)}$$

8.2 Show that an alternate version to equation (8.4.1) for the Fourier series expansion is

$$y(x) = \frac{a_0}{2} + \sum_{n=1}^{\infty} A_n \cos\left(k_n x + \varphi_n\right),$$

where $A_n = \sqrt{a_n^2 + b_n^2}$ and $\varphi_n = \tan^{-1}\left(-\dfrac{b_n}{a_n}\right)$. (This is the version used for figure 8.1.1.)

8.3 **The complex version of the Fourier expansion.** Frequently, instead of thinking of the Fourier expansion in terms of sines and cosines, it is more convenient to think of it in terms of complex exponentials. **(a)** Assume that the set of complex exponentials $e^{ik_n x}$, where $k_n = n\dfrac{2\pi}{\lambda}$ and n ranges from $-\infty$ to $+\infty$, forms a complete basis for functions $y(x)$ with periodicity λ, so long as we allow the expansion coefficients to be complex. (It is reasonable to assume that this is a complete basis, since each exponential contains some cos character and some sin character, and since we can vary the balance between the sin and cos by varying the balance between real and imaginary in the expansion coefficient.) This means that we can write

$$y(x) = \sum_{n=-\infty}^{\infty} C_n e^{ik_n x}.$$

Note: There is no "Re" in the above expression; this is intentional and correct.

Show that the basis functions in this expansion are mutually orthogonal. *Hint: Remember that for a Fourier expansion, the inner product is defined as an integral from 0 to λ.* **(b)** Show that the expansion coefficients are given by $C_n = \dfrac{1}{\lambda}\int_0^\lambda e^{-ik_n x} y(x)\,dx$. *Hint: you may wish to refer to the discussion of generic orthonormal function expansions with nonnormalized basis functions in section 8.3.* **(c)** If $y(x)$ is real (the usual case), then all the imaginary parts of the expansion above must somehow cancel out. Show in detail how this happens. *Hint: Cancellation occurs between the term at $n = m$ and the term at $n = -m$.*

8.4 **The Fourier transform.** So far, we've discussed how any function with periodicity λ can be expressed as a sum of sines and cosines with the same periodicity (plus a constant term); the difference in the wavenumber k between the longest wavelength sinusoidal component and the next-to-longest is given by $k_1 = \dfrac{2\pi}{\lambda}$. (In fact the difference between any two "adjacent" (in k-space) sinusoidal components is k_1.) It is also possible to express even nonperiodic functions in this way, so long as we allow this spacing to become infinitesimal. In this problem, you'll derive the equations which apply in this limit.

In problem 8.3, you can show that the complex exponential version of Fourier analysis is

$$y(x) = \sum_{n=-\infty}^{\infty} C_n e^{ik_n x}, \tag{1}$$

where $k_n = n\dfrac{2\pi}{\lambda}$ and $C_n = \dfrac{1}{\lambda}\int_0^\lambda e^{-ik_n x} y(x)\,dx.$

It should be fairly clear that, to evaluate the C_n's, it is only necessary to integrate over one period of $y(x)$; as long as we use the same integration interval for all the C_n's, we can choose whatever interval corresponding to one period that we like. In particular, it will make things look prettier if, instead of integrating from 0 to λ, we integrate from $-\lambda/2$ to $\lambda/2$, that is, we could just as well write:

$$C_n = \frac{1}{\lambda} \int_{-\lambda/2}^{\lambda/2} e^{-ik_n x} y(x)\, dx. \tag{2}$$

We can think of a nonperiodic $y(x)$ as a periodic function with $\lambda \to \infty$. According to the above equation, all the C_n's appear to go to zero in this limit, so we must proceed carefully. Although the C_n's do get smaller, the number of them in any interval from k to $k + dk$ gets bigger, so the net result winds up still being OK, as you'll show. As shown in the above equation (on the right), each C_n depends on the value of $k_n = nk_1$, so instead of writing them as C_n, we might write them as $C(k_n)$. In the limit $\lambda \to \infty$, the spacing (in k-space) between the C's becomes infinitesimal. In this limit, we can write them as $C(k)$ instead of $C(k_n)$. Unless something pathological is happening, all the C's in the interval k to $k + dk$ will have the same value, if we're taking the limit $dk \to 0$. A representative one of the C's in this interval would have the value

$$C(k) = \frac{1}{\lambda} \int_{-\lambda/2}^{\lambda/2} e^{-ik x} y(x)\, dx. \tag{3}$$

To form $y(x)$ from these C's, we must form an integral over k instead of the sum shown in (1). This integral (in the limit $\lambda \to \infty$) would be given by

$$y(x) = \int_{-\infty}^{\infty} (\text{\# of } C\text{'s in the interval } k \text{ to } k + dk)(\text{value of a typical } C(k)$$

$$\text{in this range}) e^{ikx},$$

where the (value of a typical $C(k)$ in this range) would be given by (3)

(a) Show that the (# of allowed C's in the interval k to $k + dk$) is given by $\dfrac{\lambda}{2\pi} dk$. (This is almost trivial.)

(b) Use this result to show that

$$y(x) = \frac{1}{\sqrt{2\pi}} \int_{-\infty}^{\infty} Y(k) e^{ikx} dk, \quad \text{where } Y(k) = \frac{1}{\sqrt{2\pi}} \int_{-\infty}^{\infty} y(x) e^{-ikx} dx.$$

Note: $Y(k)$ and $y(x)$ are referred to as a "Fourier transform pair." They contain the same information, but one version, $y(x)$, is expressed in regular space, while the other, $Y(k)$, is expressed in k-space.

8.5 What is the Fourier series representation for the triangle wave, shown in figure 8.P.1?

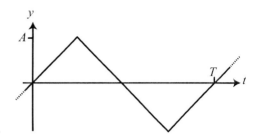

Figure 8.P.1 The triangle wave.

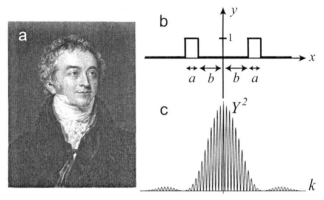

Figure 8.P.2 a: Thomas Young 1773–1829. b: Model for two-slit apparatus. c: The Fourier transform of the model shown in part b gives the interference pattern produced when light goes through the two-slit apparatus.

8.6 Interference as a Fourier Transform. As you will recall from a previous course, Thomas Young (figure 8.P.2a) showed in 1800 that light has wave character. He did this by sending a beam of light through a pair of slits, and observing the resulting interference pattern. We can model his two-slit apparatus with the following function:

$$y(x) = \begin{cases} 1 & -b-a < x < -b \quad \text{and} \quad b < x < b+a \\ 0 & \text{elsewhere} \end{cases}$$

as shown in figure 8.P.2b. Surprisingly, Fourier transforms turn out to be very handy in understanding the interference pattern. Show that the Fourier transform of the above $y(x)$ is given by $Y(k) = \sqrt{\dfrac{2}{\pi}} \dfrac{\sin[k(a+b)] - \sin(kb)}{k}$.

You are now done with the actual problem. However, it turns out that the Fourier transform you just calculated is equal to the amplitude of the electric field as the interference pattern impinges on the screen. Since the observed intensity is proportional to E^2, if we plot Y^2, we get the interference pattern. The plot of your Y^2 is shown in figure 8.P.2c.

You can see the short wavelength wiggles, which are due to the interference between the two slits, and also the more slowly varying structure which is due to the single slit diffraction pattern.

In fact, one can show that the Fourier transform of the "aperture function" always gives the amplitude of the electric field, so that the square of the Fourier transform gives the interference pattern.

8.7.   (a) Find the Fourier transform of $y = Ae^{-a|x|}$, where $a > 0$.

(b) Find the FWHM (full width at half maximum) of y, and set this FWHM equal to W.

(c) Express the FWHM of the Fourier transform of y in terms of W.

8.8. In problem 8.4, you showed that we can express *any* function $y(x)$ in the form

$$y(x) = \frac{1}{\sqrt{2\pi}} \int_{-\infty}^{\infty} Y(k)e^{ikx}dk. \tag{1}$$

You also showed that, if we write $y(x)$ this way, then its "Fourier transform" $Y(k)$ can be found by

$$Y(k) = \frac{1}{\sqrt{2\pi}} \int_{-\infty}^{\infty} y(x)e^{-ikx}dx. \tag{2}$$

The above equations are my favorite way of writing the Fourier transform, since they emphasize the symmetry between $y(x)$ and $Y(k)$. Although about half the world uses the above way of writing the Fourier transform, unfortunately the other half uses a slightly different formulation. In this version, we instead express $y(x)$ as

$$y(x) = \frac{1}{2\pi} \int_{-\infty}^{\infty} Y(k)e^{ikx}dk. \tag{3}$$

For this way of expressing $y(x)$, the expression for $Y(k)$ (equation 2) needs to be adjusted slightly to be correct. What would be the correct expression for $Y(k)$ to go with equation 3? **Explain your answer thoroughly.** *Hint: There is almost no additional math required for this problem. It will probably help you to think in terms of basis functions/vectors in Hilbert space, although there are other good ways to do this problem.*

8.9 Time scaling of the Fourier transform. Consider a function $y(t)$ with Fourier transform $Y(\omega)$. The function $y'(t) = y(at)$ is the same as $y(t)$, except that it is compressed by the factor a along the time axis. For example, if there is a peak in $y(t)$ at $t = 1$s, and $a = 2$, then the same peak appears in $y'(t) = y(at)$ at $t = 0.5$ s. Assuming $a > 0$ show that the Fourier transform of $y'(t)$ is $Y'(\omega) = \frac{1}{a}Y\left(\frac{\omega}{a}\right)$. (This means that $Y'(\omega)$ is *expanded* by the factor a along the ω axis.)

8.10 Fourier transform of the Dirac delta function

(a) Explain briefly why the Kronecker delta function has the property that it "picks out" the "matching" term in a series, that is, $\sum_{n=-\infty}^{\infty} f(n)\,\delta_{mn} = f(m)$, where f is a function.

(b) The Kronecker delta function only works for discrete variables, such as n in the example above or k_n for a finite-length string stretched between two walls. However, it is often handy to have a similar function which works for continuous variables. This is called the "Dirac delta function," and is written $\delta(x - a)$. It has the property that for $x \neq a$, $\delta(x - a) = 0$ (this is analogous to the Kronecker delta function). However, for $x = a$, $\delta(x - a) =$ "∞", where the infinity is in quotes because the "extent" of the infinity is carefully defined below. However, it is correct to picture the graph of $\delta(x - a)$ versus x as being zero everywhere except for an infinitely tall and infinitely narrow spike at $x = a$. Here's the final part of the definition of the Dirac delta function: the height of the spike is the "right amount of infinity" so that

$$\int_{-\infty}^{\infty} f(x)\,\delta(x - a)\,dx = f(a),$$

which is analogous to the equation in part a) of this problem, that is, the Dirac delta function "picks out" the "matching" term in the integral. Find the Fourier transform of the Dirac delta function for the case $a = 0$.

(c) Now that you know the answer, explain how you might have predicted it qualitatively based on the example "Fourier Transform of a Gaussian" in section 8.5.

(d) What is the "inverse Fourier transform" of $\delta(k - k_0)$ for the case $k_0 = 0$? (The "inverse Fourier transform" means the function $y(x)$ whose Fourier transform is $\delta(k - k_0)$.)

(e) When a (or k_0) is not equal to zero, things are slightly more complicated, though in a philosophical sense they are still about the same. What is the inverse Fourier transform of $\delta(k - k_0)$ for k_0 not necessarily equal to zero?

(f) Now that you know the answer to the above, explain why it makes qualitative sense.

8.11 More about the Dirac delta function. Read the first two parts of problem 8.10. (You need not actually do them before you do this problem). In part b of that problem, you can show that the Fourier transform of $\delta(x)$ is $1/\sqrt{2\pi}$. This, of course, means that the Fourier transform of $\delta(t)$ is also $1/\sqrt{2\pi}$. **(a)** Use this as a starting point for this problem, and show that $\int_{-\infty}^{\infty} e^{i\omega t}\,d\omega = 2\pi\,\delta(t)$.

(b) Explain why, therefore, $\int_{-\infty}^{\infty} e^{-i\omega t}\,d\omega = 2\pi\,\delta(t)$.

8.12 The left column of figure 8.P.3 shows several different functions of time. The right column shows the magnitudes of the Fourier amplitudes for these functions, though the order has been scrambled, for example plot g does not show the Fourier amplitudes for the function $y(t)$ in part a. "Fourier amplitude" means the amplitude A_n in the Fourier expansion expressed

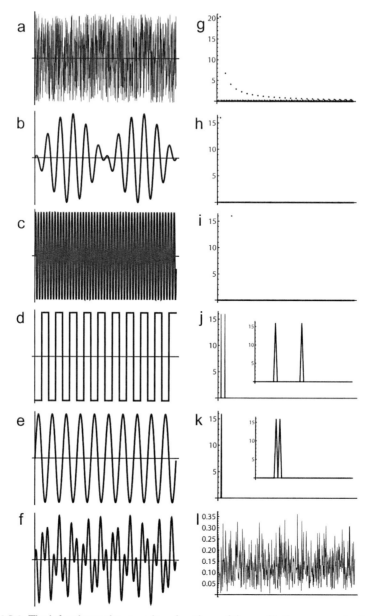

Figure 8.P.3 The left column shows various functions of time, with the same vertical and horizontal scale for each plot. The right column shows the magnitude of the Fourier amplitude as a function of ω, with the same horizontal scale for each plot. The insets for j and k show magnified views of the low-ω region; the horizontal scale on the two insets is the same. Note that parts h and i each show just a single dot, near the top left.

as (8.4.5): $y(t) = \dfrac{a_0}{2} + \sum\limits_{n=1}^{\infty} A_n \cos\left(\omega_n x + \varphi_n\right)$. The Fourier amplitudes are plotted as a function of ω. Which entry from the right column goes with each of the entries from the left column? Explain each of your choices briefly.

8.13 **Parseval's theorem.** We have seen that energy is proportional to the square of amplitude. For example, the potential energy of a simple harmonic oscillator is $\frac{1}{2}kA^2$. Therefore, it is reasonable to expect that the sums of squares of amplitudes in the time domain are proportional to the sums of squares of amplitudes in the frequency domain, since both sums should be proportional to the energy. **(a)** We begin with a simple illustration of this idea. Let $y(t) = A_1 \cos \omega_1 t + A_2 \cos \omega_2 t$, where $\omega_1 = \dfrac{2\pi}{T}$ and $\omega_2 = 2\omega_1$. The sum of squares of amplitudes in the frequency domain would simply be $A_1^2 + A_2^2$. To find the "sum of squares of amplitudes in the time domain," we must integrate: $\int_0^T [y(t)]^2 \, dt$. Show that this is proportional to $A_1^2 + A_2^2$.

(b) In part a, we considered Fourier synthesis of a periodic function, as described in section 8.4. Now, we consider Fourier transforms. As you'll recall from section 8.5, these involve complex exponentials. Therefore, instead of discussing the "sum of squares of amplitudes," we discuss the "sum of squares of magnitudes." Prove Parseval's theorem, which states that $\int_{-\infty}^{\infty} |y(t)|^2 \, dt = \int_{-\infty}^{\infty} |Y(\omega)|^2 \, d\omega$, where $y(t)$ and $Y(\omega)$ are a Fourier transform pair, as defined by (8.5.4). *Hint: use the result of problem 8.11 ; since this is a mathematical result, you can interchange the symbols ω and t and it will still be correct.*

8.14 State whether each of the following is true or false. If true, explain why briefly. If false, explain why and provide a corrected version that is not simply a negation of the original statement. Important: assume that $y(t)$ is real.

    (a) In the Fourier transform (8.5.4), the imaginary part of $Y(\omega)$ is just a mathematical convenience. We could just as well write

$$y(t) = \frac{1}{\sqrt{2\pi}} \int_{-\infty}^{\infty} Y(\omega) e^{i\omega t} \, d\omega,$$

$$\text{where } Y(\omega) = \text{Re}\left[\frac{1}{\sqrt{2\pi}} \int_{-\infty}^{\infty} y(t) e^{-i\omega t} \, dt \right].$$

    (b) In the Fourier transform (8.5.4), the negative frequencies are just a mathematical convenience. We could just as well write

$$y(t) = 2\text{Re}\left[\frac{1}{\sqrt{2\pi}} \int_{0}^{\infty} Y(\omega) e^{i\omega t} \, d\omega \right],$$

$$\text{where } Y(\omega) = \frac{1}{\sqrt{2\pi}} \int_{-\infty}^{\infty} y(t) e^{-i\omega t} \, dt.$$

8.15 **Aliasing.** We have discussed why the maximum angular frequency that can be represented by samples taken at time intervals spaced by Δ is $\omega_{N/2} = \dfrac{2\pi}{2\Delta} = \dfrac{\pi}{\Delta}$. (The corresponding frequency, $f_c = \dfrac{\omega_{N/2}}{2\pi}$ is called the

"Nyquist frequency.") What happens if we sample a continuous waveform with a higher angular frequency than this? The resulting effect is called "aliasing," and can cause serious problems in digital signal processing.

(a) Sketch several periods of a continuous sinusoidal waveform. Put dots on it indicated the times at which samples are taken, with Δ slightly less that the period of your waveform.

(b) Based on your sketch, explain why the apparent frequency based just on the sampled points is much lower than the actual frequency $1/T_w$ of your waveform.

(c) Based on your explanation from part b), for what value of $\dfrac{\Delta}{T_w}$ does the apparent frequency equal zero?

(d) Now, we will explore the same ideas more quantitatively. A continuous waveform $A \cos \omega_\alpha t$ is sampled, where $\omega_\alpha = \alpha \dfrac{2\pi}{N\Delta}$, Δ is the time interval between samples, and N is the number of samples. For simplicity, we'll assume α is an integer, but similar things happen if it isn't. Explain why, if $\omega_\alpha > \omega_{N/2}$, then peaks appear in the Discrete Fourier Transform (DFT) at ω_α and at $\omega_{N-\alpha}$, where $\omega_\alpha > \omega_{N-\alpha} > 0$.

(e) Explain how your result from part d) is consistent with your result from part c).

(f) Explain why a waveform $A \cos \omega_{N-\alpha}$ would produce peaks at ω_α and $\omega_{N-\alpha}$. (This means that DFT of the waveform with angular frequency ω_α is indistinguishable from the DFT for the waveform with angular frequency $\omega_{N-\alpha}$.)

8.16 **Windowing.** (Mathematica or other symbolic algebra program is required for this problem. Further instructions for how to implement this problem in Mathematica are available on the website for this text, under the entry for this problem.) In this problem, you will make sets of $N = 1,024$ samples taken at intervals of $\Delta = 1$ ms, so that the total sampling time is $T = 1.024$ s. Recall that the Discrete Fourier Transform (DFT) algorithm assumes that the input waveform $y(t)$ has the periodicity T. In this problem, you'll explore what happens when this isn't true.

(a) Create a dataset of 1,024 samples of the continuous wave $y = \sin \dfrac{2\pi t}{T_w}$, with a sample interval $\Delta = 1$ ms, and $T_w = 10.24$ ms, so that exactly 100 periods fit into the sampling interval. Plot the magnitude of the DFT of this dataset as a function of ω using a logarithmic scale for the vertical axis, and comment briefly on it.

(b) Now create a new dataset of 1,024 samples, with T_w adjusted so that 100.5 periods fit into the sampling interval. Because the DFT assumes periodicity T, this creates a "glitch" at the border between one interval of length T and the next, *for example*, at the border between the interval $t = 0$ to $t = T$ and the assumed repeat of the function y that occurs in the interval $t = T$ to $t = 2T$. Sketch by hand what the assumed y looks like near this border.

(**c**) Plot the magnitude DFT of this new dataset (again with a logarithmic scale for the vertical axis), and comment briefly on it.

(**d**) Now create a "windowed" version of the dataset from part b, using the Hanning window described in figure 8.6.2. This eliminates the "glitch," but of course distorts the input waveform in other ways. Again, plot the magnitude of the DFT, and comment briefly on how this plot compares to those from parts (a) and (c).

8.17 Samples are taken of a continuous waveform at 1 ms intervals. Only $N = 4$ samples are taken. (**a**) For this part, the samples have the following values: A, $-A, A, -A$. Compute by hand the Discrete Fourier Transform of this dataset, and explain how your results fit with expectations. (**b**) For this part, the samples have the following values: $A, 0, -A, 0$. Compute by hand the Discrete Fourier Transform of this dataset, and explain how your results fit with expectations.

8.18 The Heisenberg Uncertainty Principle and expectation values. As we've discussed at various points in this text, the "probability density" for a quantum mechanical particle is given by $|\Psi|^2$, where $\Psi(x, t)$ is the wavefunction. So far, I've only said that $|\Psi|^2$ is proportional to the probability of finding the particle near a particular position. The exact definition is that the probability of finding the particle somewhere between $x = a$ and $x = b$ is

$$\frac{\int\limits_{a}^{b} |\Psi|^2 \, dx}{\int\limits_{-\infty}^{\infty} |\Psi|^2 \, dx}.$$

You might recognize the term on the bottom as a normalizing factor; the numerator is the integrated probability density over the range of interest, whereas the denominator is the integrated probability density over all space. If the wavefunction is properly normalized (the same idea as normalizing an eigenfunction), then the denominator equals 1.

In this problem, we will focus on what happens at $t = 0$. Let us define $\psi(x) \equiv \Psi(x, t = 0)$, so that the above probability of finding the particle somewhere between $x = a$ and $x = b$ can be written as

$$\frac{\int\limits_{a}^{b} |\psi|^2 \, dx}{\int\limits_{-\infty}^{\infty} |\psi|^2 \, dx}.$$

Now, we are equipped to understand the concept of "expectation value." Let's say we measure some function of the position of the particle, $f(x)$. This means that if the particle is at $x = a$, then a measurement of f yields $f(a)$. If we imagine preparing a large number of particles in identical initial states, and measuring f for each of them, then sometimes we'll get $f(a)$, sometimes $f(b)$, sometimes something else. If $|\psi|^2$ is large near $x = a$ and small near $x = b$, we are more likely to get the result $f(a)$ than the result $f(b)$. The appropriately weighted average of many measurements of f on a large number of initially identical particles is called the "expectation value of f," and is computed like this:

$$\langle f \rangle = \frac{\int\limits_{-\infty}^{\infty} |\psi|^2 f(x) \, dx}{\int\limits_{-\infty}^{\infty} |\psi|^2 \, dx}.$$

For example, if $f(x) = x^2$, then

$$\langle x^2 \rangle = \frac{\int\limits_{-\infty}^{\infty} |\psi|^2 x^2 \, dx}{\int\limits_{-\infty}^{\infty} |\psi|^2 \, dx}.$$

The "standard deviation" of x is a measure of the width (in x) of the of the graph of $|\psi|^2$ *versus* x, and is defined as

$$\Delta x \equiv \sqrt{\left\langle (x - \langle x \rangle)^2 \right\rangle}.$$

The quantity $(x - \langle x \rangle)$ is the difference between x and the average value of x. By squaring this, we get a number that is positive whether or not x is greater than $\langle x \rangle$. In the above equation, we take the average of this square, then take the square root of the average, so as to get back to something that has the same units as x. Thus, Δx is the Root of the Mean of the Square of the difference between x and $\langle x \rangle$. I capitalized the words "root," "mean," and "square" because you will encounter this same idea in chapter 9; at that point it will be referred to as the "rms" amplitude. The Δx defined in this way is the "more careful definition" of the width that was referred to in our initial discussion of the Heisenberg uncertainty principle, back in section 1.12.

(a) Assume that $\psi = Ae^{-x^2/2\sigma^2}$, where A and σ are constants. Show that $\Delta x = \dfrac{\sigma}{\sqrt{2}}$. *Hint: because ψ is symmetrical about $x = 0$, we have right away that $\langle x \rangle = 0$.*

(b) We can write any wavefunction ψ as a Fourier sum: $\psi(x) = \dfrac{1}{\sqrt{2\pi}} \int\limits_{-\infty}^{\infty} Y(k) e^{ikx} dx$. The Fourier transform $Y(k)$ plays the same role in k-space that $\psi(x)$ plays in real space. For example, the probability of measuring a k for the particle that lies between

k_a and k_b is

$$\frac{\int\limits_{k_a}^{k_b} |Y|^2 \, dk}{\int\limits_{-\infty}^{\infty} |Y|^2 \, dk}.$$

We can define the width (in k) of the graph of $|Y|^2$ *versus* k:

$$\Delta k \equiv \sqrt{\left\langle (k - \langle k \rangle)^2 \right\rangle},$$

where, for example,

$$\left\langle k^2 \right\rangle = \frac{\int\limits_{-\infty}^{\infty} |Y|^2 \, k^2 \, dk}{\int\limits_{-\infty}^{\infty} |Y|^2 \, dk}.$$

For the ψ assumed in part (a), show that $\Delta k = \dfrac{1}{\sigma\sqrt{2}}$.

(c) Combine the results of parts (a) and (b) to show that $\Delta x \, \Delta k = \dfrac{1}{2}$, as claimed in section 1.12.

(d) For a quantum mechanical particle, the momentum is given by $p = \hbar k$, where $\hbar$ is Planck's constant. Show that therefore, for the ψ assumed in part (b), $\Delta x \, \Delta p = \dfrac{\hbar}{2}$. (In fact, the ψ assumed in part (b) gives the minimum value for $\Delta x \, \Delta p$, so that in general we have $\Delta x \, \Delta p \geq \dfrac{\hbar}{2}$, which is Heisenberg's uncertainty relation.)

9 Traveling Waves

Only for one does my heart resonate
Like a column of air in the wind
I seek the one whose frequency suits me
To release the sound within.
— Anonymous

9.1 Introduction

So far, we've examined oscillations limited to a finite region of space, such as the standing waves on a rope stretched between two walls. Such situations are analogous to the quantum mechanical problem of an electron confined to a finite region of space. For example, the electron might be confined by its attraction to a nucleus, as suggested in figure 9.1.1. These "bound states" will make up at least 75% of your work in an introductory quantum mechanics course, and will lead, for example, to the structure of the periodic table. However, we can also make waves that move.

One of the joys of physics is in discovering hidden connections. In this chapter, we will explore electromagnetic waves (such as light) in vacuum and in matter, waves on ropes, sound waves, and waves on transmission lines. We will find that the mathematical structure for all of these waves is identical, once we find the appropriate variables to study. In each case, we will find that there are two components to the wave; in the example of electromagnetic waves in vacuum, these are the oscillating electric and magnetic fields.

9.2 The wave equation

Let's begin by looking at traveling waves on a string, because they're easy to visualize. (Later, we'll examine other types of traveling waves, such as electromagnetic waves.) We begin with the beaded string. Recall from section 7.1 that bead j experiences tension forces from the the string connecting to bead $j - 1$ (on the left) and from the string

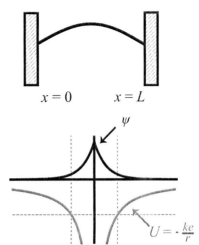

Figure 9.1.1 Standing waves on a rope (top) are analogous to states of an electron for which it is confined, such as that shown in the bottom part of the figure; these are called "bound states."

connecting to bead $j + 1$ (on the right, leading to

$$(7.1.2): \ddot{y}_j = -\frac{T}{ma}\left(-y_{j-1} + 2y_j - y_{j+1}\right),$$

where y_j is the y-position of bead j, T is the tension in the string, m is the mass of a bead, and a is the spacing between beads. We are interested in a continuous string, so we allow the spacing of the beads to become very small. We can then think of a continuous function $y(x, t)$ which describes the string. For notational convenience, we will stop bothering to indicate explicitly that y is a function of t as well as x, so that we simply write $y_j = y(x_j)$. Since a is small, we can approximate y_{j-1} and y_{j+1} using a second-order Taylor series:

$$y_{j-1} = y\left(x_j - a\right) \cong y\left(x_j\right) - a \left.\frac{\partial y}{\partial x}\right|_{x_j} + \frac{a^2}{2} \left.\frac{\partial^2 y}{\partial x^2}\right|_{x_j}, \qquad (9.2.1a)$$

$$y_{j+1} = y\left(x_j + a\right) \cong y\left(x_j\right) + a \left.\frac{\partial y}{\partial x}\right|_{x_j} + \frac{a^2}{2} \left.\frac{\partial^2 y}{\partial x^2}\right|_{x_j}. \qquad (9.2.1b)$$

The notation $\dfrac{\partial y}{\partial x}$ means "partial derivative of y with respect to x." This simply means "take the derivative of y with respect to x, treating t as a constant." For example, if $y = ax^2t^3$, then $\dfrac{\partial y}{\partial x} = 2axt^3$. Similarly, the notation $\dfrac{\partial^2 y}{\partial x^2}$ means "second partial derivative of y with respect to x," which means "take the second derivative of y with respect to x, treating t as a constant." For example, if $y = ax^2t^3$, then $\dfrac{\partial^2 y}{\partial x^2} = 2at^3$. You should not get too worried about this notation; it is needed because y is a function of both x and t, but there is really nothing mysterious about it.

> **Your turn:** Substitute equation (9.2.1) into 7.1.2 above, and show that the result is
>
> $$\ddot{y}\left(x_j\right) \cong \frac{Ta}{m} \left.\frac{\partial^2 y}{\partial x^2}\right|_{x_j}. \tag{9.2.2}$$

In the limit that $a \to 0$, the second-order Taylor series approximation becomes exact, so that the equality above becomes exact. As we take this limit, we imagine simultaneously shrinking the mass of the beads, so that the mass per unit $\mu \equiv \dfrac{m}{a}$ remains constant. We can then rewrite equation (9.2.2)[1] as $\left[\dfrac{\partial^2 y}{\partial t^2} = \dfrac{T}{\mu} \dfrac{\partial^2 y}{\partial x^2}\right]_{x_j}$, where the subscript in square brackets indicates that both sides are evaluated at $x = x_j$. Since we can do this same analysis for any bead j, and since the beads are infinitesimally close together, we see that the equation holds for any x, so we may as well write it just as

$$\frac{\partial^2 y}{\partial t^2} = \frac{T}{\mu} \frac{\partial^2 y}{\partial x^2}. \tag{9.2.3}$$

> **Your turn:** Show that the units of $\dfrac{T}{\mu}$ are $\left(\dfrac{m}{s}\right)^2$.

Given the above, we can define something with the units of velocity:

$$v_{\mathrm{p}} \equiv \sqrt{\frac{T}{\mu}} \tag{9.2.4}$$

(We'll soon see just what this is the velocity of.) With this, we can rewrite equation (9.2.3) as

$$\boxed{\frac{\partial^2 y}{\partial t^2} = v_{\mathrm{p}}^2 \frac{\partial^2 y}{\partial x^2}.} \tag{9.2.5}$$

The Wave Equation

This is the "wave equation." It says that the second derivative of y with respect to t is the same as the second derivative with respect to x, except for the additional factor of v_{p}^2.

Claim: Any function of the form $y(x - v_{\mathrm{p}}t)$, that is, a function that has as its argument the factor $(x - v_{\mathrm{p}}t)$ is a solution to the wave equation.

---

1. Note that $\ddot{y} = \dfrac{\partial^2 y}{\partial t^2}$, meaning the second partial derivative of y with respect to t. Again, this notation is nothing to worry about; it simply means, "take the second derivative of y with respect to t, treating x as a constant." For example, if $y = ax^2 t^3$, then $\dfrac{\partial^2 y}{\partial t^2} = 6ax^2 t$.

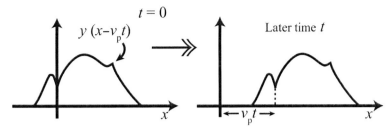

Figure 9.2.1 The function $y(x - v_\text{p}t)$ represents a rigid shape moving to the right at speed v_p.

Demonstration by example: Let's try $y = (x - v_\text{p}t)^b$. To test whether this works, we need to evaluate the derivatives:

$$\frac{\partial^2}{\partial x^2}\left(x - v_\text{p}t\right)^b = \frac{\partial}{\partial x}\left[\frac{\partial}{\partial x}\left(x - v_\text{p}t\right)^b\right] = \frac{\partial}{\partial x}\left[b\left(x - v_\text{p}t\right)^{b-1}\right] = b(b-1)\left(x - v_\text{p}t\right)^{b-2}$$

and

$$\frac{\partial^2}{\partial t^2}\left(x - v_\text{p}t\right)^b = \frac{\partial}{\partial t}\left[\frac{\partial}{\partial t}\left(x - v_\text{p}t\right)^b\right] = \frac{\partial}{\partial t}\left[b\left(x - v_\text{p}t\right)^{b-1}\left(-v_\text{p}\right)\right]$$

$$= b\,(b-1)\left(x - v_\text{p}t\right)^{b-2} v_\text{p}^2.$$

Comparing these, we see that $y = (x - v_\text{p}t)^b$ is indeed a solution to the wave equation. In fact, we can now see why any function of the form $y(x - v_\text{p}t)$ is a solution: the process of taking a derivative with respect to t is identical to taking one with respect to x, except that each t derivative also brings out (because of the chain rule) a factor of $(-v_\text{p})$, so that taking two time derivatives brings out the extra factor v_p^2, which is just what we need for a solution to the wave equation.

> **Your turn:** Show that any function of the form $y(x + v_\text{p}t)$ is also a solution to the wave equation.

You know from your study of functions in high school that the function $y(x - b)$, where b is a constant, looks just the same as $y(x)$, but shifted to the *right* by the amount b. Therefore, a function of the form $y(x - v_\text{p}t)$ looks just the same as $y(x)$, but shifted right by the amount $v_\text{p}t$. Since this shift increases linearly in time, at the rate v_p, we see that $y(x - v_\text{p}t)$ represents a rigid shape moving to the right at speed v_p, as shown in figure 9.2.1. Similarly, a function of the form $y(x + v_\text{p}t)$ represents a rigid shape moving left at speed v_p.

> **Conclusion:** The solutions to the wave equation are of the form $y(x - v_\text{p}t)$ (representing a rigid shape moving right at speed v_p) or of the form $y(x + v_\text{p}t)$ (representing a rigid shape moving left at speed v_p).

So, we see that the string can indeed sustain traveling waves, with speed $v_{\mathrm{p}} = \sqrt{\dfrac{T}{\mu}}$. Note that each infinitesimal segment of the string moves straight up and down (in the $+$ or $-y$ direction) as the wave passes, even though the wave is moving right or left (in the $+$ or $-x$ direction).

There is something quite remarkable about the conclusion in the box above: the speed of the wave, v_{p}, doesn't depend on the shape. In particular, for sinusoidal waves *the speed doesn't depend on the wavelength or the amplitude!*

9.3   Traveling sinusoidal waves

The most important example of a traveling wave is a sinusoid; We know from Fourier analysis that any function can be formed by summing up sinusoids. Recall from chapter 7 that a sine wave of wavenumber $k = \dfrac{2\pi}{\lambda}$ (where λ is the wavelength) and amplitude A is written $A \sin kx$. To make this travel to the right, we simply replace x by $\left(x - v_{\mathrm{p}}t\right)$:

$$y = A \sin\left[k\left(x - v_{\mathrm{p}}t\right)\right] = A \sin\left(kx - kv_{\mathrm{p}}t\right). \qquad (9.3.1)$$

When t advances from 0 to $\dfrac{2\pi}{k\,v_{\mathrm{p}}}$, the argument of the sin increases by 2π. Therefore, the period of the wave is $T = \dfrac{2\pi}{k\,v_{\mathrm{p}}}$. Since the angular frequency is given by $\omega \equiv \dfrac{2\pi}{T}$, we see that $\omega = kv_{\mathrm{p}} \Leftrightarrow$

$$\boxed{v_{\mathrm{p}} = \frac{\omega}{k}.} \qquad (9.3.2)$$

This is really nothing new. You've known since you were a toddler that the speed of a sinusoidal wave is given by $v = \dfrac{\lambda}{T}$. We can reexpress this in terms of k and ω:

$$v = \frac{\lambda}{T} = \frac{2\pi/k}{2\pi/\omega} = \frac{\omega}{k}.$$

Using equation (9.3.2), we can rewrite equation (9.3.1), to express a right-traveling sinusoidal wave as

$$y = A \sin\left(kx - \omega t\right),$$

which is the most common way of writing it. The speed of the wave is the speed at which the crest (corresponding to a phase of $90°$) or the trough (corresponding to a phase of $270°$) advances, so it is called the "phase velocity," hence the use of "p" as a subscript in v_{p}.

The superposition principle for traveling waves

In section 4.4, we found that when an oscillator is driven by multiple drive forces, the resulting response is simply the sum of the responses to each drive force on its own. This is a consequence of the linearity of the differential equation governing the oscillator. In this section, we explore a related consequence of linearity, but this time for a system that is not driven. In particular, we will explore whether we can combine left- and right-moving waves, and what happens when they "collide."

Claim: For any differential equation that is linear (i.e., there are no terms proportional to y^2 or to $y\dot{y}$, etc.) and homogeneous (i.e., there are no terms that are constant), if y_A is a solution to the differential equation, and y_B is a different solution, then the linear combination $Ay_A + By_B$ (where A and B are constants) is also a solution.

Proof for the wave equation: To say that y_A is a solution to the wave equation simply means that

$$\frac{\partial^2 y_A}{\partial t^2} = v_p^2 \frac{\partial^2 y_A}{\partial x^2}. \tag{9.4.1}$$

Similarly, to say that y_B is a solution simply means that

$$\frac{\partial^2 y_B}{\partial t^2} = v_p^2 \frac{\partial^2 y_B}{\partial x^2}. \tag{9.4.2}$$

Now, we multiply equation (9.4.1) by A, multiply equation (9.4.2) by B and add them together:

$$A\frac{\partial^2 y_A}{\partial t^2} + B\frac{\partial^2 y_B}{\partial t^2} = v_p^2\left(A\frac{\partial^2 y_A}{\partial x^2} + B\frac{\partial^2 y_B}{\partial x^2}\right) \Leftrightarrow \frac{\partial^2\left(Ay_A + By_B\right)}{\partial t^2} = v_p^2\frac{\partial^2}{\partial x^2}\left(Ay_A + By_B\right),$$

which shows that $Ay_A + By_B$ is a solution to the wave equation. (Perhaps you can see how you could do this same type of proof for *any* linear, homogeneous differential equation.)

This principle of superposing (i.e., adding together) solutions is a *very powerful one*, and one that you will use a *lot* in quantum mechanics, which is governed by the Schrodinger equation

$$-\frac{\hbar^2}{2m}\frac{\partial^2\Psi}{\partial x^2} + U\Psi = i\hbar\frac{\partial\Psi}{\partial t}, \tag{9.4.3}$$

where m is the mass of the electron, U is the potential energy of the electron, and $\Psi(x, t)$ is the electron wavefunction, which is analogous to $y(x, t)$. You can see that this differential equation is linear (i.e., it contains no terms proportional to Ψ^2 or to $\Psi\dot{\Psi}$, etc.) and homogeneous (i.e., it contains no constant terms), so it must obey the Principle of Superposition.

As applied to waves that obey the wave equation, the Principle means, for example, that we can have left- and right-traveling waves propagating at the same time. As they move into the same area, they simply add up, as shown in figure 9.4.1, but they don't alter each other in any profound way – each just keeps going, and eventually they pass through each other with no change!

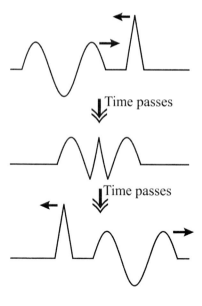

Figure 9.4.1 For waves governed by a linear differential equation, left- and right-traveling waves simply add together, and pass through each other without other effect; this is the Principle of Superposition.

Recall from chapter 7 that when a string is stretched between two walls, that is, when we impose the boundary conditions that $y = 0$ at $x = 0$ and $x = L$, the solutions are standing waves. The basic physics of the string that we used in chapter 7 are exactly the same as the physics we used to derive the wave equation. We know that the left- and right-traveling wave solutions are the only solutions to the wave equation, so it must be possible to express standing waves as a superposition of traveling waves. For example, a pure normal mode n of a string between two walls is

$$y = C_n \sin k_n x \, \cos \omega_n t. \tag{9.4.4}$$

(Note how, for a standing wave such as this, the time and space dependence appear in the arguments of two different functions, whereas for a traveling wave they appear together in the argument of one function.) Can we express this as a superposition of traveling waves?

Your turn: Use the basic trig identity $\sin (A + B) = \sin A \, \cos B + \cos A \, \sin B$ to show that

$$\frac{C_n}{2} \sin \left(k_n x + \omega_n t \right) + \frac{C_n}{2} \sin \left(k_n x - \omega_n t \right) = C_n \sin k_n x \cos \omega_n t,$$

that is, that **the sum of two equal-amplitude waves traveling in opposite directions produces a standing wave!**

This result is important in a number of contexts. For example, in solid-state physics the "energy gaps" which are responsible for the electronic properties of semiconductors arise as the result of a standing wave which forms when electron waves of certain special wavelengths undergo diffraction from the lattice of atoms in the semiconductor crystal, creating a "backscattered" wave which travels in the opposite direction from the original, and has the same amplitude.

Note that standing waves are a special case, because of the imposition of the boundary conditions. Therefore, although we can superpose the more general solutions (traveling waves) to create the special case solution (standing waves), we can't go the other way, that is, we can't superpose standing waves to create traveling waves.

9.5 Electromagnetic waves in vacuum

The phenomena of electricity and magnetism are governed by Maxwell's four equations:

Gauss's Law: $\oint \mathbf{E} \cdot \hat{\mathbf{n}}\, dA = \dfrac{Q_{\text{net, enclosed}}}{\varepsilon_0}$ (9.5.1)

"Electric field lines can only begin on $+$ charges and can only end on $-$ charges. The electric field due to a point charge is given by $E = \dfrac{1}{4\pi\varepsilon_0}\dfrac{Q}{r^2}$"

Gauss's Law for magnetic fields: $\oint \mathbf{B} \cdot \hat{\mathbf{n}}\, dA = 0$ (9.5.2)

"There are no magnetic monopoles."

Faraday's Law: $\oint \mathbf{E} \cdot d\vec{\ell} = -\dfrac{d}{dt}\int \mathbf{B} \cdot \hat{\mathbf{n}}\, dA$ (9.5.3)

"A time-varying **B** creates a spatially varying **E**."

Ampère's Law: $\oint \mathbf{B} \cdot d\vec{\ell} = \mu_0 I_{\substack{\text{net} \\ \text{threading}}} + \mu_0\varepsilon_0\dfrac{d}{dt}\int \mathbf{E} \cdot \hat{\mathbf{n}}\, dA$ (9.5.4)

"Magnetic fields are created by currents or by time-varying **E**."

Maxwell's equations restrict which combinations of electric fields, magnetic fields, and charge are allowed. For example, the magnetic field pattern shown in figure 9.5.1 is not allowed by equation (9.5.2). Let us ask whether "plane waves" are allowed in

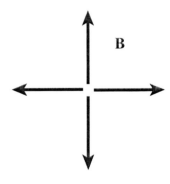

Figure 9.5.1 This magnetic field is not consistent with equation (9.5.2).

vacuum, that is, do Maxwell's equations allow the following combination of electric and magnetic fields in vacuum:

$$\mathbf{E} = E(x, t)\,\hat{\mathbf{j}} \quad \text{and} \quad \mathbf{B} = B(x, t)\,\hat{\mathbf{k}}, \tag{9.5.5}$$

where $\hat{\mathbf{j}}$ is the unit vector in the y-direction and $\hat{\mathbf{k}}$ is the unit vector in the z-direction. The configuration described by equation (9.5.5) is called a "plane wave" because there is no dependence of the electric or magnetic field strength on y or z. Therefore, in any plane perpendicular to the x-axis, the $\mathbf{E}$ and $\mathbf{B}$ are uniform (because x has the same value throughout the plane). The electric field component of one particular type of plane wave is illustrated in figure 9.5.2, using the convention that the length of the vector indicates the strength of the field. (Note: the planes shown could be infinite in extent, but to conserve paper we have only shown a portion of each plane.)

It's pretty clear that the plane wave does satisfy both the electric and magnetic versions of Gauss's Law, since the flux of either $\mathbf{E}$ or $\mathbf{B}$ would be zero through any closed surface (as required, since there are no enclosed charges in a vacuum). Next, let's check whether this combination of electric and magnetic fields satisfies Faraday's Law.

We apply Faraday's Law to the rectangular loop shown in figure 9.5.3. We'll choose to have $\vec{d\ell}$ point counterclockwise around this loop when evaluating the left side of Faraday's law: $\oint \mathbf{E} \cdot \vec{d\ell}$. The closed loop integral is equal to the sum of the four integrals along the four segments that make up the path:

$$\oint \mathbf{E} \cdot \vec{d\ell} = \int_1 + \int_2 + \int_3 + \int_4 = 0 + E(x_2)\,\Delta y + 0 - E(x_1)\,\Delta y, \tag{9.5.6}$$

where the 0's arise because $\mathbf{E}$ is perpendicular to $\vec{d\ell}$ along segments 1 and 3, and the minus sign arises because $\mathbf{E}$ is anti-parallel to $\vec{d\ell}$ along segment 4. We can simplify this by taking the limit $\Delta x \to 0$, and Taylor expanding $E(x)$ around $E(x_1)$:

$$E(x_2) \cong E(x_1) + \Delta x\,\frac{\partial E}{\partial x}.$$

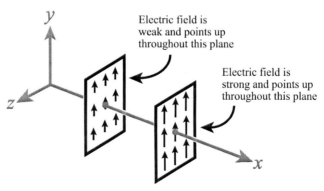

Figure 9.5.2 A plane wave in the electric field; the field strength and direction depends on x, but not on y or z.

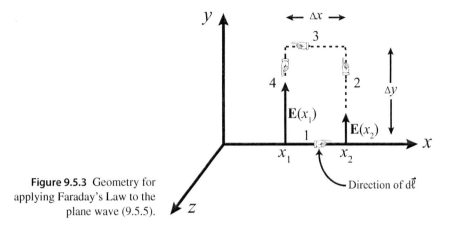

Figure 9.5.3 Geometry for applying Faraday's Law to the plane wave (9.5.5).

Plugging this into equation (9.5.6) gives

$$\oint \mathbf{E} \cdot d\vec{\ell} \cong \frac{\partial E}{\partial x} \Delta x \, \Delta y. \tag{9.5.7}$$

The equality becomes exact in the limit $\Delta x \to 0$.

Now we turn our attention to the right side of Faraday's Law: $-\dfrac{d}{dt} \int \mathbf{B} \cdot \hat{\mathbf{n}} \, dA$. The direction of $\hat{\mathbf{n}}$ is determined by our choice of the directionality of $d\vec{\ell}$: curl the fingers of your right hand in the direction of $d\vec{\ell}$, and your thumb points in the direction of $\hat{\mathbf{n}}$, along $\hat{\mathbf{k}}$ in this case. The form of the magnetic field we're investigating is (9.5.5): $\mathbf{B} = B(x, t)\, \hat{\mathbf{k}}$, therefore $\mathbf{B} \cdot \hat{\mathbf{n}} = B(x, t)$. Since Δx is infinitesimal, and since B doesn't depend on y, B is uniform over the area bounded by the loop. Therefore, $\mathbf{B} \cdot \hat{\mathbf{n}} = B \, \Delta x \, \Delta y$. Plugging this and equation (9.5.7) into Faraday's Law (9.5.3) gives

$$\frac{\partial E}{\partial x} \Delta x \, \Delta y = -\frac{\partial B}{\partial t} \Delta x \, \Delta y \Rightarrow$$

$$\frac{\partial E}{\partial x} = -\frac{\partial B}{\partial t}. \tag{9.5.8}$$

So, fields of the form $\mathbf{E} = E(x, t)\, \hat{\mathbf{y}}$ and $\mathbf{B} = B(x, t)\, \hat{\mathbf{z}}$ are allowed by Faraday's law, so long as E and B are related as shown by equation (9.5.8).

There is one more Maxwell equation that we must check, Ampère's law:

$$\oint B \cdot d\vec{\ell} = \mu_0 I_{\substack{\text{net} \\ \text{threading}}} + \mu_0 \varepsilon_0 \frac{d}{dt} \int \mathbf{E} \cdot \hat{\mathbf{n}} \, dA.$$

Since we are considering fields in a vacuum, $I_{\substack{\text{net} \\ \text{threading}}} = 0$. The reasoning following from this is almost exactly the same as the reasoning for Faraday's law. You can show in problem 9.6 that the result is

$$\frac{\partial B}{\partial x} = -\mu_0 \varepsilon_0 \frac{\partial E}{\partial t}. \tag{9.5.9}$$

Let's combine equations (9.5.9) with (9.5.8). We begin by taking $\dfrac{\partial}{\partial x}$ of both sides of equation (9.5.8), giving

$$\frac{\partial^2 E}{\partial^2 x} = -\frac{\partial}{\partial x}\frac{\partial B}{\partial t} = -\frac{\partial}{\partial t}\frac{\partial B}{\partial x}.$$

Now, we use equation (9.5.9) to substitute for $\dfrac{\partial B}{\partial x}$, giving

$$\frac{\partial^2 E}{\partial^2 x} = -\frac{\partial}{\partial t}\left(-\mu_0\varepsilon_0\frac{\partial E}{\partial t}\right) \Rightarrow$$

$$\frac{\partial^2 E}{\partial^2 x} = \mu_0\varepsilon_0\frac{\partial^2 E}{\partial t^2}.$$

This is the wave equation, (9.2.5): $\dfrac{\partial^2 y}{\partial t^2} = v_{\text{p}}^2\dfrac{\partial^2 y}{\partial x^2}$, with E playing the role of y, and phase velocity

$$\boxed{c = \frac{1}{\sqrt{\varepsilon_0\mu_0}}} \qquad (9.5.10)$$

If you plug in the numbers on this, you find a speed of 2.998×10^8 m/s, exactly equal to the speed of light!

What about the magnetic field? You can make any wave by adding up sinusoids. So, for simplicity, let's consider

$$\mathbf{E} = E_0 \sin k\,(x - ct)\hat{\mathbf{j}}.$$

Since the phase velocity $c = \omega/k$, we can rewrite this as $\mathbf{E} = E\hat{\mathbf{j}}$ with

$$E = E_0 \sin(kx - \omega t). \qquad (9.5.11)$$

Plugging this into equation (9.5.8): $\dfrac{\partial E}{\partial x} = -\dfrac{\partial B}{\partial t}$ gives

$$\frac{\partial B}{\partial t} = -E_0 k \cos(kx - \omega t) \xrightarrow[\text{respect to } t]{\text{integrate with}} B = E_0\frac{k}{\omega}\sin(kx - \omega t) + \text{const.}$$

Because we are interested in waves, we'll set the constant equal to zero. (The presence of the constant means that we're allowed to have a uniform, constant magnetic field in addition to the electromagnetic wave.) Since $c = \omega/k$, this gives

$$B = E_0\frac{1}{c}\sin(kx - \omega t) = \frac{E}{c} \Rightarrow$$

$$\boxed{E = cB.} \qquad (9.5.12)$$

Relation between amplitudes of E and B for an electromagnetic wave in vacuum

Because of this direct proportionality, we see that the electric and magnetic components of the wave are in phase. This means that the magnetic field also propagates as a wave, moving at speed c.

We have just shown that plane waves are allowed by all four of Maxwell's equations, but only if they travel at speed c and have $E = cB$. These electromagnetic waves (which include light, radio waves, X-rays, microwaves, gamma rays, etc.) consist of a self-sustaining oscillation, in which the changing magnetic field creates a changing electric field, which in turn creates a changing magnetic field, etc. The whole thing can only keep going if it travels at exactly c. The realization that the speed of light is a consequence of Maxwell's equations led directly to Einstein's theory of special relativity: If we assume that the laws of physics are the same in all reference frames traveling at constant velocity, then Maxwell's equations must apply equally well in all such frames, and the speed of light must be exactly the same in all such frames. This is the basic postulate of special relativity, and immediately leads to such counterintuitive results as the speed of a light beam emanating from the flash light being *exactly* the same as perceived in the frame in which the flashlight is at rest as it is in a frame moving at speed $0.999\ c$ in the direction the beam is traveling!

Your turn: On the previous page, we showed that $E(x, t) = cB(x, t)$ for a wave moving in the positive x-direction, meaning that the E and B waves are in phase. Now, show that $E(x, t) = -cB(x, t)$ for a wave moving along the negative x-direction, meaning that the E and B waves are out of phase.

For any electromagnetic (em) wave, the propagation direction is given by the "Poynting vector":

$$\mathbf{S} = \frac{1}{\mu_0}\mathbf{E} \times \mathbf{B}. \tag{9.5.13}$$

(The fortuitously named John Henry Poynting was a student of Maxwell.) In section 9.10, we'll see that this vector not only points in the direction of propagation, but also has a magnitude equal to the intensity (power per area) of the wave.

At the risk of being repetitive, let us reinforce the true nature of a plane wave, the type of wave we've been discussing. Recall that we began with the assumption (9.5.5) that E and B depend on x, but not on y or z. Some students misinterpret this to mean that the wave exists only as a ray along the x-axis. This incorrect thinking is reinforced by the usual graphical way of portraying a plane wave, shown in figure 9.5.4a. This shows the magnitudes of $\mathbf{E}$ and $\mathbf{B}$ for points along the x-axis, using the convention that the length of the vector represents the strength of the field. However, this picture must *not* be taken to mean that $\mathbf{E}$ and $\mathbf{B}$ are zero at points that aren't on the x-axis. Instead, since E and B depend only on x, and not on y or z, the picture would be the same for any line parallel to the x-axis, as shown in figure 9.5.4b. Figure 9.5.5 shows the correspondence between the two ways of portraying the strength of the electric field. In the foreground, the strength is indicated by the spacing between field lines; this emphasizes that the field extends throughout space, and is has the same strength and direction throughout any plane perpendicular to the x-axis. In the background, the strength of the field is indicated by the length of the vectors.

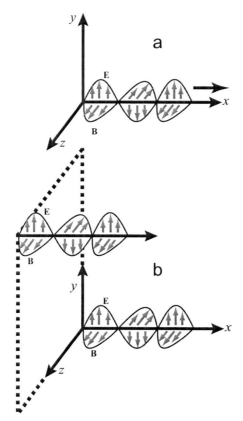

Figure 9.5.4 a: The conventional way of portraying an electromagnetic wave, showing the magnitudes of the electric and magnetic fields along the x-axis. b: Another portrayal, emphasizing that the magnitudes of the fields are equal at equivalent points along any line parallel to the x-axis.

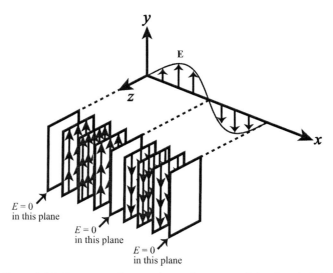

Figure 9.5.5 Two ways of portraying a plane wave of the electric field.

Key equations for isomorphisms. Over the rest of this chapter, we will develop isomorphisms (exact analogies) between em waves in vacuum and various other types of waves. The essential relations are:

$$(9.5.8): \frac{\partial E}{\partial x} = -\frac{\partial B}{\partial t} \quad \text{and} \quad (9.5.9): \frac{\partial B}{\partial x} = -\mu_0 \varepsilon_0 \frac{\partial E}{\partial t}.$$

All the other results follow from these two relations, including the fact that the two components of the wave (in this case E and B) each obey the wave equation, the relation between the magnitude of the two components ($E = cB$), and the speed of the wave $c = \dfrac{1}{\sqrt{\varepsilon_0 \mu_0}}$. So, if we can find equations relating the two components of a wave that are isomorphic to equations (9.5.8) and (9.5.9), then there is a complete isomorphism. It will sometimes make things a bit easier to substitute $c = \dfrac{1}{\sqrt{\varepsilon_0 \mu_0}}$ into equation (9.5.9), so that the two essential relations become

$$(9.5.8): \frac{\partial E}{\partial x} = -\frac{\partial B}{\partial t}$$

$$\text{and} \quad \frac{\partial B}{\partial x} = -\frac{1}{c^2}\frac{\partial E}{\partial t}. \tag{9.5.14}$$

Isomorphism with rope waves. As a guide to developing the isomorphism between rope waves and em waves in vacuum, we refer back to the isomorphism between the mechanical oscillator and the electrical oscillator. From section 1.5, we have:

Table 9.5.1. Isomorphism between mechanical and electrical oscillators

Mass and spring	Electrical oscillator
Position relative to equilibrium x	Charge q on capacitor
Mass m	Inductance L
Spring constant k	Inverse capacitance $1/C$

For the mechanical oscillator, the oscillations are a result of the restoring force which tends to bring the system back toward equilibrium and the momentum which tends to make the system overshoot the equilibrium point. The restoring force is associated with the spring constant k which is isomorphic to $1/C$. The capacitor is associated with electric fields. Therefore, we may expect some type of connection between the restoring force in a rope and the E for the em wave. The momentum is associated with the mass m which is isomorphic to L. The inductor is associated with magnetic fields. Therefore, we may expect some type of connection between the momentum of a rope wave and the B for the em wave.

Now, let's get more quantitative. For a rope wave, we speak of the mass per length, μ, rather than simply the mass (which would be infinite for an infinitely long rope). Thus, the transverse momentum per unit length is given by $\mu \dfrac{\partial y}{\partial t}$. For simplicity, we will refer to this as the "momentum wave," and use the symbol $p_y \equiv \mu \dfrac{\partial y}{\partial t}$, bearing in mind that this is really the transverse momentum per unit length. We will hope that this is isomorphic to B, and search for the quantity (related to the restoring force) that is isomorphic to E.

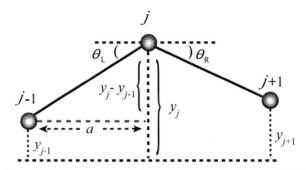

Figure 9.5.6 Geometry needed for calculating the force on bead j.

For a right-traveling em wave in vacuum, E and B are in phase, so the quantity we are searching for in the rope case should be in phase with the momentum wave. From equation (9.2.2), which describes waves on a beaded rope, we have that the acceleration of bead j is given by $\ddot{y}\left(x_j\right) \cong \dfrac{Ta}{m}\left.\dfrac{\partial^2 y}{\partial x^2}\right|_{x_j}$, where T is the tension, a is the spacing between beads, and m is the mass of a bead. Therefore, the total force in the y-direction on the bead, which equals $m\ddot{y}$, is proportional to $\dfrac{\partial^2 y}{\partial x^2}$. If we consider a sinusoidal wave $y = A\cos(kx - \omega t)$, we have that $\dfrac{\partial^2 y}{\partial x^2} = -Ak^2\cos(kx - \omega t)$, which is *not* in phase with the momentum wave $p_y = \mu\dfrac{\partial y}{\partial t} = \mu A\omega\sin(kx - \omega t)$.

However, from figure 7.1.1, reproduced here as figure 9.5.6, we see that the y-component of the force exerted *from the left side* on bead j is given by

$$-T\sin\theta_L = -T\frac{y_j - y_{j-1}}{\sqrt{a^2 + \left(y_j - y_{j-1}\right)^2}} \cong -\frac{T}{a}\left(y_j - y_{j-1}\right),$$

where the last step follows because we assume $a \gg y_j - y_{j-1}$. We denote this force as $F_L = -\dfrac{T}{a}\left(y_j - y_{j-1}\right)$.

> $F_L \equiv$ **the y-component of force exerted on a bead or piece of rope from the left**

(We will be using this quantity quite a bit in the rest of this section and in chapter 10, so please make sure you understand this definition.)

Since $a = \Delta x$, we then have

$$F_L = -T\frac{\Delta y}{\Delta x}.$$

In the limit of a continuous rope (for which we decrease a and m toward zero while keeping the mass per length, $\mu = \dfrac{m}{a}$, constant), this becomes

$$F_L = -T\frac{\partial y}{\partial x}, \tag{9.5.15}$$

where now we interpret this as the y-component of the force exerted from the left side on a tiny segment of the rope. To check whether this is in phase with the momentum wave, we again consider the sinusoidal wave $y = A\cos(kx - \omega t)$. For this, $F_L = -T\dfrac{\partial y}{\partial x} = TAk\sin(kx - \omega t)$, which *is* in phase with the momentum wave $p_y = \mu\dfrac{\partial y}{\partial t} = \mu A\omega\sin(kx - \omega t)$.

It may seem odd that the y-component of the force exerted from the left should be so important. However, if you hold the left end of a long rope with your hand, you must exert a y-component of force (from the left) on this end to start a wave propagating down the rope. Saying the same thing in a different way: for right-traveling waves, it is the y-component of force exerted from the left which is responsible for the propagation of a wavefront to the right on a rope that is initially at equilibrium; the force from the left is what pulls an initially quiescent piece of rope away from equilibrium. We will see in chapter 10 that this idea of the y-component of force exerted from the left is quite important for the behavior of right-traveling waves when they encounter an interface between two ropes with different values of μ.

So, are the momentum wave $p_y \equiv \mu\dfrac{\partial y}{\partial t}$ and the y-force-from-the-left wave $F_L = -T\dfrac{\partial y}{\partial x}$ related in the same ways as B and E? From equation (9.2.2), we have for the beaded rope

$$\frac{\partial^2 y\left(x_j\right)}{\partial t^2} = \frac{Ta}{m}\left.\frac{\partial^2 y}{\partial x^2}\right|_{x_j} \Leftrightarrow \left.T\frac{\partial^2 y}{\partial x^2}\right|_{x_j} = \frac{m}{a}\frac{\partial^2 y\left(x_j\right)}{\partial t^2} \Rightarrow -\frac{\partial}{\partial x}\left[-T\frac{\partial y}{\partial x}\right] = \frac{\partial}{\partial t}\left[\frac{m}{a}\frac{\partial y}{\partial t}\right] \Rightarrow$$

$$\frac{\partial}{\partial x}F_L = -\frac{\partial}{\partial t}\left[\mu\frac{\partial y}{\partial t}\right] \xleftrightarrow{\text{isomorphic with}} (9.5.8)\colon \frac{\partial E}{\partial x} = -\frac{\partial B}{\partial t}.$$

We see that, in the isomorphism for this equation, F_L plays the role of E, and $p_y \equiv \mu\dfrac{\partial y}{\partial t}$ plays the role of B, as expected.

To complete the isomorphism between rope waves and em waves in a vacuum, we need to find the equation for a rope that is isomorphic with equation (9.5.14): $\dfrac{\partial B}{\partial x} = -\dfrac{1}{c^2}\dfrac{\partial E}{\partial t}$. Because the order of derivatives doesn't matter, we have

$$\frac{\partial}{\partial x}\frac{\partial y}{\partial t} = \frac{\partial}{\partial t}\frac{\partial y}{\partial x}.$$

From equation (9.2.4), $v_p \equiv \sqrt{\dfrac{T}{\mu}} \Rightarrow \mu = \dfrac{T}{v_p^2}$. Combining this with the above gives

$$\mu\frac{\partial}{\partial x}\frac{\partial y}{\partial t} = \frac{T}{v_p^2}\frac{\partial}{\partial t}\frac{\partial y}{\partial x} \Rightarrow \frac{\partial}{\partial x}\left[\mu\frac{\partial y}{\partial t}\right] = -\frac{1}{v_p^2}\frac{\partial}{\partial t}\left[-T\frac{\partial y}{\partial x}\right] \Rightarrow$$

$$\frac{\partial}{\partial x}p_y = -\frac{1}{v_p^2}\frac{\partial}{\partial t}F_L \xleftrightarrow{\text{isomorphic with}} (9.5.14)\colon \frac{\partial B}{\partial x} = -\frac{1}{c^2}\frac{\partial E}{\partial t}.$$

So, there is a complete isomorphism between rope waves and em waves in a vacuum. Therefore, there is a isomorphism between the phase velocity for rope waves, $v_p = \sqrt{\dfrac{T}{\mu}}$,

Table 9.5.2. Isomorphism between rope waves and electromagnetic waves in a vacuum

Rope waves	Electromagnetic waves in vacuum
Transverse velocity $\dot{y}$	B/μ_0
y-component of force from the left	Electric field E
$F_L = -T\dfrac{\partial y}{\partial x}$	
Mass per length μ	Permeability of free space μ_0
Tension T	Inverse of permittivity of free space: $1/\varepsilon_0$

and the phase velocity for em waves in a vacuum, $c = \dfrac{1}{\sqrt{\mu_0 \varepsilon_0}}$. However, it is not clear whether we can push things further than this. For example, is T isomorphic to $1/\mu_0$ and μ isomorphic to ε_0, or instead is T isomorphic to $1/\varepsilon_0$ and μ isomorphic to μ_0? Since μ is associated with momentum (which is isomorphic to B) and μ_0 is associated with B, one might expect that μ might be isomorphic to μ_0. Further, since T is associated with the force from the left (which is isomorphic to E) and ε_0 is associated with E, one might expect that T is isomorphic to $1/\varepsilon_0$. In problem 9.5, you can show that these are indeed the correct isomorphisms. Since $p_y \equiv \mu\dfrac{\partial y}{\partial t}$ is isomorphic to B and μ is isomorphic to μ_0, we see that $\dot{y} = \dfrac{\partial y}{\partial t}$ is isomorphic to B/μ_0, and this version of the isomorphism turns out to be somewhat more useful.

The remaining isomorphisms with other types of waves will take a lot less effort to develop!

Self-test: Use this isomorphism and $E = cB$ first to show that the F_L wave is in phase with the $\dot{y}$ wave, and then to show that $\dfrac{\partial y}{\partial t} = -v_p\dfrac{\partial y}{\partial x}$ for a rope wave. (You should find this helpful for problem 9.2. You can derive this expression using a different method in problem 9.1.)

Since $E = cB$ for right-moving em waves in vacuum, the E and B waves are in phase. By the isomorphism, this means that the F_L and $\dot{y}$ waves are in phase for right-moving rope waves. For left-moving waves, $E = -cB$ (as you showed earlier in this section), so E and B are out of phase. By the isomorphism, this means that the F_L and $\dot{y}$ waves are out of phase for left-moving rope waves.

9.6   Electromagnetic waves in matter

A full description of electric and magnetic fields in matter is well beyond this text. For a deeper understanding, you should consult an introductory electrodynamics text.[2]

---

2. The best treatment at the undergraduate level is *Introduction to Electrodynamics, 3rd Ed.*, by David J. Griffiths (Prentice-Hall, Upper Saddle River, NJ, 1999).

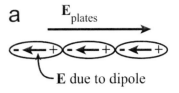

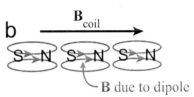

Figure 9.6.1 Part a: when electric dipoles align with a field, the field they create opposes the applied field. Part b: when magnetic dipoles align with a field, the field they create enhances the applied field.

What follows is just the basics for simple geometries and the most common materials.

When the space between the plates of a parallel plate capacitor is filled with an insulating material, the atoms and molecules within the material become polarized by the electric field; each atom or molecule becomes an electric dipole, as shown schematically in figure 9.6.1a. (Note that the polarization is caused not only by the field due to the plates, but instead by the combined field of the plates and all the other dipoles in the material.) If we consider a line of such dipoles within the material, we see that the electric field due to the dipoles opposes the electric field due to the plates, so that the total field between the plates is reduced. The factor of reduction is called the dielectric constant:

$$\kappa = \frac{E_{\text{plates}}}{E_{\text{total}}}. \tag{9.6.1}$$

The parallel plate capacitor provides a particularly simple geometry; in a more complicated geometry, the relationship between the total field and the field due to the "free charge" (the charge on the plates) is more complicated. For almost all materials, the dielectric constant really is constant, meaning that it doesn't depend on the strength of the electric field (at least until the field gets quite strong). Another way of saying this is that the degree of polarization is linearly proportional to the total electric field. *We will focus exclusively on these "linear materials,"* though you should be aware that there are some materials for which the behavior is more complicated.

An insulating material is also called a "dielectric," derived from the Greek "dia" meaning "through," therefore indicating that an electric field can penetrate through a dielectric. This is a bit of a misnomer, since the dielectric constant for typical insulating solids is from about 2–8, meaning that the field in a parallel plate capacitor is reduced by a factor of 2–8 by the presence of the dielectric. We define the permittivity of the dielectric material to be

$$\varepsilon = \kappa \varepsilon_0. \tag{9.6.2}$$

Thus, ε is always greater than or equal to ε_0.

Similarly, when the space inside a solenoid is filled with matter, the magnetic dipoles associated with the spin of the electrons and with the orbital motion of the electrons are affected by the magnetic field. The effect on the electron spins is shown schematically in figure 9.6.1b. Again, the dipoles associated with the electron spin align "with" the field of the solenoid coil (meaning that the north pole of each dipole is on the right, and the south pole on the left), but for magnetic dipoles the field due to the dipole *within* the dipole points from south to north, in the same direction as the field from the coil. (This is opposite to what happens for an electric dipole; inside the dipole, the field due to the dipole points from positive to negative.) Therefore, the field of the spin dipoles enhances the field due to the coils somewhat. A material with this type of behavior is therefore called "paramagnetic," from the Greek "para" meaning "alongside." However, in other materials, the magnetic response is dominated by the interaction with the orbital motion of the electron, which is more complicated. The net effect of this interaction is to make the magnetic moment associated with the orbital motion align *opposite* the field of the coils. Thus, if the solenoid is filled with a material dominated by this type of response, the total field is smaller than without the material. Such materials are called "diamagnetic." Again, *we restrict ourselves to linear materials* for which the degree of polarization is proportional to the total field. We define the permeability of the material to be

$$\mu = \mu_0 \left(1 + \chi_m\right), \tag{9.6.3}$$

where χ_m (Greek letter "chi"-sub-m) is the "magnetic susceptibility." For diamagnetic materials, the susceptibility is negative, and typically ranges from 10^{-6} to 10^{-5}. For paramagnetic materials, the susceptibility is positive, and typically ranges from 10^{-5} to 10^{-4}. In any case, the effect on the total field is quite small, in strong contrast to the case for dielectrics. (Note that we are *not* discussing ferromagnetic materials such as iron, which dramatically affect the magnetic field.) Thus, for linear materials, μ is close to μ_0; it is slightly less for diamagnetic materials, and slightly more for paramagnetic materials.

It is convenient to define a new field called "H":

$$\mathbf{H} = \frac{1}{\mu}\mathbf{B}. \tag{9.6.4}$$

Even for linear materials there is no simple physical interpretation for **H** other than what is evident from the above equation. Within the material, it is still the magnetic field **B** that exerts the force $\mathbf{F} = q\mathbf{v} \times \mathbf{B}$ on moving charges. However, **H** is calculationally convenient. (It turns out that, *for symmetrical geometries*, you can calculate **H** just from the currents in coils, without referring to the complexities of the diamagnetic or paramagnetic materials.)

The standard form of Maxwell's equations is still correct within any material, whether linear or not. However, it is sometimes convenient to separate the effects of the "free charges" (such as the charges on capacitor plates and the charges flowing through a solenoid coil) from those of "bound charges" (the charges which rearrange slightly when atoms and molecules are polarized by an electric field) and "bound currents" (the currents associated with electron spin and orbital motion). To create the alternate set of

Maxwell's equations which is valid in a linear material, we simply replace ε_0 by ε, μ_0 by μ, and all charges by just the free charges. The two equations we care about most are Faraday's law and Ampère's Law. Applying the recipe to Ampère's Law, we get

$$\underbrace{\oint \mathbf{B} \cdot d\vec{\ell} = \mu_0 I_{\underset{\text{threading}}{\text{net}}} + \mu_0 \varepsilon_0 \frac{d}{dt} \int \mathbf{E} \cdot \hat{\mathbf{n}}\, dA \rightarrow}_{\text{Ampere's Law for any circumstance}}$$

$$\underbrace{\oint \mathbf{B} \cdot d\vec{\ell} = \mu I_{\underset{\underset{\text{threading}}{\text{net}}}{\text{free}}} + \mu \varepsilon \frac{d}{dt} \int \mathbf{E} \cdot \hat{\mathbf{n}}\, dA.}_{\substack{\text{An alternate version of Ampere's Law} \\ \text{for an isotropic linear material}}}$$

We can rearrange this a bit to obtain

$$\oint \frac{1}{\mu} \mathbf{B} \cdot d\vec{\ell} = I_{\underset{\underset{\text{threading}}{\text{net}}}{\text{free}}} + \varepsilon \frac{d}{dt} \int \mathbf{E} \cdot \hat{\mathbf{n}}\, dA.$$

If we restrict ourselves to materials that are not electrically conducting (so that $I_{\underset{\underset{\text{threading}}{\text{net}}}{\text{free}}} = 0$), and make use of $\mathbf{H} = \dfrac{1}{\mu} \mathbf{B}$, we get

$$\oint \mathbf{H} \cdot d\vec{\ell} = \varepsilon \frac{d}{dt} \int \mathbf{E} \cdot \hat{\mathbf{n}}\, dA . \tag{9.6.5}$$

Recall that the original version of Ampère's Law in vacuum, $\oint \mathbf{B} \cdot d\vec{\ell} = \mu_0 \varepsilon_0 \dfrac{d}{dt} \int \mathbf{E} \cdot \hat{\mathbf{n}}\, dA$, when applied to a plane wave led to equation (9.5.9), $\dfrac{\partial B}{\partial x} = -\mu_0 \varepsilon_0 \dfrac{\partial E}{\partial t}$. So, we can see that equation (9.6.5) leads to

$$\frac{\partial H}{\partial x} = -\varepsilon \frac{\partial E}{\partial t} \Rightarrow$$

$$\frac{\partial (\mu H)}{\partial x} = -\varepsilon \mu \frac{\partial E}{\partial t} \xleftrightarrow{\text{isomorphic with}} (9.5.14) : \frac{\partial B}{\partial x} = -\frac{1}{c^2} \frac{\partial E}{\partial t}.$$

We see that μH plays the role of B, E plays the role of E, and $\varepsilon \mu$ plays the role of $1/c^2$.

To complete the isomorphism, we must find the equation for em fields in matter that is isomorphic to equation (9.5.8): $\dfrac{\partial E}{\partial x} = -\dfrac{\partial B}{\partial t}$. If we apply the recipe of replacing ε_0 by ε, μ_0 by μ, and all charges by just the free charges to Faraday's Law, there is no change:

$$\underbrace{\oint \mathbf{E} \cdot d\vec{\ell} = -\frac{d}{dt} \int \mathbf{B} \cdot \hat{\mathbf{n}}\, dA}_{\text{Faraday's Law for any circumstance}} \rightarrow \underbrace{\oint \mathbf{E} \cdot d\vec{\ell} = -\frac{d}{dt} \int \mathbf{B} \cdot \hat{\mathbf{n}}\, dA.}_{\substack{\text{An alternate version of Faraday's Law} \\ \text{for linear materials}}}$$

However, since $\mathbf{B} = \mu\,\mathbf{H}$, we can rewrite this as

$$\oint \mathbf{E} \cdot \mathrm{d}\vec{\ell} = -\frac{\mathrm{d}}{\mathrm{d}t} \int \mu\,\mathbf{H} \cdot \hat{\mathbf{n}}\,\mathrm{d}A. \qquad (9.6.6)$$

Recall that the original version of Faraday's law when applied to a plane wave led to equation (9.5.8), $\dfrac{\partial E}{\partial x} = -\dfrac{\partial B}{\partial t}$, so we can see that equation (9.6.6) leads to

$$\frac{\partial E}{\partial x} = -\frac{\partial\,(\mu H)}{\partial t} \xleftrightarrow{\text{isomorphic with}} (9.5.8): \frac{\partial E}{\partial x} = -\frac{\partial B}{\partial t}.$$

Thus, the isomorphism between em waves in a vacuum and em waves in linear materials is complete. As for the case with rope waves, it will be convenient to write the isomorphism in terms of B/μ_0, which is isomorphic to $\mu H/\mu = H$:

Table 9.6.1. Isomorphism between electromagnetic waves in linear materials and in a vacuum

Electromagnetic waves in linear materials	Electromagnetic waves in vacuum
H	B/μ_0
Electric Field E	Electric Field E
Permeability μ	Permeability of free space μ_0
Permittivity ε	Permittivity of free space ε_0

Since $B = \mu H$, you may well ask, "Why bother to write H in the isomorphism rather than just B/μ"? Indeed, we could well have done so. We will see that when we consider reflections at a boundary between two different media (with different values of μ), it is somewhat easier to phrase things in terms of H.

Concept test (answer below[3]): What is the speed of em waves in linear materials?

Concept test (answer below[4]): Recall that in section 9.5 I stated that the intensity (power per area) for em waves in vacuum is $\mathbf{S} = \dfrac{1}{\mu_0}\mathbf{E} \times \mathbf{B}$. Use this to explain why the intensity of em waves in linear materials is given by the magnitude of

$$\boxed{\mathbf{S} = \mathbf{E} \times \mathbf{H}.} \qquad (9.6.7)$$

---

3. In vacuum, $c = \dfrac{1}{\sqrt{\mu_0\,\varepsilon_0}}$, so in linear media $v_{\mathrm{p}} = \dfrac{1}{\sqrt{\mu\,\varepsilon}}$.

4. From equation (9.5.13), in vacuum $\mathbf{S} = \dfrac{1}{\mu_0}\mathbf{E} \times \mathbf{B} = \mathbf{E} \times \left(\dfrac{\mathbf{B}}{\mu_0}\right)$. Using the isomorphism, this translates into $\mathbf{S} = \mathbf{E} \times \mathbf{H}$.

> **Concept test (answer below[5]):** Explain why the intensity of em waves in linear materials is given by
>
> $$S = \sqrt{\frac{\varepsilon}{\mu}} E^2. \tag{9.6.8}$$

9.7   Waves on transmission lines

One of the two most important applications for em waves is the transmission of information. For example, we use em waves for radio and television broadcasts. However, we frequently want to have better control of where information goes, and for these applications we use cables, such as those used for telephone, cable TV, computer networks, speaker wires, and so on. Such cables are called "transmission lines" by physicists. The information is always transmitted through a pair of conductors. For a telephone line, there are two wires twisted together. For cable TV, there is a central wire surrounded by a cylindrical outer conductor, with plastic insulation between. (This is called a "coaxial cable.") For transmission of information on circuit boards, the transmission line often consists of a wire above a sheet of metal called a "ground plane"; the ground plane acts as the second conductor. Sometimes, the metal chassis of an apparatus, or the ground itself, is used as one of the two conductors needed for a transmission line.

Depending on the system, the information is either represented by the time-varying voltage difference between the two conductors, or by the time-varying current traveling through them (into a target device on one wire, and back out on the other). No matter whether the information is transmitted by applying a controlled voltage to one end of the transmission line or by applying a controlled current, it propagates as a linked wave of current and voltage travelling along the line. The current creates a magnetic field, and the voltage difference between the two conductors is associated with an electric field between them, so that the wave is actually a self-sustaining em phenomenon; the mathematics are analogous to those for an em wave in a vacuum, with the current analogous to B and the voltage analogous to E. We will see that, in a standard coaxial cable such as you may have used in the laboratory to connect electrical signals to an oscilloscope, the wave travels at $^2\!/_3\, c$.

Our reasoning about these waves begins with two points: (i) Any two conductors that aren't infinitely far apart have a capacitance between them. (ii) Any length of wire, even if it's straight, has some inductance. One way to see this is that when you run a current through a straight wire, it sets up a magnetic field. It takes energy to create this

---

5. From (9.5.12), $E = cB = \dfrac{1}{\sqrt{\mu_0 \varepsilon_0}} \mu_0 \left(\dfrac{B}{\mu_0} \right) = \sqrt{\dfrac{\mu_0}{\varepsilon_0}} \left(\dfrac{B}{\mu_0} \right)$. This translates into $E = \sqrt{\dfrac{\mu}{\varepsilon}} H \Leftrightarrow$

$H = \sqrt{\dfrac{\varepsilon}{\mu}} E$. Substituting this into equation (9.6.7) gives equation (9.6.8).

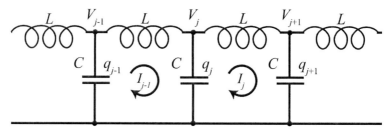

Figure 9.7.1 Model for a transmission line.

field (recall from introductory electricity and magnetism that there is an energy density associated with the magnetic field), so you know by a Lenz's law type of argument that this means it will be "harder" to start current flowing through the wire, which is exactly the characteristic of an inductor: $V = L\dfrac{dI}{dt}$ is the voltage produced across an inductor when the current through it is changed, in the same way that $V = IR$ is the voltage produced across a resistor when the current I flows through it.

Using (i) and (ii), you should now find it reasonable that any transmission line can be modeled as a series of inductors along one wire (the top wire in figure 9.7.1) and capacitors to the other wire, as shown. For a coaxial cable, the top wire would be the inner conductor and the lower wire would be the cylindrical outer conductor. For the single wire running above a ground plane, the top wire would be the wire running above the ground and the lower wire would be the ground plane. We begin with the "lumped circuit element" model shown; very soon, we will take the limit where the size of each cell in the model becomes infinitesimally small, corresponding to the limit of a continuous transmission line. For mathematical simplicity, we assume the lower wire is grounded, that is, that it is at zero voltage. We define the current circulating within each cell to be positive when it is moving clockwise, as shown, that is, moving to the right on the top wire and to the left on the bottom wire.

We will find an isomorphism between this system and em waves in vacuum.

Your turn: (a) Explain briefly why $I_{j-1} - I_j = \dfrac{dq_j}{dt}$, and why this equals $C\dfrac{dV_j}{dt}$.

(b) Explain briefly why $L\dfrac{dI_j}{dt} = V_j - V_{j+1}$

From the result of part (b), we have

$$L\frac{dI_j}{dt} = V_j - V_{j+1} = -\left(V_{j+1} - V_j\right) \equiv -\Delta V.$$

Dividing both sides of this by the cell size a gives

$$\frac{L}{a}\frac{dI_j}{dt} = -\frac{\Delta V}{a} \Leftrightarrow \frac{\Delta V}{a} = -\frac{L}{a}\frac{dI_j}{dt}.$$

We define L_0 to be the inductance per unit length. Therefore $L/a \xrightarrow{\lim a \to 0} L_0$. So, in the limit that the cell size becomes infinitesimal, the above becomes

$$\frac{\partial V}{\partial x} = -L_0 \frac{\partial I}{\partial t} \Leftrightarrow$$

$$\frac{\partial V}{\partial x} = -\frac{\partial (L_0 I)}{\partial t} \xleftarrow{\text{isomorphic with}} (9.5.8): \frac{\partial E}{\partial x} = -\frac{\partial B}{\partial t}.$$

In the isomorphism, V plays the role of E, and $L_0 I$ plays the role of B.

To show that the isomorphism is complete, we must find the equation that is isomorphic to equation (9.5.9): $\frac{\partial B}{\partial x} = -\mu_0 \varepsilon_0 \frac{\partial E}{\partial t}$. From part (a) of "Your turn" given earlier, we have

$$C \frac{dV_j}{dt} = I_{j-1} - I_j = -\left(I_j - I_{j-1} \right) \equiv -\Delta I.$$

Dividing both sides of this by the cell size a gives

$$\frac{\Delta I}{a} = -\frac{C}{a} \frac{dV_j}{dt}.$$

We define C_0 to be the capacitance per unit length. Therefore $C/a \xrightarrow{\lim a \to 0} C_0$. So, in the limit that the cell size becomes infinitesimal, the above becomes

$$\frac{\partial I}{\partial x} = -C_0 \frac{\partial V}{\partial t} \Leftrightarrow$$

$$\frac{\partial (L_0 I)}{\partial x} = -L_0 C_0 \frac{\partial V}{\partial t} \xleftarrow{\text{isomorphic with}} (9.5.9): \frac{\partial B}{\partial x} = -\mu_0 \varepsilon_0 \frac{\partial E}{\partial t}.$$

Again, V plays the role of E and $L_0 I$ plays the role of B. We also see that the combination $L_0 C_0$ plays the role of the combination $\mu_0 \varepsilon_0$. As for the isomorphism between rope waves and em waves in vacuum, it is not clear whether L_0 is isomorphic with μ_0 and C_0 with ε_0, or instead whether L_0 is isomorphic with ε_0 and C_0 with μ_0. However, since L_0 is associated with the current (which is isomorphic to B), and μ_0 is associated with B, one might expect that L_0 might be isomorphic to μ_0. Further, since C_0 is associated with the voltage (which is isomorphic to E), and ε_0 is associated with E, one might expect that C_0 is isomorphic to ε_0. By considering the power transmitted in the wave, you can show that this hunch is correct; see problem 9.14.

So, the isomorphism between waves on transmission lines and em waves in vacuum is complete. For convenience in comparing with other isomorphisms, we say that $1/C_0$ is isomorphic to $1/\varepsilon_0$, instead of saying that C_0 is isomorphic to ε_0 (which is equivalent). Also, we write the isomorphism in terms of $L_0 I/L_0 = I$, which is isomorphic to B/μ_0. See table 9.7.1.

We can use this isomorphism to quickly obtain some important results for waves on transmission lines:

1. Em waves in vacuum are a self-sustaining combination of a wave in **E** and a wave in **B**. By the isomorphism, waves on a transmission line are a self-sustaining combination of a wave in V and a wave in I.

Table 9.7.1. Isomorphism between waves on a transmission line and electromagnetic waves in a vacuum

Waves on a transmission line	Electromagnetic waves in vacuum
I	B/μ_0
V	Electric Field E
Inductance per length L_0	Permeability of free space μ_0
Inverse of capacitance per length: $1/C_0$	Inverse of permittivity of free space: $1/\varepsilon_0$

2. The speed of em waves in vacuum is $c = \dfrac{1}{\sqrt{\mu_0 \varepsilon_0}}$, so the phase velocity of waves on transmission lines is

$$v_{\mathrm{p}} = \frac{1}{\sqrt{L_0 C_0}}. \tag{9.7.1}$$

Standard RG58 coaxial cable (the type used in laboratories, and the type that has a BNC coaxial connector at the end) has $C_0 = 100$ pF/m and $L_0 = 250$ nH/m; plugging in these numbers gives $v_{\mathrm{p}} = \dfrac{1}{\sqrt{L_0 C_0}} = 2.00 \times 10^8$ m/s, or almost exactly 2/3 c. The coaxial cable typically used for cable television, RG6, has $C_0 = 53.1$pF/m and $L_0 = 348$ nH/m; plugging in these numbers gives $v_{\mathrm{p}} = \dfrac{1}{\sqrt{L_0 C_0}} = 2.33 \times 10^8$ m/s, which is a bit more than $3/4$ c.

3. For em waves in vacuum travelling in the positive x-direction, $E(x, t) = c B(x, t) = \dfrac{1}{\sqrt{\mu_0 \varepsilon_0}} B(x, t) = \sqrt{\dfrac{\mu_0}{\varepsilon_0}} \dfrac{B(x, t)}{\mu_0}$. Using the isomorphism, this means that for waves on transmission lines,

$$V(x, t) = \sqrt{\frac{L_0}{C_0}}\, I(x, t). \tag{9.7.2}$$

(Wave traveling in $+ x$-direction)

This means that the current flowing to the right on the top wire is in phase with the voltage; the current on the bottom wire is always opposite that on the top wire.

As for em waves in vacuum, for a transmission line wave traveling in the $-x$-direction, there is a negative sign added to the relation between V and I, so that

$$V(x, t) = -\sqrt{\frac{L_0}{C_0}}\, I(x, t). \tag{9.7.3}$$

(Wave traveling in $- x$-direction)

9.8   Sound waves

For most of us, sound is second only to light as the most important type of wave. Sounds fill our lives, inform us about our surroundings, and are the main medium for inter-personal relations. Sound is a traveling wave of fluctuations in the pressure and density of the air. (All the arguments we will make will work just as well for sound waves in water or any other medium.) As shown in figure 9.8.1, the fluctuations in pressure P and density ρ are small compared to the background pressure and density. In fact, the fluctuations shown in the picture are greatly exaggerated; in a typical sound wave, the pressure and density only change by about 0.1% of the background value. We define P' to be the *change* in pressure relative to the background, and ρ' to be the change in density relative to the background, so that

$$P(t) = P_0 + P'(t) \qquad\qquad (9.8.1)$$

$$\text{and } \rho(t) = \rho_0 + \rho'(t). \qquad\qquad (9.8.2)$$

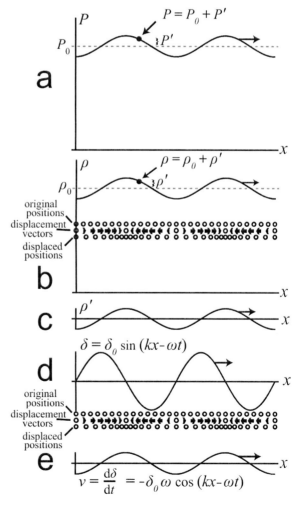

Figure 9.8.1 A sound wave consists of small variations in the pressure and density, which are in phase, as shown in parts (a) and (b). Parts (c)–(e) shown the relationships between the variation in pressure ρ', the displacement δ, and velocity $v = d\delta/dt$. The displacement (δ) and variation in pressure (ρ') waves are 90° out of phase, but the velocity wave is in phase with the density wave. The uniformly spaced dots underneath parts b and d represent the original positions of packets of air molecules. The arrows represent the displacement imparted to these packets by the sound wave, with positive displacement δ corresponding to right-pointing arrows. The bottom row of dots shows the displaced positions of the air packets. From this part of the figure, you can see how the displacement wave gives rise to the variation in density.

Higher density means higher pressure, so the pressure and density waves are in phase, as shown. As we did for em waves, we will focus on plane waves in our study of sound, that is, waves in which the pressure and density only depend on x, not on y or z. This means that all points in a plane perpendicular to the x-axis have the same pressure and density.

The density fluctuations are caused by displacements of the air molecules from their evenly spaced positions, as suggested by the dots in the figure which are closer together at the peaks in the density and by the arrows below the dots which indicate the displacements required to achieve this clumping. The displacements point toward the peaks in the density and away from the valleys. As we did in our study of longitudinal standing waves, we define the displacement of an air molecule (in the x-direction) relative to its original position to be $\delta(x, t)$.

We will show below that the pressure variations are governed by the wave equation, so that we get traveling waves. This means, for example, that we can have a sinusoidal traveling wave in the displacement, $\delta = \delta_0 \sin(kx - \omega t)$. Since the displacement is in the x-direction, there is a corresponding wave in the velocity: $v \equiv \dfrac{d\delta}{dt} = -\delta_0 \omega \cos(kx - \omega t)$, so that the velocity wave is in phase with the density wave, as shown. Because this velocity is due to the relatively small variations in P and ρ, the velocity itself is also small.

The air or other medium must obey two basic equations: the continuity equation (which is really a statement of the conservation of mass) and Euler's equation (a different Euler's equation from $e^{i\theta} = \cos\theta + i\sin\theta$, and one that is really a statement of $F = ma$). Below, we derive these two equations, and then combine them to show that the pressure variations are described by the wave equation.

The continuity equation

We consider a region of square cross-section, with length along the x-axis, as shown in figure 9.8.2. Let's think about the mass of air in the shaded section of infinitesimal thickness that lies between x and $x + dx$. At a particular instant t, the mass flowing into this section per unit time is

$$\frac{\text{mass}}{\text{time}} = \frac{\text{mass}}{\text{volume}} \cdot \frac{\text{volume}}{\text{time}} = \rho(x, t) \frac{A \cdot \text{distance}}{\text{time}} = \rho(x, t)\, A\, v(x, t).$$

Similarly, the mass flowing out of the section per unit time is $\rho(x + dx, t)\, A\, v(x + dx, t)$. Therefore, the change in mass contained in the section per unit time is

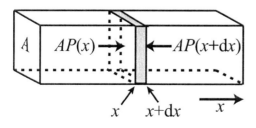

Figure 9.8.2 Geometry for the derivation of the Continuity Equation and Euler's equation.

$$\frac{dm}{dt} = \{\rho(x) A v(x)\} - \{\rho(x + dx) A v(x + dx)\}.$$

We can use a first-order Taylor series for $\rho(x + dx)$ and $v(x + dx)$:

$$\rho(x + dx) = \rho(x) + \frac{\partial \rho}{\partial x}dx \quad \text{and} \quad v(x + dx) = v(x) + \frac{\partial v}{\partial x}dx.$$

Because dx is infinitesimal, these expressions are exact. Plugging them into the expression above gives

$$\frac{dm}{dt} = \{\rho(x) A v(x)\} - \left\{\left[\rho(x) + \frac{\partial \rho}{\partial x}dx\right] A \left[v(x) + \frac{\partial v}{\partial x}dx\right]\right\}.$$

$$= -A \left[\frac{\partial \rho}{\partial x}dx\, v(x) + \rho(x)\frac{\partial v}{\partial x}dx + \frac{\partial \rho}{\partial x}dx\frac{\partial v}{\partial x}dx\right].$$

Since the last term is of order dx^2, it is negligible compared to the others, so

$$\frac{1}{A}\frac{dm}{dt} = -\left[\frac{\partial \rho}{\partial x}dx\, v(x) + \rho(x)\frac{\partial v}{\partial x}dx\right] \Leftrightarrow$$

$$\frac{1}{A\,dx}\frac{dm}{dt} = -\left[\frac{\partial \rho}{\partial x}v(x) + \rho(x)\frac{\partial v}{\partial x}\right]. \tag{9.8.3}$$

For the shaded section in figure 9.8.2, $\rho = \frac{m}{A\,dx}$. We also have that $\frac{\partial(v\rho)}{\partial x} = \frac{\partial \rho}{\partial x}v(x) + \rho(x)\frac{\partial v}{\partial x}$. Substituting these into equation (9.8.3) gives the continuity equation:

$$\boxed{\frac{\partial \rho}{\partial t} + \frac{\partial(v\rho)}{\partial x} = 0.} \tag{9.8.4}$$

The continuity equation for a system with variation in the x-direction only.

Next, we will do some massaging of this which, together with a massaged version of Euler's equation, will lead to the wave equation for the pressure. Plugging equation (9.8.2): $\rho(t) = \rho_0 + \rho'(t)$ into the above gives

$$\underbrace{\frac{\partial \rho_0}{\partial t}}_{=0} + \frac{\partial \rho'}{\partial t} + \frac{\partial}{\partial x}\left[v\left(\rho_0 + \rho'\right)\right] = 0 \Rightarrow$$

$$\frac{\partial \rho'}{\partial t} + \rho_0\frac{\partial v}{\partial x} + \frac{\partial(v\rho')}{\partial x} = 0.$$

Since both v and ρ' are small, the last term is negligible compared to the other two, giving

$$\frac{\partial \rho'}{\partial t} + \rho_0\frac{\partial v}{\partial x} = 0. \tag{9.8.5}$$

Pressure is a function of density. Expanding the pressure in a Taylor series gives

$$P(\rho) = P(\rho_0 + \rho') = \underbrace{P(\rho_0)}_{P_0} + \rho'\frac{\partial P}{\partial \rho} + \left(\text{terms of order } \rho'^2 \text{ and higher}\right).$$

Since ρ' is small, we ignore the higher order terms. Also, we define

$$v_s^2 \equiv \frac{\partial P}{\partial \rho}. \tag{9.8.6}$$

(Soon, you will understand the reason why we chose the symbol v_s^2 to represent this derivative.) So,

$$P(\rho, t) = P_0 + \rho' v_s^2 \Rightarrow \frac{\partial P}{\partial t} = v_s^2 \frac{\partial \rho'}{\partial t} \Leftrightarrow \frac{\partial \rho'}{\partial t} = \frac{1}{v_s^2} \frac{\partial P}{\partial t}.$$

Since $P(t) = P_0 + P'(t)$, we have that $\dfrac{\partial P}{\partial t} = \dfrac{\partial P'}{\partial t}$. Plugging this into the above gives

$$\frac{\partial \rho'}{\partial t} = \frac{1}{v_s^2} \frac{\partial P'}{\partial t}.$$

Now, we plug this into equation (9.8.5), giving

$$\frac{1}{v_s^2} \frac{\partial P'}{\partial t} + \rho_0 \frac{\partial v}{\partial x} = 0. \tag{9.8.7}$$

Taking $\dfrac{\partial}{\partial t}$ of both sides yields

$$\frac{1}{v_s^2} \frac{\partial^2 P'}{\partial t^2} + \rho_0 \frac{\partial}{\partial t} \frac{\partial v}{\partial x} = 0.$$

The order of partial derivatives doesn't matter, so we could write this as

$$\rho_0 \frac{\partial}{\partial x} \frac{\partial v}{\partial t} = -\frac{1}{v_s^2} \frac{\partial^2 P'}{\partial t^2}. \tag{9.8.8}$$

This is as far as we need to go with massaging the continuity equation. We'll use this result a bit later to get the wave equation for the pressure. You should examine figure 9.8.3 at this point.

Figure 9.8.3 This surfer is probably thinking about partial derivatives, since they are so important for the understanding of waves. Image © Quincy Dein/Dreamstime.com

Euler's equation. The shaded section in figure 9.8.2 experiences a force $AP(x)$ from the left side (pushing to the right), and a force of magnitude $AP(x + dx)$ from the right side (pushing to the left). So, the net force is $F_{net} = A[P(x) - P(x + dx)]$. Applying $F = ma$ to the section then gives

$$F_{net} = A[P(x) - P(x + dx)] = (\rho A \, dx)\frac{dv}{dt}.$$

Again, we use a first-order Taylor series: $P(x + dx) = P(x) + \dfrac{\partial P}{\partial x}dx$, so that

$$-\frac{\partial P}{\partial x}dx = \rho \, dx \frac{dv}{dt} \Rightarrow$$

$$-\frac{\partial P}{\partial x} = \rho \frac{dv}{dt}. \tag{9.8.9}$$

Recall that v is a function of both x and t. Therefore, the full derivative $\dfrac{dv}{dt}$ can be expressed in terms of the partial derivatives via

$$\frac{dv}{dt} = \frac{\partial v}{\partial t} + \frac{\partial v}{\partial x}\frac{dx}{dt} = \frac{\partial v}{\partial t} + \frac{\partial v}{\partial x}v.$$

Plugging this into equation (9.8.9) gives

$$-\frac{\partial P}{\partial x} = \rho\left(\frac{\partial v}{\partial t} + v\frac{\partial v}{\partial x}\right). \tag{9.8.10}$$

Euler's equation for a system with variation in the x − direction only.

Again, we will now do a bit of massaging of this, in our pursuit of the wave equation for the pressure.

Your turn: Use equations (9.8.10), (9.8.1): $P(t) = P_0 + P'(t)$, (9.8.2): $\rho(t) = \rho_0 + \rho'(t)$, and the fact that both v and ρ' are small to show that

$$-\frac{\partial P'}{\partial x} = \rho_0\frac{\partial v}{\partial t}. \tag{9.8.11}$$

Taking $\dfrac{\partial}{\partial x}$ of both sides gives

$$-\frac{\partial^2 P'}{\partial x^2} = \rho_0\frac{\partial}{\partial x}\frac{\partial v}{\partial t}.$$

Finally, we use equation (9.8.8), $\rho_0\dfrac{\partial}{\partial x}\dfrac{\partial v}{\partial t} = -\dfrac{1}{v_s^2}\dfrac{\partial^2 P'}{\partial t^2}$ to substitute for the right side, giving

$$-\frac{\partial^2 P'}{\partial x^2} = -\frac{1}{v_s^2}\frac{\partial^2 P'}{\partial t^2} \Leftrightarrow$$

$$\frac{\partial^2 P'}{\partial t^2} = v_s^2\frac{\partial^2 P'}{\partial x^2}. \tag{9.8.12}$$

This is the wave equation, equation (9.2.5): $\dfrac{\partial^2 y}{\partial t^2} = v_p^2 \dfrac{\partial^2 y}{\partial x^2}$. So, sound behaves as a linear wave, with speed given by equation (9.8.6): $v_s^2 \equiv \dfrac{\partial P}{\partial \rho}$!

> **Concept test (answer below[6]):** What is the difference between v_s and v?

The speed of sound. To find the speed of sound $v_s \equiv \sqrt{\dfrac{\partial P}{\partial \rho}}$, we must ask what should be held constant when evaluating the partial derivative. Clearly, the pressure does depend on the absolute temperature T, so perhaps it is T that should be held constant. However, in fact the compressions and expansions of the air in a sound wave are fast enough that the temperature is not constant; it is higher in regions where the gas is compressed and lower in regions where it is expanded. So, instead of holding T constant, it is a better approximation to hold the total energy of the system constant. In other words, when we think of the air being compressed in part of the sound wave, it is better to assume that in this compression all the work done on the gas goes to increase its energy (so that its temperature goes up), rather than assuming that the compression is done at constant temperature (which would require that we allow energy to leak out as the gas is compressed). Note that the work done on the part of the gas being compressed is done *by* other parts of the gas, so that the total energy is constant. This type of compression, in which all the energy used to compress the gas stays in the gas and none leaves, is called an "adiabatic" compression, from the Greek *a dia batos*, "not through to pass," meaning that no energy leaks out of (or into) the system. So,

$$v_s = \sqrt{\left. \frac{\partial P}{\partial \rho} \right|_{\text{adiabatic}}} . \tag{9.8.13}$$

The density is $\rho = \dfrac{Nm}{V} \Leftrightarrow V = \dfrac{Nm}{\rho}$, where m is the mass of a gas molecule and N is the number of molecules in volume V. Therefore, we have that

$$\frac{\partial P}{\partial \rho} = \frac{\partial P}{\partial V}\frac{dV}{d\rho} = \frac{\partial P}{\partial V}\left(-\frac{Nm}{\rho^2}\right). \tag{9.8.14}$$

We can model air using the ideal gas law:

$$PV = Nk_B T = nRT, \tag{9.8.15}$$

where $n = N/N_A$ is the number of moles of gas, $N_A = 6.022 \times 10^{23}$ is Avogadro's number, $k_B = 1.381 \times 10^{23}$ J/K is Boltzmann's constant, and $R = N_A k_B = 8.314$ J mol^{-1} K^{-1} is the Universal Gas Constant. So, we see that, if T were held constant, we

---

6. The quantity v_s is the phase velocity of the wave, that is, the speed at which the crests of the sound wave move through space. It is constant in time. The quantity v is defined as $d\delta/dt$, that is, the time derivative of the displacement of the air molecules relative to their equilibrium positions. As shown in figure 9.8.1e, v varies sinusoidally in space. So, as the wave propagates, v also varies sinusoidally in time.

would have $P = \dfrac{\text{constant}}{V}$, where the constant would be $N k_B T$. However, in fact as the gas is compressed adiabatically, the temperature increases, so that the pressure increases more quickly with a decrease in volume. One can show that, for an adiabatic compression,

$$P = \frac{C}{V^\gamma}, \tag{9.8.16}$$

where C is a constant and the "adiabatic index" γ depends on the type of gas; for air, $\gamma = 1.40$. Therefore, using equation (9.8.14),

$$\frac{\partial P}{\partial \rho} = \frac{\partial P}{\partial V}\left(-\frac{Nm}{\rho^2}\right) = \frac{\gamma\, C}{V^{\gamma+1}}\frac{Nm}{\rho^2}.$$

Since $C = PV^\gamma$, we have

$$\frac{\partial P}{\partial \rho} = \frac{\gamma\, P}{V}\frac{Nm}{\rho^2} = \frac{\gamma\, P}{\rho}.$$

So, $v_s \equiv \sqrt{\dfrac{\partial P}{\partial \rho}} \Rightarrow$

$$\boxed{v_s = \sqrt{\frac{\gamma\, P_0}{\rho_0}}}, \tag{9.8.17}$$

Speed of sound in a gas

where we have explicitly indicated that one uses the equilibrium values of the pressure and density to calculate the speed of sound, since the speed of sound is an average property of the gas, not something that varies on the scale of the wavelength of the sound. For air under standard conditions, $\rho_0 = 1.2 \text{ kg/m}^3$ and $P_0 = 1.01 \times 10^5$ Pa. (Recall that 1 Pa, pronounced "one Pascal," equals 1 N/m^2.) Plugging in these numbers gives $v_s = 343$ m/s, which matches very well with experimental values.

We are also interested in the speed of sound in liquids. Recall from section 2.3 that Young's modulus E was defined by equation (2.3.3): $\dfrac{F_{\text{applied}}}{A} = E\dfrac{x}{\ell}$, where F_{applied} is the force applied to one face of a solid with cross-sectional area A and length ℓ and $x = -\Delta\ell$ is the magnitude of the resulting change in length of the solid. Since F_{applied}/A is pressure, we could rewrite this as $P = -E\dfrac{\Delta\ell}{\ell}$, so that

$$\frac{\partial P}{\partial \ell} = -\frac{E}{\ell} \Leftrightarrow E = -\ell\frac{\partial P}{\partial \ell}.$$

If the pressure is instead applied "hydrostatically" (meaning that it is applied to all faces), then we usually think about the change in the volume of the solid, instead of the change in the length. We define the bulk modulus by analogy with Young's modulus to be

$$B \equiv -V\frac{\partial P}{\partial V}. \tag{9.8.18}$$

The same definition works equally well for liquids, which also get compressed when pressure is applied hydrostatically. Rearranging equation (9.8.18) gives

$$\frac{\partial P}{\partial V} = -B/V.$$

Substituting this into equation (9.8.14) gives

$$\frac{\partial P}{\partial \rho} = \frac{B}{V}\frac{Nm}{\rho^2} = \frac{B}{\rho}.$$

Since $v_s \equiv \sqrt{\dfrac{\partial P}{\partial \rho}}$, we then have

$$\boxed{v_s = \sqrt{\frac{B}{\rho_0}},}$$

(9.8.19)

Speed of sound in a liquid.

where again we have explicitly indicated that we use the equilibrium value of the density to calculate the speed of sound. Now, as discussed earlier, for sound waves the compressions are close to adiabatic. However, the bulk modulus is usually measured under constant temperature ("isothermal") conditions. Luckily, for a liquid the difference between adiabatic compressions and isothermal compressions is much smaller than for a gas, because the pressure increase when the volume is decreased is determined more by changes in the interaction between molecules due to the reduction in the average distance between them, so that the change in the temperature of the molecules is not as important in determining the change in pressure. So, using typical tabulated values of B for liquids works well with equation (9.8.19). For example, experimental values for the bulk modulus of fresh water at $20°$C range from 2.19 to 2.22 GPa, and the density is 1,000 kg/m^3. Thus, equation (9.8.19) predicts a speed of sound in water of 1,480–1,490 m/s, which is fairly close to the experimental value of 1,498 m/s.

Finally, we are also interested in the speed of sound in solids. When part of a solid is compressed, it bulges out. Unlike in a liquid, the neighboring parts of the solid resist this bulging, effectively increasing the bulk modulus for the part being compressed. Furthermore, the parts of the solid which are outside the region of the sound wave exert shear stress on the parts that are inside the region, further increasing the effective stiffness. A full discussion of these effects is beyond the level of this book. However, one can show[7] that for typical solids,

$$\boxed{v_s \approx \sqrt{\frac{1.5\,B}{\rho_0}}.}$$

(9.8.20)

Approximation for speed of sound in a typical solid.

---

7. *Understanding the Properties of Matter, 2nd Ed.*, by Michael de Podesta (Taylor and Francis, London, 2002), p. 287.

For example, for aluminum the bulk modulus is 75.5 GPa, and the density is 2,698 kg/m^3, so that equation (9.8.20) predicts $v_s = 6{,}480$m/s, whereas the experimental value is 6,374 m/s.

Isomorphism with em waves in a vacuum. As with the other types of waves we've studied, we can form an isomorphism between sound waves and em waves in a vacuum. We start with equation (9.8.11):

$$-\frac{\partial P'}{\partial x} = \rho_0 \frac{\partial v}{\partial t} \Leftrightarrow$$

$$\frac{\partial P'}{\partial x} = -\frac{\partial (\rho_0 v)}{\partial t} \xrightarrow{\text{isomorphic with}} (9.5.8): \frac{\partial E}{\partial x} = -\frac{\partial B}{\partial t}.$$

In the isomorphism, the pressure variation P' plays the role of E and the momentum-per-volume $\rho_0 v$ plays the role of B.

To complete the isomorphism, we must find the equation for sound that is isomorphic with equation (9.5.14): $\dfrac{\partial B}{\partial x} = -\dfrac{1}{c^2}\dfrac{\partial E}{\partial t}$. We make use of equation (9.8.7):

$$\frac{1}{v_s^2}\frac{\partial P'}{\partial t} + \rho_0 \frac{\partial v}{\partial x} = 0 \Rightarrow$$

$$\frac{\partial (\rho_0 v)}{\partial x} = -\frac{1}{v_s^2}\frac{\partial P'}{\partial t} \xrightarrow{\text{isomorphic with}} (9.5.14): \frac{\partial B}{\partial x} = -\frac{1}{c^2}\frac{\partial E}{\partial t}.$$

Again the pressure variation P' plays the role of E and the momentum-per-volume $\rho_0 v$ plays the role of B.

For sound waves in a gas, we have equation (9.8.17): $v_s = \sqrt{\dfrac{\gamma P_0}{\rho_0}}$, which, by the above reasoning, is isomorphic to $c = \dfrac{1}{\sqrt{\mu_0 \varepsilon_0}}$. As in previous isomorphisms, it is not clear exactly how to form the isomorphisms between the quantities γ, P_0, and ρ_0 and the quantities μ_0 and ε_0. However, since ρ_0 is associated with the momentum-per-volume $\rho_0 v$ (which is isomorphic to B), and μ_0 is associated with B, one might expect that ρ_0 might be isomorphic to μ_0. Further, since P_0 is associated with the pressure variation (which is isomorphic to E), and ε_0 is associated with E, one might expect that γP_0 is isomorphic to $1/\varepsilon_0$. By considering the power transmitted in the wave, you can show that this hunch is correct; see problem 9.16. Therefore, the isomorphism is complete. Instead of writing that $\rho_0 v$ is isomorphic to B, it is more convenient to say that v is isomorphic to B/μ_0.

Table 9.8.1. Isomorphism between sound waves in gases and electromagnetic waves in a vacuum

Sound waves	Electromagnetic waves in vacuum
Velocity $v \equiv d\delta/dt$	B/μ_0
Pressure variation P'	Electric Field E
Equilibrium density ρ_0	Permeability of free space μ_0
γP_0	Inverse of permittivity of free space: $1/\varepsilon_0$

> **Concept test (answer below[8]):** What is the relationship between the magnitude of the pressure variation wave P' and the magnitude of the velocity wave $v \equiv d\delta/dt$?

9.9 Musical instruments based on tubes

All wind instruments, from the oboe to the organ, whether woodwind or brass, are based on resonance in a tube filled with air. We saw in section 9.4 that we can superpose two waves of equal amplitude traveling in opposite directions to create a standing wave. This is true for any system described by the wave equation. Since sound is described by the wave equation, we know that there can be standing waves of sound. These are similar to the standing waves we found for strings; they are the normal modes of the air-filled tube. For the string fixed at both ends, the normal modes fit an integer number of half wavelengths between the walls. As we'll see, the boundary conditions for tubes of air can be different, so that in some cases the condition instead is that we must fit an *odd* integer number of *quarter* wavelengths between the ends.

There are two types of boundary conditions. In a flute, the player blows air across the mouthpiece, which is near one end of the flute. The mouthpiece is an open hole, so that the pressure at this end of the flute is kept essentially constant at atmospheric pressure. The other end of the flute is open, so the pressure there is also essentially constant. Thus, the boundary conditions for a flute are that the variation in pressure P' must go to zero at the ends, just as for a string fixed between two walls the amplitude must go to zero at the walls. We could also say that there must be a node of the standing wave in pressure at each end of the flute. Therefore, the condition for determining the frequencies of the normal modes is that we must fit an integer number of half wavelengths into the length of the flute:

$$L = n\frac{\lambda_n}{2} \Leftrightarrow \lambda_n = \frac{2L}{n}. \tag{9.9.1}$$

The phase velocity of the wave is the speed of sound, so that

$$v_s = \frac{\omega}{k} = \frac{2\pi f}{2\pi/\lambda} = \lambda f \Rightarrow$$

$$f = \frac{v_s}{\lambda}. \tag{9.9.2}$$

Substituting from equation (9.9.1) gives

$$f_n = n\frac{v_s}{2L}. \tag{9.9.3}$$

Resonant frequencies for an air tube open at both ends.

The lowest frequency mode, $n = 1$, is called the "fundamental." There is a pressure node at each end of the flute, and an "antinode" (a point of maximum amplitude) at

---

8. We use the isomorphism to translate (9.5.12), $E = cB = c\mu_0 \dfrac{B}{\mu_0}$ into $P' = v_s \rho_0 v$.

the center. The next mode, $n = 2$, is called the "first harmonic"; this has pressure nodes at both ends and also in the middle. The $n = 3$ mode is called the "second harmonic," and so on. We can see from equation (9.9.3) that the frequency progression of the normal modes is quite simple for a tube open at both ends: $f_1, 2f_1, 3f_1$, and so on. When the system is excited with a wide range of frequencies simultaneously (as when a flute player blows across the mouthpiece), the response at the resonant frequencies is much stronger than the response at other frequencies (because the quality factor Q is high), resulting in a well-defined musical pitch. Most instruments are designed so that the fundamental mode is excited with the highest amplitude, but the excitation of the other modes in addition is critical to the musical timbre of the instrument.

It is also worthwhile to consider what the wave in displacement looks like. We saw in figure 9.8.1 that the displacement wave is $90°$ out of phase with the pressure wave. Therefore, for a standing wave:

> A node in the pressure wave corresponds to an antinode in the displacement wave, and vice-versa.

So, the standing waves for a flute look as shown in figure 9.9.1a.

> **Concept test (answer below[9]):** In figure 9.9.1a, how can you tell that the solid line for the displacement curve occurs at the same time as the solid line for the pressure curve, rather than at the same time as the dashed line for the pressure curve?

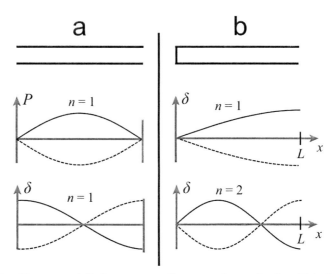

Figure 9.9.1 a: Pressure and displacement standing waves for an air tube with both ends open, for the mode $n = 1$. b: Displacement standing waves for a tube that is closed on the left end and open on the right. Top: fundamental ($n = 1$). Bottom ($n = 2$).

---

9. High pressure requires high density. For the solid line, the pressure is highest at the center, so we need positive displacements for the left half and negative displacements for the right half.

There are some types of musical instruments in which one end of the tube is closed. For example, some pipes on organs are like this. In all organ pipes, air is blown in at the bottom end, providing the constant pressure boundary condition discussed above. However, for some pipes the top end is open, and for others it is closed. (These two types produce a different quality of musical note, as we'll explore below.) There must be a node in the displacement at a closed end, since the end prevents motion along the axis of the tube.

Also, for any instrument which is excited at one end by vibrating lips (trumpet, etc.) or by a vibrating reed (clarinet, etc.), it is appropriate to count that end as closed, as we can easily see. Consider a mass on a spring that is driven at its resonant frequency by moving the support point up and down. In steady state, the amplitude of the motion of the mass is Q times the amplitude of the support point motion (assuming the damping is not too heavy). Therefore, compared to the motion of the mass, the support point is almost a fixed point. The analogy for a trumpet is that the motion of the lips at the mouthpiece is very small compared to the motion of the air molecules at the antinodes of the standing wave, so we can consider the mouthpiece end to essentially be a node of the displacement.

Let's consider a tube with the left end closed (corresponding to a node in the displacement) and the right end open (corresponding to a node in the pressure, and so an antinode in the displacement). For the longest wavelength normal mode (the fundamental), we fit a quarter wavelength into the length, as shown in the top part figure 9.9.1b. In the mode with next shorter wavelength, shown in the lower part of the figure, we fit $3/4$ of a wavelength into the length. In the next mode, we would fit $5/4$ of a wavelength. We can see that the general pattern is

$$L = (2n - 1)\frac{\lambda_n}{4} \Leftrightarrow \lambda_n = \frac{4L}{2n - 1}.$$

From equation (9.9.2), $f = \frac{v_s}{\lambda}$, so

$$f_n = (2n - 1)\frac{v_s}{4L}. \tag{9.9.4}$$

Resonant frequencies for an air tube open one end and closed at the other.

Unlike the case when both ends are open, the frequency progression is more complicated: f_1, $3f_1$, $5f_1$, and so on. Because the frequencies corresponding to even multiples of f_1 are missing, the musical timbre of an instrument with one end closed is quite different from one with both ends open.

9.10   Power carried by rope and electromagnetic waves; RMS amplitudes

Earlier, we noted that transmission of information is one of the two most important applications for waves. The other is the transmission of energy. Almost all the energy we use on our planet has been transmitted to us via em waves from the sun. (Much of it was transmitted hundreds of millions of years ago, and stored up in the form of fossil

Figure 9.10.1 A wave train propagates to the right. The part of the rope to the right of the dashed vertical line is initially stationary, but starts moving once the wave impinges on it, showing that the wave must carry energy.

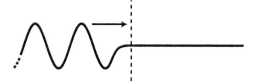

fuels.) Microwave ovens use em waves with a wavelength of about 10 cm to transmit energy into the water molecules in food.[10] Infrared lasers are used for surgery and for cutting metal. Dentists use ultraviolet light (with a wavelength of about 300 nm) to harden dental adhesives.

It is easy to see that energy is carried by a rope wave. For the wave train shown in figure 9.10.1, the pieces of the rope on the right are initially at rest, and so have no kinetic energy. However, as the wave reaches them, they begin to move, so the wave has moved energy from left to right. The further the wavefront moves to the right, the more energy has been moved from the region to the left of the dashed vertical line into the region on the right. Thus, if the wave is originally semi-infinite (i.e., it extends infinitely far to the left), it must contain an infinite amount of energy. So, we see that, instead of discussing the total energy in a wave, it is more useful to discuss the power carried by a wave, that is, the energy per unit time. For a sinusoidal rope wave, you can show in problem 9.15 that the power is given by

$$P = \tfrac{1}{2}\mu A^2 \omega^2 v_\mathrm{p},$$ (9.10.1)

where μ is the mass per unit length, A is the amplitude, ω is the angular frequency, and v_p is the speed. The important thing to note about this relation is that the power is proportional to the *square* of the amplitude; this is a universal feature of all waves.

Most applications of waves for power transmission rely on em waves. It is not difficult to find the power transmitted by em waves in vacuum, if we make use of two results that you may have encountered in introductory electricity and magnetism: there is energy density (i.e., energy per unit volume) in the electric field and in the magnetic field:

Energy density of **E**: $\quad \eta_E = \dfrac{\varepsilon_0}{2}E^2$ (9.10.2)

Energy density of **B**: $\quad \eta_B = \dfrac{1}{2\mu_0}B^2.$ (9.10.3)

10. Scientists have investigated the use of microwaves for heating homes for decades. The basic idea is that very low level microwaves would be broadcast throughout the home, heating the humans within. Because the heat is mostly absorbed by the humans and not the furniture or the air, much less energy is wasted than in conventional heating systems. However, so far these ideas have not progressed beyond the experimental stage, because of concerns over whether such a system could be successfully marketed.

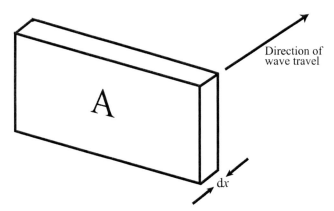

Figure 9.10.2 A box of volume $A\,dx$.

For em waves in vacuum, $E = cB$. Plugging this into equation (9.10.3) gives

$$\eta_B = \frac{1}{2\mu_0}\frac{E^2}{c^2} = \frac{1}{2\mu_0}\frac{E^2}{1/\left(\mu_0\varepsilon_0\right)} = \frac{\varepsilon_0}{2}E^2.$$

So, we see that, for em radiation in vacuum, the energy is carried equally in the magnetic and electric fields. The total energy density is thus

$$\eta_{\mathrm{Rad}} = \varepsilon_0 E^2 \qquad (9.10.4)$$

Energy density of em radiation in vacuum

To calculate the power delivered by these waves, we first calculate the energy contained in a box of cross-sectional area A, as shown in figure 9.10.2. The box has an infinitesimal length dx along the direction of wave travel, so that E is essentially constant throughout the box. The volume of the box is $A\,dx$, so the energy contained within it is $\varepsilon_0 E^2 A\,dx$. We imagine that the box moves forward with the wave, and that we have placed a perfect energy absorber right in front of the box. Then, the entire energy content of the box is deposited into the absorber in a time dx/c. The power is given by

$$P = \frac{\text{Energy}}{\text{time}} = \frac{\varepsilon_0 E^2 A\,dx}{dx/c} = c\varepsilon_0 E^2 A.$$

The wavefronts of a plane wave are infinite in the y- and z-directions, so that to calculate the power of such a wave, we would need a box with infinite A, which according to the above would be infinite. This is not a helpful notion – it is much more useful to quote the power *per unit area*, which is the definition of intensity:

$$\text{Intensity} \equiv \frac{\text{Power}}{\text{area}} \equiv S = c\varepsilon_0 E^2. \qquad (9.10.5)$$

To bring this to the standard form, we use $E = cB$, so that

$$S = c\varepsilon_0 E\left(cB\right) = c^2\varepsilon_0 EB = \left(\frac{1}{\mu_0\varepsilon_0}\right)\varepsilon_0 EB = \frac{1}{\mu_0}EB. \qquad (9.10.6)$$

Recall that we had earlier found that the Poynting vector points in the direction of propagation. We defined it as

$$\boxed{\mathbf{S} = \frac{1}{\mu_0}\mathbf{E} \times \mathbf{B},} \tag{9.10.7}$$

Instantaneous intensity of an em wave in vacuum.

and now we see that it has magnitude equal to the intensity of the wave.

Since E and B vary sinusoidally in time, this shows that the intensity of an em wave varies in time. This is seldom of any importance – usually we are far more interested in the *average* intensity. We revert temporarily to the form of the instantaneous intensity given by equation (9.10.5):

$$S = c\varepsilon_0 E^2.$$

So, the *average* intensity would be given by

$$\langle S \rangle = c\varepsilon_0 \left\langle E^2 \right\rangle, \tag{9.10.8}$$

where the angle brackets indicate an average over one period.

We define the "root mean square" or "rms" amplitude in terms of this average:

$$\boxed{E_{\text{rms}} \equiv \sqrt{\langle E^2 \rangle}.} \tag{9.10.9}$$

So, E_{rms} is the square *root* of the *mean* of the *square* of E. The concept of rms amplitude is quite important for the study of almost all time-varying phenomena.

Thus,

$$\langle S \rangle = c\varepsilon_0 E_{\text{rms}}^2.$$

Since $E = cB$, we have $E_{\text{rms}} = cB_{\text{rms}}$, so that

$$\langle S \rangle = c^2\varepsilon_0 E_{\text{rms}} B_{\text{rms}} = \frac{1}{\mu_0} E_{\text{rms}} B_{\text{rms}}.$$

Thus, we see that equation (9.10.7),

$$\boxed{\mathbf{S} = \frac{1}{\mu_0}\mathbf{E} \times \mathbf{B},}$$

Instantaneous intensity (if E and B are amplitudes) or average intensity
(if E and B are rms amplitudes) of an em wave in vacuum

can be interpreted in two ways, both of which are correct. Either it can indicate a time-varying vector with instantaneous magnitude equal to the instantaneous intensity of the wave, or (and this is the more usual interpretation) we can use the rms amplitudes of **E** and **B** to calculate the cross product, with a result that gives the *average* intensity.

Most often, rms amplitudes are used for sinusoidal waves. In this case, the relation between the amplitude (i.e., the peak value) and the rms amplitude is simple. We make use of the handy fact (which we encountered in chapter 4) that the average of a sinusoid squared over a full wavelength is half the peak value,[11] that is,

$$\text{For a sinusoid, } \left\langle E^2 \right\rangle = \frac{E_{\text{peak}}^2}{2}.$$

Plugging this into the definition of rms amplitude, (9.10.9), gives

$$E_{\text{rms}} \equiv \sqrt{\left\langle E^2 \right\rangle} = \sqrt{\frac{E_{\text{peak}}^2}{2}} = \frac{E_{\text{peak}}}{\sqrt{2}}.$$

So, *for a sinusoidal waveform*, the rms amplitude is just the (peak) amplitude over $\sqrt{2}$.

Self-test (answer below[12]): Sketch a waveform for which $E_{\text{rms}} \approx \dfrac{E_{\text{peak}}}{20}$.

9.11 Intensity of sound waves; decibels

For a sound wave, the energy is carried partly by the kinetic energy associated with the oscillating longitudinal velocity, and partly by the potential energy associated with the compressions and expansions of the medium. We first consider the density of the kinetic energy, which is given by

$$\eta_K \equiv \frac{\text{kinetic energy}}{\text{volume}} = \frac{\frac{1}{2}mv^2}{\text{volume}} = \frac{1}{2}\rho v^2.$$

Your turn: Employing the same ideas used to develop (9.10.5), show that the intensity of kinetic energy for sound waves is

$$S_K = \tfrac{1}{2} v_s \rho_0 v^2. \tag{9.11.1}$$

The above is the instantaneous kinetic energy intensity, which is highest at the velocity peaks and valleys of the sound wave. Usually, we are more interested in the intensity averaged over a wavelength. For a sinusoidal right-moving wave, we have

$$v = v_0 \sin(kx - \omega t), \tag{9.11.2}$$

where v_0 is the amplitude of the longitudinal velocity wave, and is not to be confused with the phase velocity of the sound wave, $v_s = \dfrac{\omega}{k}$. We can compute the the average

---

11. Recall, that $\cos^2 \omega_d t = \dfrac{1}{2}\left(1 + \cos 2\omega_d t\right)$ and that $\left\langle \cos 2\omega_d t \right\rangle = 0$ over a complete cycle.

12. Answer to self-test: One possibility is a train of pulses, with the waveform equal to zero between pulses, and with the repeat period equal to 20 times the pulse width.

kinetic energy intensity by plugging equation (9.11.2) into (9.11.1):

$$\langle S_K \rangle = \tfrac{1}{2} v_s \rho_0 \langle v^2 \rangle = \tfrac{1}{2} v_s \rho_0 v_0^2 \langle \sin^2 (kx - \omega t) \rangle.$$

Again, we make use of the fact that the average of the square of a sinusoid over one period is half. Therefore,

$$\langle S_K \rangle = \tfrac{1}{4} v_s \rho_0 v_0^2.$$

In problem 9.11, you can show that the potential energy averaged over a wavelength is equal to the kinetic energy averaged over a wavelength. (This is analogous to the situation for em waves in a vacuum, for which we saw that the energy is equally divided between the electric and magnetic components.) Therefore, the total average intensity is

$$\langle S \rangle = \tfrac{1}{2} v_s \rho_0 v_0^2. \tag{9.11.3}$$

It is more common to discuss sound intensity in terms of pressure. Plugging equation (9.11.2) into (9.8.11), $-\dfrac{\partial P'}{\partial x} = \rho_0 \dfrac{\partial v}{\partial t}$, gives

$$\frac{\partial P'}{\partial x} = \omega \rho_0 v_0 \cos(kx - \omega t). \tag{9.11.4}$$

We saw (figure 9.8.1) that the pressure wave is in phase with the velocity wave. Therefore, we must have

$$P' = P_m \sin (kx - \omega t) \Rightarrow$$

$$\frac{\partial P'}{\partial x} = P_m k \cos(kx - \omega t). \tag{9.11.5}$$

(Note that P_m is the amplitude of the pressure wave, while P_0 is the constant background pressure.) Comparing equations (9.11.5) and (9.11.4), we see that

$$P_m k = \omega \rho_0 v_0 \Leftrightarrow v_0 = P_m \frac{k}{\omega} \frac{1}{\rho_0} \Rightarrow$$

$$v_0 = P_m \frac{1}{v_s \rho_0}. \tag{9.11.6}$$

So, we can re-express equation (9.11.3) as

$$\langle S \rangle = \frac{P_m^2}{2 v_s \rho_0}.$$

It is more common to deal with the rms amplitude of the pressure, which, for a sinusoidal wave is

$$P_{rms} = \frac{P_m}{\sqrt{2}},$$

so that

$$\langle S \rangle = \frac{P_{rms}^2}{v_s \rho_0}. \tag{9.11.7}$$

Average intensity of a sound wave in terms of the rms amplitude P_{rms} of the pressure variation.

Sound intensity is usually quoted in decibels, abbreviated dB:

$$\text{dB} \equiv 10 \log_{10} \frac{\langle S \rangle}{\langle S_{\text{ref}} \rangle},$$

(9.11.8)

where $\langle S_{\text{ref}} \rangle$ is the intensity of a sound wave with $P_{\text{ref,rms}} = 2 \times 10^{-5}$ Pa; this corresponds to the quietest sound that can be perceived by a human. We can re-express this equation as

$$\frac{\langle S \rangle}{\langle S_{\text{ref}} \rangle} = 10^{\text{dB}/10},$$

(9.11.9)

which shows that each 10 dB increase in the intensity corresponds to a factor of 10 increase in intensity.

> **Self-test (answer below[13]):** Does a 5 dB increase in intensity correspond to a factor of 5 increase in intensity?

Since v_{s} and ρ_0 are the same for $\langle S \rangle$ and $\langle S_{\text{ref}} \rangle$, we also have

$$\text{dB} \equiv 10 \log_{10} \frac{P_{\text{rms}}^2}{P_{\text{ref,rms}}^2} \Rightarrow$$

$$\text{dB} = 20 \log_{10} \frac{P_{\text{rms}}}{P_{\text{ref,rms}}}.$$

(9.11.10)

> **Example:** What is the rms amplitude of pressure in a sound wave of intensity 0 dB?
> Answer: $\text{dB} = 20 \log_{10} \frac{P_{\text{rms}}}{P_{\text{ref,rms}}} \Rightarrow$
>
> $$P_{\text{rms}} = P_{\text{ref,rms}} 10^{\text{dB}/20}.$$
>
> (9.11.11)
>
> So, for 0 dB, $P_{\text{rms}} = P_{\text{ref,rms}} = 2 \times 10^{-5}$ Pa.

Concept test (answer below[14]): How much larger is the rms amplitude of pressure in a sound wave of 10 dB intensity than in a sound wave of -10 dB intensity?

Decibels are used in many other measurements as well. For example, sometimes voltage is quoted in decibels, and the symbol "dBV" is used, meaning decibels relative

---

13. $\dfrac{\langle S \rangle}{\langle S_{\text{ref}} \rangle} = 10^{\text{dB}/10} = 10^{5/10} = \sqrt{10} = 3.162$, so a 5 dB increase corresponds to a factor of 3.162 increase in intensity.

14. For the -10 dB wave, $P_{\text{rms}} = P_{\text{ref,rms}} 10^{-10/20} = \dfrac{1}{\sqrt{10}} P_{\text{ref,rms}}$. For the 10 dB wave, $P_{\text{rms}} = P_{\text{ref,rms}} 10^{10/20} = \sqrt{10}\, P_{\text{ref,rms}}$, so it has ten times larger amplitude than the -10 dB wave. We can see that, in general, each increase of 20 dB corresponds to a factor of 10 increase in the amplitude (and a factor of 100 increase in the intensity).

to a reference level of 1 V:

$$dBV \equiv 20 \log_{10} \frac{V}{(1\ V)}.$$

Again, each increase of 20 dB corresponds to a factor of 10 increase in Voltage. For a resistor, power is proportional to V^2, so that an increase of 20 dB corresponds to a factor of 100 increase in power, and an increase of 10 dB corresponds to a factor of 10 increase in power.

9.12   Dispersion relations and group velocity

When I first encountered the definition below as an undergraduate, I did not fully appreciate how frequently it would be important. So, be forewarned: you will hear about dispersion relations at least once a month for the rest of your physics life!

> **Dispersion relation: The relation between angular velocity ω and wavenumber k.**

Why is this called the "dispersion" relation? We have seen that the phase velocity (i.e., the velocity of the crests or, equivalently, of the troughs) is given by equation (9.3.2):

$$v_p = \frac{\omega}{k}.$$

This is one way of presenting the dispersion relation. For most of the waves we have studied, v_p is a constant, independent of k and ω. In other words, for most of the waves we have studied, including waves on a rope, em waves in vacuum, and waves on transmission lines, the speed of the wave does not depend on its wavenumber or frequency. This fact becomes extremely important for the propagation of pulses, such as that shown here in figure 9.12.1. (The vertical axis would be y for a wave on a rope, E for an em wave, or V for a wave on a transmission line.) As you learned in the section 8.5 (on Fourier Transforms), such a wave, even though it is not periodic, can be synthesized by adding together an infinite number of sinusoids. Now, consider what happens as the wave propagates. As long as all these sinusoids propagate at the same speed, the pulse maintains its shape, since all the sinusoids continue to add together in

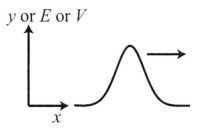

y or E or V

Figure 9.12.1 A propagating pulse. x

the same way, except for the overall motion. However, if v_p depends on wavelength, then the different Fourier components travel at different speeds, and so the pulse soon gets dispersed. Hence the name "dispersion relation."

As we'll argue below, there are many examples of wave propagation for which v_p *does* depend on ω or k, so the dispersion relation is nonlinear The phenomenon of dispersion has enormous practical consequences. In some cases, the dispersion is desirable, so as to separate different frequency components of a signal. For example, a prism takes advantage of the differences in speed within the glass for different colors (wavelengths) of light to bend them into different paths. In other cases, the dispersion is very undesirable, and must be minimized. For example, when data is transmitted as a series of pulses through fiber optic cables, it is quite important that the shape of each pulse not change too much as it propagates through the kilometers of glass between relay stations. (It turns out that there is a very fortuitous window of wavelengths, for which the glass used for fibers has a nearly linear dispersion relation, and very low absorption. This particular window happens to match almost perfectly with the wavelength produced by inexpensive solid-state lasers.)

The propagation of light through glass is one of the most important examples of dispersion. The full details of this propagation are beyond the scope of this book, so here we present a qualitative model. As the light wave moves into the material, it exerts a force, which varies sinusoidally in time, on the electrons, causing them to vibrate. (It also exerts a force on the nuclei, but they are so much more massive that they hardly move at all.) The vibrating electrons, because they are accelerated charges, emit em radiation. The total wave propagating through the glass is the superposition of the original wave with this re-radiated wave coming from each electron. The speed at which this total wave propagates depends on the details of how all these waves interfere with each other.

Since the electrons are originally in equilibrium, we can qualitatively model them as harmonic oscillators. Therefore, the phase of their response relative to the drive (provided by the incoming wave) depends on the ratio of the drive frequency (i.e., the frequency of the incoming wave) to the resonant frequency of the oscillators. Since this ratio changes as the frequency of the incoming wave changes, the phase relationship between the incoming wave and the re-radiated wave from the electrons changes, and so the speed of propagation changes. Therefore, although we can write the dispersion relation in the form (9.3.2): $v_p = \dfrac{\omega}{k}$, v_p is a function of ω, and so the dispersion relation is nonlinear.

You have already encountered another nonlinear dispersion relation: the one for a beaded string, which is the same as the relation for atoms in a crystalline solid. Recall from equation (7.2.7):

$$k_n = n\frac{\pi}{L} \qquad \omega_n = \sqrt{2}\omega_A \sin\left(n\frac{\pi a}{2L}\right) \qquad \omega_A \equiv \sqrt{\frac{2T}{am}}$$

Combining these expressions gives

$$\omega_n = \sqrt{2}\omega_A \sin\left(\frac{a}{2}k_n\right), \tag{9.12.1}$$

which is an example of a nonlinear dispersion relation.

We'll consider one more example: quantum mechanical waves. We have mentioned a few times that quantum mechanical particles, such as electrons, have a wave nature, and are described by a wavefunction $\Psi(x, t)$. As for other types of waves, this wavefunction has a frequency and a wavelength. As we discussed briefly in section 5.5, it turns out that the energy of any quantum mechanical particle (such as an electron, a photon, or a quantized vibration of a crystal called a "phonon") is given by

$$E = \hbar\omega. \tag{9.12.2}$$

(You may have seen this equation in an equivalent form, such as $E = hf$ or $E = h\nu$, where $h = 2\pi\hbar$ and f or ν ("nu") represents the frequency.) It turns out that the momentum of the particle is given by

$$p = \hbar k. \tag{9.12.3}$$

These two equations form the basis of quantum mechanics. Therefore, when you encounter them in a quantum mechanics course, be sure you understand the experimental justification for them.

For a particle with mass, such as the electron, we can express the kinetic energy in terms of the momentum:

$$KE = \tfrac{1}{2}mv_e^2 = \frac{(mv_e)^2}{2m} = \frac{p^2}{2m},$$

where v_e is the speed of the electron. Substituting from equation (9.12.3) gives

$$KE = \frac{\hbar^2 k^2}{2m}.$$

For a "free electron", that is, an electron which is simply traveling through space without any forces on it, the energy is entirely kinetic. Therefore,

$$E = \hbar\omega = KE = \frac{\hbar^2 k^2}{2m} \Leftrightarrow$$

$$\omega = \frac{\hbar k^2}{2m}. \tag{9.12.4}$$

Dispersion relation for a free electron

This is a another example of a nonlinear dispersion relation. The fact that it's nonlinear immediately tells us that quantum waves with different wavelengths travel at different speeds:

$$v_p = \frac{\omega}{k} = \frac{\hbar k^2/2m}{k} = \frac{\hbar k}{2m}, \tag{9.12.5}$$

so that higher wavenumber (smaller wavelength) quantum waves go faster.

Let's check our understanding. Plugging into the above from equation (9.12.3) gives

$$v_p = \frac{\hbar k}{2m} = \frac{p}{2m} = \frac{mv_e}{2m} = \frac{v_e}{2}.$$

The speed of the electron wave, the wave that represents the electron, is half the speed of the electron itself!

OK, let's not panic. If the electron wave has a well-defined k (as we assumed in the above discussion), that means it is mathematically represented by a pure sinusoid, which means that the wave must extend from $x \to -\infty$ to $x \to +\infty$. We discussed such a wave in section 1.11: $\Psi = \psi_0 e^{-i\omega_0 t} e^{ik_0 x}$. Because we want to emphasize that they are fixed quantities, we write the angular frequency as ω_0 (instead of simply ω) and the wavenumber as k_0 (instead of simply as k). The quantity $|\Psi|^2$ (called the "probability density") tells us the probability for finding the electron at a particular place. For this case,

$$|\Psi|^2 = \Psi\,\Psi^* = \left(\psi_0 e^{-i\omega_0 t} e^{ik_0 x}\right)\left(\psi_0 e^{i\omega_0 t} e^{-ik_0 x}\right) = \psi_0^2.$$

Since this doesn't depend on x, we say that such an electron is "completely delocalized": it is equally likely to be found anywhere between $x \to -\infty$ and $x \to +\infty$. So, perhaps it shouldn't worry us that the velocity of the electron is twice the velocity of the wave that represents it – after all, if the electron is already everywhere, what does it mean for it to have a velocity anyway?

To discuss the velocity of the electron in terms we understand better, we need to consider an electron that's in a more localized state. We can create such a state, called a "wavepacket," by multiplying the free-electron wavefunction $\Psi = \psi_0 e^{-i\omega_0 t} e^{ik_0 x}$ by an envelope function, as shown in figure 9.12.2a. The figure shows the wavepacket at $t = 0$, with the peak of the envelope at position x_{m}.

How will the wavepacket evolve in time? The pulse could be Fourier-synthesized by adding up a large number of sinusoids of the form $\Psi = \psi_0 e^{-i\omega t} e^{ikx}$, with different

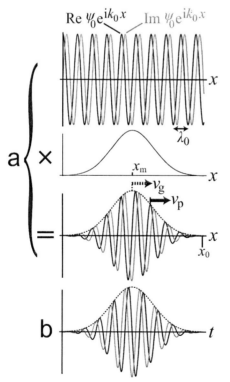

Figure 9.12.2 a: We can create a wavepacket (bottom) by multiplying a rapidly oscillating function (in this case a free-electron wavefunction, top) by an envelope function (middle). b: The wavepacket as a function of time, as it passes the point x_0.

k's (and corresponding ω's) for each sinusoid. Each of these sinusoids propagates at the speed given by equation (9.12.5), $v_{\mathrm{p}} = \dfrac{\hbar k}{2m}$, so we can propagate each sinusoid forward in time, and then add them up to see how the wavepacket has propagated. We will show that, if we use an envelope which varies slowly enough, then, over short to moderate time intervals, the envelope propagates without changing its shape, at a speed v_{g} (called the "group velocity") which matches that of the electron itself. (This means, of course, that the group velocity is different from the phase velocity v_{p}.) *This idea of group velocity is critically important for all types of waves governed by a nonlinear dispersion relation.*

Figure 9.12.2a shows the wavepacket as a function of x. If instead we consider the behavior as a function of time as the pulse passes the point x_0, the plot of Ψ *versus* time would look as shown in figure 9.12.2b. For the rest of this section, we will think primarily in terms of the dependence on time, instead of the dependence on x.

We will set aside this electron wavepacket for now, but come back to it after we've figured out how to calculate the group velocity. For now, let's consider a different example for which the ideas of dispersion and group velocity are important; we'll use this example as a basis for explicitly showing that the envelope can propagate without changing its shape, even for a system with a nonlinear dispersion relation. Our example is the important problem of sending information through a fiber optic cable. For example, figure 9.12.3 shows the voltage output from a microphone as I speak the word, "Hi!." The horizontal axis is in seconds, so that the entire word lasts about $^{1}/_{4}$ second. The first part of this is the "H" sound; if you look carefully, you can see a repeating pattern in this part, one copy of which is highlighted with a box. You can see a different repeating pattern in the last part of the word, which is the "i" sound. The highlight box is 10-ms wide, and the sharpest features in it are perhaps 0.5-ms wide. We can see that, later on in the word, there are even sharper features that are about 0.1-ms wide. So, to Fourier synthesize this waveform, we'd need sinusoids with frequencies from about $\dfrac{1}{0.25\text{ s}}$ to about $\dfrac{1}{0.1\text{ ms}}$ m, or about 4 Hz to 10 kHz. This is a typical range for audio signals. We define the function shown in figure 9.12.3 to be $f(t)$.

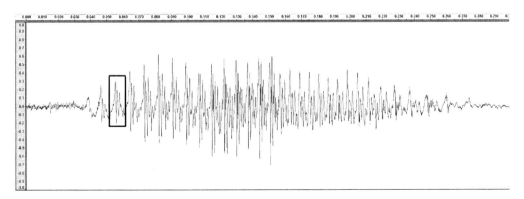

Figure 9.12.3 The word "Hi!" as recorded with a microphone. The vertical axis is proportional to the variation in air pressure relative to background, while the horizontal axis is time in s.

Say that we'd like to send this word over a fiber optic cable. We will study fiber optics in more detail in section 10.9, but just from the name you can tell that they're intended to transmit optical signals, that is, signals with frequencies from about 10^{14} to 10^{15} Hz. As we discussed earlier, the frequencies in our audio signal $f(t)$ only go up to about 10 kHz, so if we converted the signal to an em wave, it wouldn't be able to propagate through the fiber optic. One solution is to use the audio signal to "modulate" a fixed-frequency "carrier wave," that is, to use $f(t)$ as an envelope function for a pure sinusoid:

$$y(t) = \underbrace{f(t)}_{\substack{\text{envelope} \\ \text{function}}} \underbrace{\cos \omega_c t}_{\substack{\text{carrier} \\ \text{wave}}} \qquad (9.12.6)$$

where the angular frequency ω_c of the carrier *is* in the frequency range that can propagate through the fiber optic. This idea is illustrated in figure 9.12.4.

To begin with, instead of the complicated $f(t)$ that represents the "Hi!" sound, let's use a pure sinusoid as the audio signal that is to be transmitted:

$$f(t) = B \cos \omega_a t,$$

where ω_a is an audio frequency, perhaps $2\pi \cdot 10$ kHz, and B is the amplitude. Thus,

$$y(t) = B \cos \omega_a t \cos \omega_c t.$$

Recall from our discussion of beats equation (5.1.3):

$$A \cos \omega_1 t + A \cos \omega_2 t = 2A \cos \omega_e t \cos \omega_{av} t,$$

where, from equation (5.1.2),

$$\omega_1 = \omega_{av} + \omega_e \quad \text{and} \quad \omega_2 = \omega_{av} - \omega_e.$$

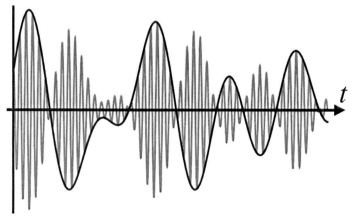

Figure 9.12.4 A part of an audio waveform (black) is used as an envelope function for a rapidly oscillating carrier wave.

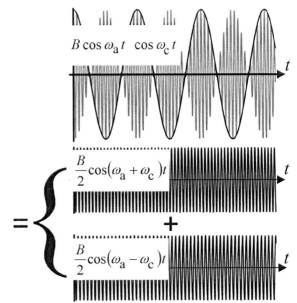

Figure 9.12.5 A rapidly oscillating carrier wave multiplied by a simple sinusoidal envelope function is equal to the sum of two sinuoids, each at an angular frequency close to that of the carrier.

To apply this to our case, we make the following subsitutions: $2A \to B$, $\omega_e \to \omega_a$, and $\omega_{av} \to \omega_c$. Therefore,

$$\omega_1 = \omega_c + \omega_a \quad \text{and} \quad \omega_2 = \omega_c - \omega_a, \tag{9.12.7}$$

$$\Rightarrow \omega_c = \frac{\omega_1 + \omega_2}{2} \quad \text{and} \quad \omega_a = \frac{\omega_1 - \omega_2}{2}, \tag{9.12.8}$$

$$\text{and} \quad y(t) = B \cos \omega_a t \, \cos \omega_c t = \frac{B}{2} \cos \omega_1 t + \frac{B}{2} \cos \omega_2 t. \tag{9.12.9}$$

This relation, shown graphically in figure 9.12.5, says that by using the audio signal as an envelope function for the carrier wave, we shift the frequency up close to ω_c. (For a fiber optic, $\omega_c \sim 2\pi \left(10^{14} \text{ Hz}\right)$, while $\omega_a \sim 2\pi \left(10^4 \text{ Hz}\right)$, so $\omega_c \pm \omega_a$ is *very* close to ω_c in percentage terms.) So, we can now transmit the signal through the fiber optic.[15]

The glass from which the fiber optic is made has a nonlinear dispersion relation, so we must inquire whether the shape of the envelope function is preserved as the wave propagates. So far, we've only been considering the wave as a function of time. To change it to a propagating wave, we replace $\omega_1 t$ by $k_1 x - \omega_1 t$, where k_1 is the wavenumber corresponding to ω_1, and we replace $\omega_2 t$ by $k_2 x - \omega_2 t$. Applying this

15. This is essentially how AM ("amplitude modulation") radio works, but the carrier is a radio wave with $\omega_c \sim 2\pi \left(10^6 \text{ Hz}\right)$.

recipe to equation (9.12.9), and using equation (9.12.8) gives

$$y(x, t) = B \cos \left[\frac{(k_1 x - \omega_1 t) - (k_2 x - \omega_2 t)}{2} \right] \cos \left[\frac{(k_1 x - \omega_1 t) + (k_2 x - \omega_2 t)}{2} \right]$$

$$= \frac{B}{2} \cos (k_1 x - \omega_1 t) + \frac{B}{2} \cos (k_2 x - \omega_2 t). \tag{9.12.10}$$

We will assume that the slope of the function $\omega(k)$ is approximately constant over the range ω_1 to ω_2, so that the wavenumber of the carrier wave is equal to the average of k_1 and k_2:

$$k_c = \frac{k_1 + k_2}{2}. \tag{9.12.11}$$

We also define

$$k_d \equiv \frac{k_1 - k_2}{2}. \tag{9.12.12}$$

Using these, together with equation (9.12.8), we can simplify equation (9.12.10) to

$$y(x, t) = \underbrace{B \cos(k_d x - \omega_a t)}_{\text{envelope function}} \underbrace{\cos(k_c x - \omega_c t)}_{\text{carrier wave}}$$

$$= \frac{B}{2} \cos(k_1 x - \omega_1 t) + \frac{B}{2} \cos(k_2 x - \omega_2 t). \tag{9.12.13}$$

We see that the envelope function travels at the speed $v_g = \dfrac{\omega_a}{k_d} = \dfrac{\omega_1 - \omega_2}{k_1 - k_2}$. Recall that we assume that the slope of the function $\omega(k)$ is approximately constant over the range ω_1 to ω_2, therefore

$$\boxed{v_g = \left. \frac{d\omega}{dk} \right|_{k_c}}. \tag{9.12.14}$$

Group velocity (the velocity of the envelope for a wave packet)

In appendix B, it is shown that these ideas all work equally well for a more complicated envelope function $f(t)$, rather than the simple sinusoid we used. If $f(t)$ can be Fourier synthesized from sinusoids with angular frequencies from 0 up to ω_m, then $y(t) = f(t) \cos \omega_c t$ is shown in the appendix to have Fourier components with angular frequencies from $\omega_c - \omega_m$ to $\omega_c + \omega_m$. The corresponding traveling wave is shown to have an envelope whose shape is determined by $f(t)$, which travels at the group velocity (9.12.14):

$$y(x, t) = \underbrace{\text{carrier wave}}_{\substack{\text{travels at} \\ v_p = \frac{\omega_c}{k_c}}} \cdot \underbrace{\text{envelope}}_{\substack{\text{shape determined} \\ \text{by } f(t), \text{ travels at} \\ v_g \equiv \left. \frac{d\omega}{dk} \right|_{k_c}}}$$

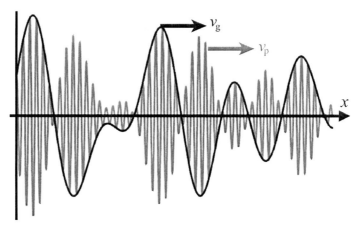

Figure 9.12.6 An example of an envelope modulating a carrier for a travelling wave.

(To obtain the above, we again require that $\dfrac{d\omega}{dk}$ be constant over the range $\omega_c - \omega_m$ to $\omega_c + \omega_m$.) An example is shown in figure 9.12.6.

To recap:

> If we start with an envelope function $f(t)$ having nonzero Fourier components from 0 to ω_m, and multiply it by a carrier wave $\cos \omega_c t$:
>
> 1. The resulting wavepacket $y(t) = f(t) \cos \omega_c t$ has Fourier components with angular frequencies from $\omega_c - \omega_m$ to $\omega_c + \omega_m$.
>
> 2. If $\dfrac{d\omega}{dk}$ is constant over the range $\omega_c - \omega_m$ to $\omega_c + \omega_m$, then the envelope of the wavepacket doesn't change shape, and propagates at the group velocity $v_g \equiv \left. \dfrac{d\omega}{dk} \right|_{k_c}$.
>
> 3. The peaks and valleys of the carrier wave (which is modulated by the envelope) propagate at $v_p = \dfrac{\omega_c}{k_c}$.

Does the idea of group velocity solve the conundrum about electron waves? Let's check to see whether the group velocity for a localized electron wavepacket equals the velocity of the electron itself:

$$\omega = \frac{\hbar k^2}{2m} \Rightarrow v_g \equiv \frac{d\omega}{dk} = \frac{2\hbar k}{2m} = \frac{p}{m} = \frac{mv_e}{m} = v_e \; \checkmark \;(\text{Phew!!})$$

In general, v_g can be equal to, greater than, or less than v_p. It even occurs in some real systems that v_g is negative, even though all the Fourier component waves have positive v_p!

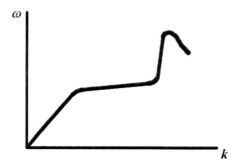

Figure 9.12.7 A hypothetical dispersion relation.

Concept test (answer below[16]): For the hypothetical dispersion relation shown in figure 9.12.7 here, identify (a) a point where $v_p = v_g$, (b) a point where $v_p < v_g$, (c) a point where $v_p > v_g > 0$, and (d) a point where $v_p > 0 > v_g$.

Concept and skill inventory for chapter 9

After reading this chapter, you should fully understand the following terms:

Traveling wave (9.1)
Partial derivative (9.2)
The wave equation (9.2)
Superposition principle for traveling waves (9.4)
Plane wave (9.5)

---

16. $v_p \equiv \omega/k$ is the slope of a straight line drawn from the origin to a point on the dispersion curve, while $v_g \equiv d\omega/dk$ is the slope of the curve itself. Therefore, as shown in figure 9.12.8, at point 1 they are equal. At point 2, $v_p > v_g > 0$. (The group velocity is only slightly positive at point 2.) At point 3, $v_p < v_g$. At point 4, $v_p > 0 > v_g$.

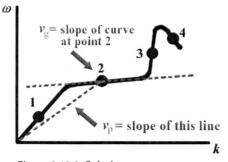

Figure 9.12.8 Solution to concept test.

Poynting vector (9.5)
Momentum wave (9.5)
Dielectric constant (9.6)
Permittivity (9.6)
Magnetic susceptibility (9.6)
Permeability (9.6)
H (9.6)
Transmission line (9.7)
Displacement wave for sound (9.8)
Velocity wave for sound (9.8)
Bulk modulus (9.8)
Adiabatic (9.8)
Continuity equation (9.8)
Euler's equation for fluids (9.8)
rms amplitude (9.10)
dB (9.11)
Dispersion relation (9.12)
Carrier wave (9.12)
Group velocity (9.12)

You should know what happens when:

Two waves "collide" (9.4)
Two equal amplitude sinusoidal traveling waves going in opposite directions are
   superposed (9.4)

You should understand the following connections:

Phase velocity, angular frequency, & wavenumber (9.3)
Traveling waves & standing waves (9.4)
Magnitudes of electric & magnetic fields for em waves in vacuum (9.5)
Directions of **E**, **B**, & **S** for em waves (9.5)
$\dot{y}$, B, H, I, & v (9.5–9.8)
F_L, E, V, & P' (9.5–9.8)
μ, μ_0, μ, L_0, & ρ_0 (9.5–9.8)
T, $1/\varepsilon_0$, $1/\varepsilon$, $1/C_0$, & γP_0 (9.5–9.8)
H & **B** (9.6)
Maxwell's equations in vacuum & in linear materials (9.6)
Density, pressure, displacement, & velocity components of a sound wave (9.8)
Fundamental frequency for a wind instrument, length of the air column, speed of sound,
   whether one or both ends are open (9.9)
Nodes in the pressure wave & antinodes in the displacement wave for a wind
   instrument (9.9)
rms amplitude & average value (9.10)
Graph of the dispersion relation, group velocity, & phase velocity (9.12)

You should understand the differences between:

The functional forms of left- & right-traveling waves (9.2)

v & v_s (9.8)

Power & intensity (9.10)

You should be familiar with the following additional concepts:

The two quantities isomorphic to E and B are in phase for right-traveling waves and out of phase for left-traveling waves. (9.5)

The energy, power, or intensity is proportional to amplitude.[2] (9.10)

You should be able to:

Take a partial derivative. (9.2)

Test a proposed solution to a differential equation involving partial dervatives. (9.2)

Recognize an isomorphism based on (9.5.8) and either (9.5.9) or (9.5.14).

Use isomorphisms to quickly copy results for em waves in vacuum to other systems (9.5–9.8)

Calculate the speed of sound for a gas, liquid, or solid (9.8)

Calculate the resonant frequencies for a wind instrument with one or both ends open (9.9)

Convert between dB and power or amplitude ratios (9.11)

Calculate the group velocity from the dispersion relation (9.12)

Explain the difference between the group velocity and the phase velocity (9.12)

In addition to all of the above, you should be able to combine the concepts you've learned to address new situations.

Problems

Note: Additional problems are available on the website for this text.

Instructor: Ratings of problem difficulty, full solutions, and important additional support materials are available on the website.

9.1 (a) Use the fact that any right-travelling wave can be expressed as $y(x - vt)$ to show that $\dot{y} = -v\dfrac{\partial y}{\partial x}$ for a right-traveling wave.

(b) Use the fact that any left-travelling wave can be expressed as $y(x + vt)$ to show that $\dot{y} = v\dfrac{\partial y}{\partial x}$ for a left-traveling wave.

9.2 The left end of a long rope with $\mu = 0.1$ kg/m and $T = 50$ N is moved in the pattern shown in figure 9.P.1, creating a pulse which then propagates to the right along the rope. The left end of the rope is at $x = 0$. Make a sketch of the rope from 0 to 20 m at $t = 0.50$ s. On your sketch, label the width and height of the pulse quantitatively, as well as the position of the leftmost point on the pulse.

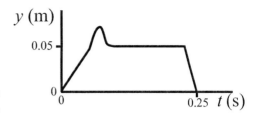

Figure 9.P.1 The left end of a rope is
moved in the pattern shown here.

9.3 In a region of outer space, the electric and magnetic fields are described by

$$\mathbf{E}(x, y, z, t) = E_0 \sin(kx - \omega t)\,\hat{\mathbf{j}} \qquad \mathbf{B}(x, y, z, t) = B_0 \sin(kx - \omega t)\,\hat{\mathbf{k}}$$

The electric field magnitude is greater than zero for $0 < x < 0.4$ m, goes to
zero at $x = 0.4$ m, and is negative for 0.4 m $< x < 0.8$ m.

(a) Using the convention that the length of the vector indicates the
strength of the field, sketch the electric and magnetic fields for points
along the x-axis from the origin to $x = 1.6$ m, at $t = 0$. Label your
sketch quantitatively.

(b) In which direction is the wave propagating? Explain how you
can tell.

(c) What is the value of k?

(d) What is the value of ω?

(e) If $E_0 = 5 \times 10^{-5}$ N/C, what is the value of B_0?

(f) Consider the following four points, with coordinates (x, y, z)
measured in meters: Point 1: $(1, 0, 0)$. Point 2: $(1, 0, 1)$. Point 3:
$(1.1, 0, 1)$. Point 4: $(1.1, 0, 0)$. At $t = 0$, rank the electric field at
these points from largest to smallest. If the field is zero at any point,
state that explicitly. If the field is equal at two or more points, state
that explicitly.

(g) Now, make a similar ranking for the magnitude of the magnetic field
at points 1-4.

(h) Now consider the following other points (again with dimensions in
meters): Point 5: $(1, 1, 1)$. Point 6: $(1, -1, 1)$. Point 7: $(1, -1, -1)$.
Point 8: $(1, 1, -1)$. Specify the strength and direction of the electric
and magnetic fields at each of these points at $t = 0$.

(i) A completely different wave is also sinusoidal, and also has
magnetic and electric fields that depend only on x and t. At $t = 0$,
the magnetic field at the origin is $B_0\,\hat{\mathbf{j}}$ and the electric field at the
origin is $E_0\hat{\mathbf{k}}$, where B_0 and E_0 are both positive. In which direction
is the wave propagating? How can you tell?

9.4 A wave pulse is traveling to the right on a rope with $\mu = 0.100$ kg/m and
$T = 20.0$ N. The pulse consists of two straight sloping sections, with a sharp
peak at the center. (To the left of the peak, the rope has a positive slope, and
to the right of the peak it has a negative slope.) Both sloping sections are at
angles of $30°$ relative to the horizontal. The peak height is 3.00 cm. Make a

sketch showing the transverse velocity (i.e., the velocity in the y-direction) along the length of the rope, and label your sketch quantitatively, including the numerical value for the maximum velocity.

9.5 From the isomorphism between rope waves and em waves in a vacuum, we have that $v_{\mathrm{p}} = \sqrt{\dfrac{T}{\mu}}$ is isomorphic to $c = \dfrac{1}{\sqrt{\mu_0 \varepsilon_0}}$. We cannot tell from this what the individual isomorphisms are between the quantities T and μ on one hand and the quantities μ_0 and ε_0 on the other. However, since μ is associated with the momentum $\mu \dot{y}$ (which is isomorphic to B) and μ_0 is associated with B, one might expect that μ might be isomorphic to μ_0. This would then require that T is isomorphic to $1/\varepsilon_0$, which makes sense, since T is associated with the force from the left (which is isomorphic to E), and ε_0 is associated with E. Note that we have not proved this isomorphism. However, check it by applying it to the Poynting vector for an em wave in vacuum. What should this translate into for a rope wave? Does it translate correctly?

9.6 For an em plane wave in vacuum, assume that $\mathbf{E}$ is parallel to $\hat{\mathbf{j}}$ and $\mathbf{B}$ is parallel to $\hat{\mathbf{k}}$. Apply Ampère's law $\oint \mathbf{B} \cdot \mathrm{d}\vec{\ell} = \mu_0 I_{\substack{\text{net} \\ \text{threading}}} + \mu_0 \varepsilon_0 \dfrac{\mathrm{d}}{\mathrm{d}t} \int \mathbf{E} \cdot \hat{\mathbf{n}} \, \mathrm{d}A$ to a rectangular loop in the $x - z$ plane to show that $\dfrac{\partial B}{\partial x} = -\mu_0 \varepsilon_0 \dfrac{\partial E}{\partial t}$. This is equation (9.5.9) that we used to derive the wave equation for em waves. *Hint: The derivation of the above equation is essentially the same as the derivation of equation (9.5.8).*

9.7 Circular polarization. **(a)** The "polarization" of an em wave refers to the direction of the electric field. Thus, for the em waves we've explicitly considered, the polarization is along the y-axis. However, this is not at all required. For example, we could have the polarization along the z-axis and still have a wave propagating in the x-direction. The electric field for such a wave would be given by $\mathbf{E} = E_0 \cos(kx - \omega t) \, \hat{\mathbf{k}}$, where $E_0 > 0$. Explain why the magnetic field for this wave would be given by $\mathbf{B} = -B_0 \cos(kx - \omega t) \, \hat{\mathbf{j}}$, where $B_0 > 0$. (Your explanation need not be very mathematical; you should include a diagram or two.)

For the rest of the problem, consider the following electric field (in vacuum):

$$\mathbf{E} = E_0 \cos(kx - \omega t) \, \hat{\mathbf{j}} + E_0 \sin(kx - \omega t) \, \hat{\mathbf{k}}$$

(b) Qualitatively describe the behavior of the above electric field, as seen by an observer at the origin. *(Hint: radiation with this type of electric field is called "circularly polarized.")* **(c)** How must k be related to ω? **(d)** What is the magnetic field that must accompany the above electric field? Be sure to specify both the magnitude (in terms of E_0 and known constants) as a function of position and time, and the direction, using the unit vectors $\hat{\mathbf{i}}$, $\hat{\mathbf{j}}$, and $\hat{\mathbf{k}}$. The best way to accomplish both these tasks will be to use a Cartesian representation, as I did for $\mathbf{E}$. Explain your reasoning briefly.

9.8 Waves in three dimensions. We have seen that waves in a one-dimensional system are described by the wave equation

$$v_p^2 \frac{\partial^2 y}{\partial x^2} = \frac{\partial^2 y}{\partial t^2},$$

where $y = y(x, t)$. The function y might describe the displacement of a rope, but we have also seen that this equation describes other types of waves, such as em waves. We have seen that the solutions of this equation that represent right-moving waves can be written as $y\left(x - v_p t\right) = y\left[\frac{1}{k}\left(kx - kv_p t\right)\right] = y\left[\frac{1}{k}(kx - \omega t)\right]$. Written in this way, we can think of the multiplication by $\frac{1}{k}$ as part of the action of the function; this means that we could *either* think of the generic form of a right moving wave as any function $y\left(x - v_p t\right)$, *or* as any function $f(kx - \omega t)$. For example, a right-moving sinusoid can be written as $f(x, t) = A \sin(kx - \omega t)$. This gives a wave with wavenumber k, wavelength $\lambda = \frac{2\pi}{k}$, angular frequency ω, and phase velocity $v_p = \frac{\omega}{k}$.

For a three-dimensional medium which is isotropic (i.e., in which all directions are equivalent), the wave equation becomes

$$v_p^2 \left(\frac{\partial^2 f}{\partial x^2} + \frac{\partial^2 f}{\partial y^2} + \frac{\partial^2 f}{\partial z^2}\right) = \frac{\partial^2 f}{\partial t^2},$$

where now the "thing that is waving" (perhaps the displacement of the rope or the electric field) is $f(x, y, z, t)$. We can write this in shorter form by introducing the "Laplacian" operator (named in honor of P. S. de Laplace):

$$\nabla^2 \equiv \frac{\partial^2}{\partial x^2} + \frac{\partial^2}{\partial y^2} + \frac{\partial^2}{\partial z^2}.$$

This is also often called "del squared," since it can be thought of as $\vec{\nabla} \cdot \vec{\nabla}$, where $\vec{\nabla}$ is del, also known as the gradient operator: $\vec{\nabla} \equiv \hat{\mathbf{i}}\frac{\partial}{\partial x} + \hat{\mathbf{j}}\frac{\partial}{\partial y} + \hat{\mathbf{k}}\frac{\partial}{\partial z}$. With this, we can write the three-dimensional wave equation for isotropic media as

$$v_p^2 \nabla^2 f = \ddot{f}.$$

    (a) Show that $f(\mathbf{k} \cdot \mathbf{r} - \omega t)$ is a solution to this equation, where $\mathbf{k} = k_x\hat{\mathbf{i}} + k_y\hat{\mathbf{j}} + k_z\hat{\mathbf{k}}$ is the three-dimensional version of the wavenumber, and is called the "wave vector." (Of course, $\mathbf{r} = x\hat{\mathbf{i}} + y\hat{\mathbf{j}} + z\hat{\mathbf{k}}$.) In the process, find the relation between k_x, k_y, k_z, ω, and v_p.

    (b) Describe qualitatively what the components of $\mathbf{k}$ represent. *Hint: Consider the limiting cases when one of the three is equal to 1 m^{-1}, and the other two equal zero.*

9.9 Phase relationships for left-traveling sound waves. In section 9.8, we showed that, for a right-traveling sound wave, the pressure, density, and velocity are all in phase, while the displacement is $90°$ out of phase with them. What are the phase relationships between these four components of the wave for a left-traveling wave?

9.10 **Standing waves of sound.** We know that we can create a standing wave by superposing left- and right-traveling waves of equal amplitudes.

(a) Superpose two traveling displacement waves of amplitude $\dfrac{\delta_0}{2}$ to show that the displacement and velocity components of a standing wave of sound are given by

$$\delta = \delta_0 \sin kx \, \cos \omega t \quad \text{and} \quad v \equiv \frac{\partial \delta}{\partial t} = -\delta_0 \omega \, \sin kx \, \sin \omega t.$$

(b) For a standing wave that is one wavelength long, sketch δ at $t = 0$, and indicate the displacement vectors in a way similar to the vectors shown in figure 9.8.1. From your sketch, explain why the density wave must be $\rho' = -\rho_{\mathrm{m}} \cos kx \, \cos \omega t$.

(c) Recalling that the potential energy is associated with the pressure wave, which is always in phase with the density wave, explain why your results mean that at the time when the kinetic energy of the standing wave is maximized, the potential energy is zero, and vice versa.

9.11 **Potential energy in sound waves**. The potential energy of a sound wave depends on the pressure. At the time of maximum pressure amplitude for a standing wave, the distribution of pressure over a wavelength is exactly the same as it is over a wavelength of a traveling wave. Therefore, the potential energy of one wavelength's worth of a standing wave is the same as the potential energy of one wavelength's worth of a traveling wave. In problem 9.10, you can show that the kinetic and potential energies for a standing wave of sound are out of phase, that is, that at the time when the kinetic energy of the standing wave is maximized, the potential energy is zero, and vice versa. By conservation of energy, this means that the maximum kinetic energy of one wavelength's worth of a standing wave equals the maximum potential energy, which by the above reasoning equals one wavelength's worth of a traveling wave's potential energy. **(a)** For an air tube of length L and cross-sectional area A that is closed at both ends, the standing wave in the velocity would be (see problem 9.10) $v = -\delta_0 \omega \sin kx \sin \omega t = -v_0 \sin kx \sin \omega t$, where v_0 is the amplitude of the velocity wave. Use this to show that, for the mode with $\lambda = L$, the maximum kinetic energy of the air tube is $K_{\max} = \frac{1}{4}\rho_0 A v_0^2 L$. Hint: Start by dividing the tube into slices of infinitesimal thickness dx along the length of the tube, and finding the maximum kinetic energy $dK_{\max}$ of a slice. **(b)** Explain why this means that the average potential energy intensity of a traveling sound wave is $\langle S_U \rangle = \frac{1}{4}v_s \rho_0 v_0^2$.

9.12 Make yourself a mug of hot cocoa. Holding the mug by the handle, stir the cocoa vigorously, then use your spoon to tap–tap–tap on the rim of the mug. You should hear a musical pitch that changes over time. You can stir the cocoa again and repeat the experiment. (If you are unable to do this yourself, you can watch a video of me doing it by going to the entry for this problem on the website for this text. However, it's a lot more fun to do it yourself.) Explain qualitatively what's going on.

9.13 Is each of the following statements true or false? If true, explain briefly. If false, briefly explain why, and provide a corrected version that is not simply a negation of the original.

    (a) In a vacuum, em radiation is carried in both the magnetic and electric fields. The amount of energy carried in the electric field is sometimes more (and therefore sometimes less) than the amount carried in the magnetic field.

    (b) If the electric field points along the $-y$ axis and the magnetic field points along $+z$ axis, then the Poynting vector (which shows the direction of wave propagation) points along the $-x$ axis.

    (c) Since we can superpose a left-moving wave with a right-moving wave to create an expression for a standing wave, we can always express any traveling wave as the sum of two or more standing waves.

    (d) An em wave is traveling in a linear material. When it enters a different material that has the same permeability but a larger permittivity, its velocity increases.

9.14 **Power in transmission line waves**. From the isomorphism between transmission line waves and em waves in vacuum, we have that $L_0 C_0$ is isomorphic to $\mu_0 \varepsilon_0$. We cannot tell from this what the individual isomorphisms are between the quantities L_0 and C_0 on one hand and the quantities μ_0 and ε_0 on the other. However, since L_0 is associated with the current (which is isomorphic to B/μ_0), and μ_0 is associated with B, one might expect that L_0 might be isomorphic to μ_0. This would then require C_0 is isomorphic to ε_0, which makes sense since C_0 is associated with the voltage (which is isomorphic to E), and ε_0 is associated with E. Check this proposed isomorphism by applying it to the Poynting vector for an em wave in vacuum. What should this translate into for a waves on a transmission line? Does it translate correctly?

9.15 **Power in a rope wave**. A traveling rope wave carries both kinetic and potential energy. In this problem, you will find the total power carried by the wave. Consider a sinusoidal wave traveling to the right $y = A \cos(kx - \omega t)$. The rope has tension T and mass per unit length μ.

    (a) Show that the kinetic energy in one wavelength's worth of this wave is $\frac{1}{4}\lambda \mu A^2 \omega^2$. Hint: you may as well make the calculation at $t = 0$. Also, recall that the average value of the square of a sinusoid is $1/2$.

    (b) Calculating the potential energy for a traveling wave is more difficult, so we will use a trick. The potential energy is determined by the shape, and the shape of one wavelength's worth of a traveling wave is the same as one wavelength's worth of a standing wave (at a moment when the wave is at its maximum amplitude). Therefore, the potential energy of one wavelength's worth of a traveling wave is the same as the maximum potential energy of one wavelength's worth of a standing wave. Explain why this is the same as the maximum kinetic energy of one wavelength's worth of a standing wave.

   (c) Show that the maximum kinetic energy in one wavelength's worth of a standing rope wave of amplitude A is $\frac{1}{4}\lambda\mu A^2\omega^2$.

   (d) Show that the power in a traveling rope wave is $P = \frac{1}{2}\mu A^2\omega^2 v_\mathrm{p}$.

9.16 Completing the isomorphism between sound waves and em waves in vacuum. For sound waves in a gas, we have $v_s = \sqrt{\dfrac{\gamma P_0}{\rho_0}}$, which is isomorphic to $c = \dfrac{1}{\sqrt{\mu_0\,\varepsilon_0}}$. We cannot tell from this what the individual isomorphisms are between the quantities γ, P_0, and C_0 on one hand and the quantities μ_0 and ε_0 on the other. However, since ρ_0 is associated with the momentum-per-volume $\rho_0 v$ (which is isomorphic to B), and μ_0 is associated with B, one might expect that ρ_0 might be isomorphic to μ_0. This would then require that γP_0 is isomorphic to $1/\varepsilon_0$, which is reasonable since P_0 is associated with the pressure variation (which is isomorphic to E), and ε_0 is associated with E. By considering the power transmitted in the wave, show that this hunch is correct. *Hint: Note that in (9.11.3), $\langle S\rangle = \frac{1}{2}v_\mathrm{s}\rho_0 v_0^2$, the quantity given is the* average *intensity.*

9.17 The 287th Annual Solar System Olympics is being held on Venus, where conditions are rather different from here. On Earth, the atmospheric pressure is 101.3 kPa. The 100-m dash is held in Sandworm's Borough Stadium on the surface of Venus, where the atmospheric pressure is 92 times that of Earth and the atmospheric density is 65 kg/m^3. The atmosphere is mostly CO_2, and is at a temperature of $400°$C. Assume the adiabatic index (γ) of this atmosphere is 1.235. The starter, standing at the start line, fires the starting pistol.

   (a) How long will it take for the sound of the gun going off to reach a timer standing at the finish line?

   (b) What is the rms amplitude of pressure in the sound wave that comes out of the barrel of the gun, if the sound intensity measured right next to the muzzle is 140 dB?

   (c) After a blistering 8.69 s, a human crosses the finish line ahead of all of the other competitors. We are now at the medal ceremony, and the Aphrodite Orchestra is playing Joe Diffie's "Third Rock from the Sun," the planetary anthem of Earth. The flautist, who happens to be from Pluto, has placed an extension onto the end of her flute to take into account the atmospheric conditions on Venus. Assuming the conditions from part a still apply, how long must the flute become if the flute is tuned so that its first harmonic occurs at 880 Hz? How long must a clarinet be if it were to fit the same characteristics?

9.18 A wave $y_1 = A \cos(k_1 x - \omega_1 t)$ is superposed with a wave $y_2 = A \cos(k_2 x - \omega_2 t)$, where k_1 is fairly close to k_2. **(a)** What is the group velocity of the resulting waveform, that is, the velocity of the envelope? **(b)** What is the wavelength of the rapidly oscillating function contained within the envelope? **(c)** What is the speed of the crests of the rapidly oscillating function?

9.19 A Gaussian wavepacket of light travels through a sheet of glass which has a nonlinear dispersion. Explain what is wrong with the following statement: "When the wavepacket emerges from the glass, the high-frequency components are shifted to the front of the packet, and the low-frequency components are shifted to the back."

9.20 For waves in deep water (depth at least as large as the wavelength), with amplitude much less than the wavelength, the dispersion relation is $\omega^2 = gk + \dfrac{\gamma k^3}{\rho}$. The gk part of this dominates at small k (i.e., at large wavelength), and describes the situation when the restoring force is provided by gravity (g is the acceleration of gravity). Waves in this regime are called "gravity waves." The $\dfrac{\gamma k^3}{\rho}$ dominates at large k, and describes the situation when the restoring force is provided by surface tension; γ is the surface tension (7.2×10^{-2} N/m for pure water), and ρ is the density (1,000 kg/m^3). Waves in this regime are called "capillary waves." **(a)** At what wavelength do these two terms make equal contributions? **(b)** Use a graphing program to make a plot of ω as a function of k for wavelengths ranging from 5 mm to 20 cm. On the same graph show separate curves for pure gravity waves and pure capillary waves as dashed lines, and label each of the three curves. **(c)** For gravity waves (waves with a wavelength long enough that surface tension effects are negligible), what is the phase velocity in terms of g and k? **(d)** For gravity waves, show that the group velocity for waves having wavenumbers in a range centered on k is half the phase velocity. **(e)** For capillary waves (which have a wavelength short enough that the gk term can be ignored), what is the phase velocity in terms of k, γ, and ρ? **(f)** For capillary waves, show that the group velocity for waves having wavenumbers in a range centered on k is 3/2 the phase velocity.

9.21 **Em waves in the ionosphere**. (This is an example of a problem that looks much more intimidating than it really is.) So far, we have considered em waves in vacuum. One of the layers of the Earth's atmosphere is made of mostly ionized gas, and is called the "ionosphere." It is responsible for long-range transmission of AM radio, since the radio waves bounce off this layer and then back down to Earth. Suppose that em waves in the ionosphere are governed by the following differential equation:

$$\frac{\partial^2 E}{\partial t^2} = c^2 \frac{\partial^2 E}{\partial x^2} - \omega_0^2 E,$$

where c and ω_0 are known constants. You are told that all possible solutions of this equation can be expressed in the form

$$E(x, t) = \int_{-\infty}^{\infty} \varepsilon(k) e^{i(kx - \omega t)} dk,$$

where $\varepsilon(k)$ is a function of k.

(a) A naïve student might think that the above could only represent right-moving waves. Explain why this is incorrect.

(b) Given that the above is a solution to the differential equation, what must be the dispersion relation? *Hint: rearrange the differential equation so that you have zero on one side. Then, bear in mind that integrating with respect to k commutes with taking the derivative with respect to x or t. Finally, if you can make the integrand zero, then the integral will definitely be zero.*

(c) A burst of sine waves with a small range of angular frequencies centered on ω_1 is emitted from the point $x = 0$. The burst has a duration τ. Approximately how long will it take the burst to reach the point $x = L$?

10 Waves at Interfaces

Come with me if you want to live!
— Arnold Schwarzenegger as the Terminator in *Terminator 2: Judgment Day*

So far, we have discussed waves traveling through an unchanging medium. However, interesting and surprising things happen at the interface between two different mediums, leading to a wealth of important science and applications. In this chapter, we begin by studying the behavior of wave pulses on ropes when they encounter a boundary at which the mass/length of the rope suddenly changes. Although this may not sound very exciting, bear in mind that waves on a rope are exactly analogous to waves in any elastic medium. Therefore, for example, by understanding how transverse waves on a rope behave, we immediately discover the behavior for torsional waves in solids. For learning purposes, waves on ropes have the distinct advantage that they are easily visualized and relatively easy to understand intuitively. Furthermore, the methods we use for determining the behavior of rope waves at interfaces are almost exactly the same as those you will use in a future course to understand the behavior of quantum waves at interfaces, a most important problem indeed. We will also explicitly discuss the behavior at interfaces of sound waves, waves on transmission lines, and electromagnetic waves in a vacuum. The work we do with rope waves will serve us well when we reach these other, perhaps more interesting, types of waves, since we will again find isomorphisms for the behavior at interfaces.

10.1   Reflections and the idea of boundary conditions

In section 1.11, we discussed the simplest type of quantum mechanical wave, $\Psi(x, t) = \psi_0 e^{-i\omega t} e^{ikx}$. We showed that an electron described by such a wave function is completely delocalized, that is, it is equally probable to be found at any point in space. Often, we are interested in electrons which are more localized; a possible wavefunction for such an electron is shown schematically in figure 10.1.1a. (Recall that the quantum mechanical wavefunction is inherently complex; only the real part is shown in the figure.) This "wave packet" is travelling in a region of constant potential energy U, but is about to encounter a place where U suddenly increases. We will not treat this

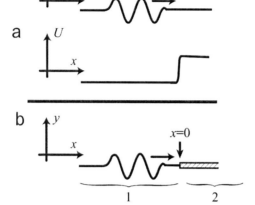

Figure 10.1.1 A Quantum mechanical wavepacket approaching a place where the potential energy suddenly changes (**a**) is analogous to a rope wavepacket approaching a place where the mass/length suddenly changes (**b**).

problem quantitatively in this book, but it is quite analogous to the problem of a wave packet on a rope wave shown in figure 10.1.1b.

At the moment shown, the wave packet (or pulse) is traveling on a rope with low mass/length μ_1, and is about to encounter an interface at which the mass/length suddenly increases to a larger value μ_2. (The tension is the same in the two parts of the rope.) For now, we assume that any conversion of the energy in the wave into thermal energy is negligible. We then recognize that some of the energy in the pulse might be transmitted into the heavier part of the rope as a pulse traveling further to the right, but some might also be reflected as a pulse traveling to the left in the lighter part of the rope. Our job is as follows: given the shape of the incident pulse, find the shape of the transmitted and reflected pulses. We define the following:

Table 10.1.1. Functions for rope waves

$R_1(x - v_1 t)$	Function describing the incident pulse, which travels to the right in medium 1 (the left part of the rope).
$L_1(x + v_1 t)$	Function describing the reflected pulse, which travels to the left in medium 1.
$R_2(x - v_2 t)$	Function describing the transmitted pulse, which travels to the right in medium 2 (the right part of the rope).

We want to describe things for all times, including the time before the pulse encounters the interface, the time during the encounter, and the time after the encounter. During the encounter, the left part of the rope has both the tail end of the pulse (moving right) and also the front end of the reflection (moving left). So, in general, the motion of the rope to the left of the interface (medium 1) is

$$y_1(x, t) = R_1(x - v_1 t) + L_1(x + v_1 t), \tag{10.1.1}$$

$\underbrace{\qquad}_{\substack{\text{motion of} \\ \text{medium 1}}}$ $\underbrace{\qquad}_{\substack{\text{incident,} \\ \text{right-moving} \\ \text{wave}}}$ $\underbrace{\qquad}_{\substack{\text{reflected,} \\ \text{left-moving} \\ \text{wave}}}$

while the motion of the rope to the right of the interface (medium 2) is

$$\underbrace{y_2\,(x,\,t)}_{\substack{\text{motion of}\\\text{medium 2}}} = \underbrace{R_2\,(x - v_2 t)}_{\substack{\text{transmitted,}\\\text{right-moving}\\\text{wave}}}. \tag{10.1.2}$$

So, given the function R_1, our task is to find L_1 and R_2. In this section, we'll find L_1.

The properties of medium 1 determine how fast pulses travel, but place no constraint on the shape of the pulse. Similarly, the properties of medium 2 determine only the speed of waves in that medium. So, the details of L_1 and R_2 must be determined by the properties of the interface itself, combined, of course, with the details of R_1. There are two important properties of the interface; these are referred to as "boundary conditions":

1. At the interface, the rope must be continuous. We set $x = 0$ at the interface, so this condition is

$$y_1\,(0,\,t) = y_2\,(0,\,t) \tag{10.1.3}$$

$$\Rightarrow \left[R_1 + L_1\right]_{x=0} = R_2\big|_{x=0}.$$

   Taking the time derivative of this gives

$$\left[\frac{\partial R_1}{\partial t} + \frac{\partial L_1}{\partial t}\right]_{x=0} = \frac{\partial R_2}{\partial t}\bigg|_{x=0}. \tag{10.1.4}$$

2. Figure 10.1.2 shows the interface at a time when the pulse is passing through it. We consider an infinitesimal segment of the rope, at the interface. This segment has infinitesimal mass. Therefore, to avoid infinite acceleration, the net force applied to it must be zero. It experiences a force $\mathbf{T}_2$ from the right part of the rope and a force $\mathbf{T}_1$ from the left part. For these forces to cancel, they must point in opposite directions. Since the tension force in a rope points along the rope, this means that

$$\frac{\partial y_1}{\partial x}\bigg|_{x=0} = \frac{\partial y_2}{\partial x}\bigg|_{x=0} \tag{10.1.5}[1]$$

$$\Rightarrow \left[\frac{\partial R_1\,(x - v_1 t)}{\partial x} + \frac{\partial L_1\,(x + v_1 t)}{\partial x}\right]_{x=0} = \frac{\partial R_2\,(x - v_2 t)}{\partial x}\bigg|_{x=0}. \tag{10.1.6}$$

Since R_1 is a function of the combination $(x - v_1 t)$, we have that

$$\frac{\partial R_1\,(x - v_1 t)}{\partial x} = \frac{\partial R_1\,(x - v_1 t)}{\partial\,(-v_1 t)}. \tag{10.1.7}$$

---

1. The same combination of boundary conditions, that is, continuity of the wavefunction (10.1.3) and continuity of its derivative (10.1.5) are used to solve the equivalent problem in quantum mechanics.

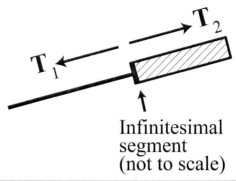

Infinitesimal
segment
(not to scale)

Figure 10.1.2 Top: Forces on an infinitesimal segment of rope at the interface must cancel, to avoid infinite acceleration. Bottom: As we'll see later in this chapter, reflection of light off the surface of water is closely analogous to reflection of rope waves off the interface between a light rope and a heavy one. This shows the 3rd Avenue bridge in Minneapolis. Image © Geoffrey Kuchera | Dreamstime.com

Now, we use the chain rule:

$$\frac{\partial R_1\left(x - v_1 t\right)}{\partial\left(-v_1 t\right)} = \frac{\partial R_1\left(x - v_1 t\right)}{\partial t}\frac{dt}{d\left(-v_1 t\right)} = \frac{\partial R_1\left(x - v_1 t\right)}{\partial t}\left(\frac{d\left(-v_1 t\right)}{dt}\right)^{-1}$$

$$= -\frac{1}{v_1}\frac{\partial R_1\left(x - v_1 t\right)}{\partial t}.$$

Combining this with equation (10.1.7) gives

$$\frac{\partial R_1\left(x - v_1 t\right)}{\partial x} = -\frac{1}{v_1}\frac{\partial R_1\left(x - v_1 t\right)}{\partial t}.$$

Applying this, and the analogous expressions for R_2 and L_1 to equation (10.1.6) yields

$$\Rightarrow \left[-\frac{1}{v_1}\frac{\partial R_1}{\partial t} + \frac{1}{v_1}\frac{\partial L_1}{\partial t}\right]_{x=0} = -\frac{1}{v_2}\frac{\partial R_2}{\partial t}\bigg|_{x=0}.$$

Using equation (10.1.4) to substitute for $\left.\dfrac{\partial R_2}{\partial t}\right|_{x=0}$ in the above, gives

$$\left[-\frac{1}{v_1}\frac{\partial R_1}{\partial t} + \frac{1}{v_1}\frac{\partial L_1}{\partial t}\right]_{x=0} = -\frac{1}{v_2}\left[\frac{\partial R_1}{\partial t} + \frac{\partial L_1}{\partial t}\right]_{x=0}.$$

Integrating with respect to time, and multiplying by $-v_1 v_2$, we obtain

$$\left[v_2 R_1 - v_2 L_1 = v_1 R_1 + v_1 L_1 + \text{const.}\right]_{x=0} \Leftrightarrow$$

$$\left[\left(v_2 - v_1\right) R_1 = \left(v_2 + v_1\right) L_1 + \text{const}\right]_{x=0}.$$

This equation must hold true under all conditions. In particular, it must be true when the rope is at equilibrium, that is, when $R_1 = L_1 = 0$. Therefore, the constant must equal zero. So, finally, we get

$$\boxed{\left[L_1 = R_1 \frac{v_2 - v_1}{v_2 + v_1}\right]_{x=0}}.$$ (10.1.8)

Core example: Applying the reflection equation.
We consider the case $v_1 = 3v_2$, so that $\left[L_1 = -\dfrac{1}{2}R_1\right]_{x=0}$. Each part of the incident pulse causes the corresponding part of the reflected pulse, so that the sequence before and after the encounter with the interface is as shown in figure 10.1.3a. Note that only the behavior for $x < 0$ is shown; in section 10.3, we'll figure out the behavior for $x > 0$.

Concept test 1 (answer below[2]): For the situation shown in the core example above, make a sketch of R_1, L_1, and y_1 for a time *during* the encounter of the pulse with the interface, so that the sketch of R_1 looks as shown in figure 10.1.3b. At this time, only about the first quarter of the pulse has encountered the interface.

Concept test 2 (answer below[3]): If we attach the end of a rope to a wall, and launch a pulse at this interface (as shown in figure 10.1.5), what does the reflection look like well after the encounter of the pulse with the interface? *Hint: Think about the wall as if it were a very heavy rope. What is the speed in such a rope?*

The opposite limit, of $v_2 \to \infty$, is more difficult to contrive. We must attach the left part of the rope to something that has the same tension, but zero mass. The only way to do this is to use a massless ring, which slides without friction on a pole, as shown in figure 10.1.7. This arrangement is obviously a bit ridiculous, but the analogs for other types of waves are important, and we'll discuss them later in this chapter.

---

2. The answer to Concept test 1 is shown in figure 10.1.4.

3. The speed of a rope wave is $\sqrt{T/\mu}$. We think of the wall as a rope with infinite μ, so that $v_2 = 0$. Using equation (10.1.8) this gives $\left[L_1 = -R_1\right]_{x=0}$, so the reflection is the same size as the incident pulse, but inverted, as shown in figure 10.1.6.

Before pulse reaches interface:

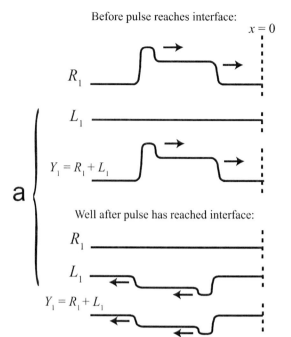

$$x = 0$$

$$R_1$$

$$L_1$$

$$Y_1 = R_1 + L_1$$

Well after pulse has reached interface:

$$R_1$$

$$L_1$$

$$Y_1 = R_1 + L_1$$

b During encounter of pulse with interface:

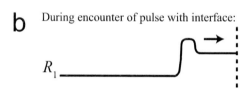

$$R_1$$

Figure 10.1.3 a: The behavior of the part of the rope to the left of the interface for the case $L_1 = -R_1/2$. b: R_1 for a time early in the encounter of the pulse with the interface. Don't look yet at the next figure.

During encounter of pulse with interface:

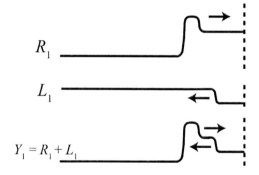

$$R_1$$

$$L_1$$

$$Y_1 = R_1 + L_1$$

Figure 10.1.4 The reflection of a pulse during the encounter with the interface.

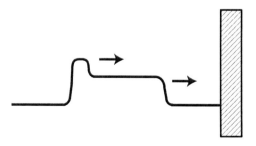

Figure 10.1.5 A pulse about to hit a brick wall.

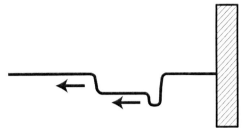

Figure 10.1.6 The reflected pulse well after the encounter with the brick wall.

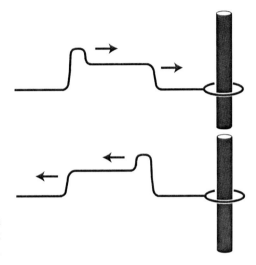

Figure 10.1.7 Reflection of a pulse off a massless ring which slides without friction on a pole.

In this limit, equation (10.1.8) gives $[L_1 = R_1]_{x=0}$, so that the reflected pulse has the same amplitude as the incident pulse, and is not inverted.

10.2 Transmitted waves

Imagine you are a dog holding the left end of a rope which is under tension, as shown in figure 10.2.1. Suddenly, you move your head up, launching a right-moving wavefront, as shown.

The same thing happens at the interface, for a rope wave; the motion of the point at the interface (the left end of the heavy part of the rope) launches the transmitted pulse into the right part of the rope.

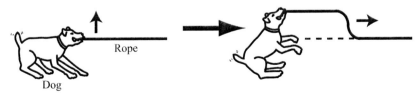

Figure 10.2.1 If you suddenly move your head up, you will launch a right-moving wavefront.

In this case, we only need the boundary condition that the rope is continuous at the interface:

$$y_1(0, t) = y_2(0, t)$$

$$\Rightarrow [R_1 + L_1]_{x=0} = R_2|_{x=0}.$$

Your turn: Use this, and the results of section 10.1, to show that

$$\left[R_2 = R_1 \frac{2v_2}{v_2 + v_1} \right]_{x=0}. \qquad (10.2.1)$$

Thus, the transmitted wave has an amplitude relative to the incident wave of $\frac{2v_2}{v_2 + v_1}$. Because it travels at speed v_2, it is compressed or expanded in the horizontal direction by the ratio v_2/v_1.

Core example: Shape of the transmitted wave pulse.
Again, we consider the case $v_1 = 3v_2$. From equation (10.2.1), we see that $R_2 = R_1/2$. We launch the same pulse as before toward the interface. As shown in figure 10.2.2a, the transmitted pulse is scaled by the factor $\frac{2v_2}{v_2 + v_1}$ in the in the y-direction, and by the factor v_2/v_1 in the x-direction.

Concept test (answer below[4]): Sketch $R_1, R_2, L_1, y_1,$ and y_2 at the point in time when the first quarter of the pulse has encountered the interface, that is, at the time shown in figure 10.2.2b. Make your sketch as quantitative as possible.

Self-test (answer below[5]): Now, let $v_1 = \frac{1}{3}v_2$, so that the left side of the rope is heavier. Use the same incident pulse as is shown in figure 10.2.2. Sketch $R_1, R_2, L_1, y_1,$ and y_2 for a point in time well after the pulse has encountered the interface. Make your sketch as quantitative as possible.

Sinusoidal waves are particularly important, partly because any wave can be expressed as a sum of sinusoids, as we saw in chapter 8. What happens to a sinusoidal

---

4. The answer to the Concept Test is shown in figure 10.2.3.
5. Answer to Self-test: Using equation (10.1.8), the amplitude of the reflected pulse is half that of the incident pulse, and the reflection is noninverted (i.e., the amplitude is positive). Using equation (10.2.1), the amplitude of the transmitted pulse is 3/2 that of the incident pulse. Because v_2 is three times v_1, the transmitted pulse is stretched in the horizontal direction by a factor of three. Putting this all together gives figure 10.2.4. It might at first appear that this violates conservation of energy, since the transmitted pulse is higher and wider than the incident pulse. However, since medium 2 is lighter, it takes less energy to create pulses in it.

Before pulse encounters the interface:

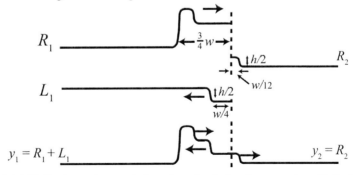

Figure 10.2.2 a: An incident pulse (top) is partly transmitted and partly reflected (bottom). Note that the reflected pulse travels three times as fast as the transmitted pulse. b: One quarter of the way through the encounter. Don't look yet at the next figure.

During encounter of pulse with interface:

Figure 10.2.3 Reflection and transmission of a pulse during the encounter with the interface.

wave when it encounters an interface? As for all other waves, the horizontal scale for the transmitted wave is changed by the factor v_2/v_1, so that the wavelength in medium 2 is

$$\lambda_2 = \frac{v_2}{v_1}\lambda_1. \qquad (10.2.2)$$

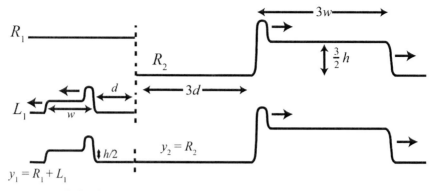

Figure 10.2.4 Reflection and transmission for a case where the speed is three times greater on the right side of the interface.

Does the angular frequency change?

$$v_2 = \frac{\omega_2}{k_2} \Leftrightarrow \omega_2 = v_2 k_2 = \frac{2\pi v_2}{\lambda_2} = \frac{2\pi v_2}{\frac{v_2}{v_1}\lambda_1} = \frac{2\pi v_1}{\lambda_1} = v_1 k_1 = \omega_1.$$

So:

> **When a sinusoidal wave moves from one medium to another, the wavelength changes, but the frequency doesn't change.**

We can also see this by considering that each crest of the transmitted wave is launched by a crest in the incident wave.

10.3 Characteristic impedances for mechanical systems

> **Concept test (answer below[6]):** We saw in section 10.1 that, when a rope wave encounters a brick wall, it is reflected with full amplitude but inverted. If it instead encounters a massless ring that slides frictionlessly on a pole, it is reflected with full amplitude but noninverted. One might expect that there would be some middle ground where there is no reflection at all, that is, a condition halfway between an inverted and a noninverted reflection. Perhaps if we attach the rope to a ring that has some carefully chosen mass, and that slides frictionlessly on a pole, we could suppress the reflection. Explain, using fundamental physical principles, why this couldn't work.

As we saw in chapter 9, transmission lines (such as coaxial cables) can support electromagnetic waves, which behave in many ways like waves on ropes. Thus, we

---

6. The wave pulse carries energy. Since the ring slides frictionlessly, it cannot dissipate the energy, so the only way for energy to be conserved is by generating a reflection.

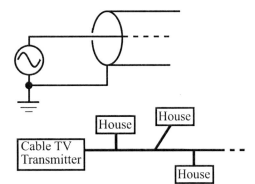

Figure 10.3.1 Top: model for a cable TV transmitter attached to a coaxial cable. Bottom: schematic for a cable TV network; the lines represent coaxial cables.

can expect that, when a wave pulse travelling along a coaxial cable reaches the end of the cable, reflections will be generated (just as reflections are generated when a pulse reaches the end of a rope attached to a wall or a ring). This could create a very serious problem for the operator of a cable television system. To broadcast on such a system, waves are launched into the cable, as shown schematically in the top part of figure 10.3.1. This signal must be sent to many houses, as shown in the bottom part; if reflections are generated at each one, there will be terrible interference problems, dramatically reducing picture quality. What is needed is some way to end the cable (as we must do at each house) but suppress the reflections. The device attached to the end of the cable that accomplishes this suppression is called a "terminator."

This is analogous to the problem of suppressing reflections for a rope wave, and the basic solution is the same in both cases. If instead of ending the cable, we attached an infinite length of cable of the same type, clearly no reflections would be generated. So, how can we make a compact device that "feels" like an infinite length of cable? For a rope, how can we make a compact device that feels like an infinite length of rope under tension?

For the rope, we will borrow an idea from our earlier study of ac circuits, in section 1.10. There, we encountered the generalized version of the resistance, called the impedance. This is defined as $Z \equiv \tilde{V}/\tilde{I}$, where $\tilde{V}$ is the complex version of the voltage and $\tilde{I}$ is the complex version of the current. For a resistor, we have simply $Z = R$; because this is real, the current and voltage are in phase for a resistor. However, for a capacitor we found that $Z = \dfrac{1}{i\omega C}$; because this is purely imaginary, the current and voltage are 90° out of phase for a capacitor. One might say that the impedance characterizes the way a circuit element "feels." We will define the equivalent quantity for a mechanical system, and then show that, if we construct a compact object with the same impedance it does indeed suppress reflections.

What is the mechanical equivalent of the impedance $Z \equiv \tilde{V}/\tilde{I}$? Recall the isomorphism between the damped driven mechanical and electrical oscillators:

$$(4.1.3a): m\ddot{x} + b\dot{x} + kx = F_0 \cos \omega_d t$$

$$(4.6.1): L\ddot{q} + R\dot{q} + \frac{q}{C} = V_0 \cos \omega_d t$$

So, the voltage is isomorphic with the applied force, and the current (which equals $\dot{q}$) is isomorphic with the velocity $\dot{x}$. By analogy with $Z \equiv \tilde{V}/\tilde{I}$, this suggests the following definition of the mechanical impedance:

$$Z \equiv \frac{\tilde{F}_{\text{applied}}}{\dot{z}},$$ (10.3.1)

Definition of mechanical impedance for a compact object

where $\tilde{F}_{\text{applied}}$ is the complex version of the applied force (e.g., for the damped driven oscillator, $\tilde{F}_{\text{applied}} = F_0 e^{i\omega_d t}$) and $\dot{z}$ is the complex version of the velocity $\dot{y}$ (e.g., for the damped driven oscillator in a steady-state, $\dot{z} = \dfrac{d}{dt}\left[A e^{i(\omega_d t - \delta)}\right]$). If F_{applied} is in phase with the velocity $\dot{y}$, then Z is real, just as the impedance is real for an electrical circuit if the voltage is in phase with the current. However, if Z is complex, then F_{applied} is not in phase with $\dot{y}$. Since Z may generally be complex, we don't bother to include the tilde above it.

Self-test (answer below[7]): For a damped driven harmonic oscillator, what is the impedance Z? Express your answer in terms of F_0, A, δ, and ω_d.

For an extended object such as a rope, we must be a bit more careful with our definition. Imagine you are holding the left end of a rope in your hand. You can launch waves down the rope by moving your hand up and down, exerting a force on the tiny piece of the rope you are actually touching. However, this is not the only force this piece of rope feels; it also feels a force from the rest of the rope, off to its right. So, because our definition involves the applied force, we could also write it (for the case of right-moving waves) as the force coming from the left side of a piece of rope:

$$Z \equiv \frac{\tilde{F}_{\text{L}}}{\dot{z}},$$

Definition of mechanical impedance for right-traveling waves

where $\tilde{F}_{\text{L}}$ is the complex version of F_{L} (the y-component of the force exerted on a piece of rope from its left end), and $\dot{z}(x, t)$ is the complex version of the velocity in the y-direction, that is, $\dot{y} = \text{Re}\,\dot{z}$. We already know from the arguments of section 9.5 that F_{L} is in phase with $\dot{y}$ for a right-moving wave, so Z is real, and we can simply write

$$Z = \frac{F_{\text{L}}}{\dot{y}}$$ (10.3.2)

Mechanical impedance for right-traveling waves.

---

7. Answer to self-test: $Z = \dfrac{\tilde{F}_{\text{applied}}}{\dot{z}} = \dfrac{F_0 e^{i\omega_d t}}{\dfrac{d}{dt}\left[A e^{i(\omega_d t - \delta)}\right]} = -i e^{i\delta} \dfrac{F_0}{\omega_d A} = e^{i(\delta - \pi/2)} \dfrac{F_0}{\omega_d A}$. So, when $\delta = \pi/2$ (which occurs when $\omega_d = \omega_0$), the impedance is real, meaning that the drive force is in phase with the velocity, as we already knew.

It is not difficult to calculate Z, since we already know from equation (9.5.15) that $F_L = -T\frac{\partial y}{\partial x}$. So, we need to express $\dot{y}$ in terms of $\frac{\partial y}{\partial x}$, so that when we take the ratio we will get a constant. For a right moving wave, the solution to the wave equation is of the form $y(x - vt)$. Therefore, by the chain rule, $\dot{y} = \frac{\partial y}{\partial t} = -v\frac{\partial y}{\partial x}$. (For example, if $y = A\sin[k(x - vt)]$, then $\frac{\partial y}{\partial t} = -kvA\cos[k(x - vt)]$, while $\frac{\partial y}{\partial x} = kA\cos[k(x - vt)]$.) Therefore,

$$Z = \frac{-T\dfrac{\partial y}{\partial x}}{-v\dfrac{\partial y}{\partial x}} = \frac{T}{v}.$$

From equation (9.2.4), $v = \sqrt{\dfrac{T}{\mu}}$, so that

$$\boxed{Z_{\text{rope}} = \sqrt{T\mu}} \tag{10.3.3}$$

For a left-traveling wave, the solution to the wave equation is of the form $y(x + vt)$, so that $\dot{y} = \frac{\partial y}{\partial t} = +v\frac{\partial y}{\partial x}$. Since the impedance should be a property of the rope, and should not depend on the direction of wave travel, we therefore see that we must alter the definition (10.3.2) for the case of left-moving waves:

$$Z = -\frac{F_L}{\dot{y}}.$$

Mechanical impedance for left-traveling waves.

Just to check: $Z = -\dfrac{-T\dfrac{\partial y}{\partial x}}{v\dfrac{\partial y}{\partial x}} = \dfrac{T}{v} = \sqrt{T\mu}$, as desired. You may be wondering why, for left-traveling waves, we don't instead define $Z = +F_r/\dot{y}$, where F_r is the y-component of force applied to a segment of rope from the right. This would be correct, but it is not as convenient. We wish to build on the isomorphisms developed in chapter 9, which are all based on F_L, so it is better to stick with F_L in this chapter as well. Furthermore, it is easier to discuss boundary conditions when we use the same quantity F_L to describe left- and right-traveling waves.

Let us return now to right-traveling waves. If you hold the left end of the rope, the force you apply to it is F_L. Because the *velocity* $\dot{y}$ is in phase with F_L, it is in phase with the force you apply. Thus, the rope "feels" different from a small rock, for which the *acceleration* is in phase with the force you apply. The difference comes because the end of the rope feels not only the force you are exerting now, but also the force from the part of the rope just to its right. That part of the rope is at a position that is determined by the propagating wave, that is, that is determined by the force you exerted on the end of the rope a short time ago.

So, how can we terminate the rope in such a way as to suppress reflections? In other words, which compact object has the same impedance as an infinite length of rope, so that it feels the same as an infinite length of rope? In other words, which compact system has a velocity that is in phase with the applied force? The answer

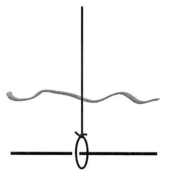

Figure 10.3.2 By connecting a rope to a massless ring immersed in a suitably chosen damping fluid, we can suppress reflections.

is a massless ring, sliding on a frictionless pole, and immersed in a viscous damping medium. Because the ring is massless, the net force applied to it must be zero (otherwise its acceleration would be infinite). Therefore, the force applied to it (in this case by the rope) must be opposite the damping force, that is,

$$F_{\text{applied}} = -(-b\dot{y}) = b\dot{y}, \qquad (10.3.4)$$

so that, as we need, the y-velocity is in phase with the applied force. The impedance of such a ring is the ratio of the force to the velocity, so that $Z_{\substack{\text{damped} \\ \text{massless} \\ \text{ring}}} = b$, and to suppress reflections of waves traveling down the rope, we just match impedances (so that the ring feels like an infinite rope), that is, we choose the size of the ring and the viscosity of the damping medium so that

$$b = \sqrt{T\mu}.$$

One way to accomplish this would be to have the rope be vertical, and the ring be immersed in a pool of liquid, as suggested in figure 10.3.2. Note that the damping provides a way to absorb the energy of incident pulses.

Now, unless you plan to create a telephone company based on tin cans tied together with strings, the previous discussion might seem rather contrived. However, in section 10.4 we will develop the ideas needed to determine reflection and transmission for all the types of waves we have studied, making use of the isomorphisms uncovered in chapter 9. Then, in section 10.5, we apply these ideas to the more practical considerations of how to suppress reflections in transmission lines (such as coaxial cables).

10.4   "Universal" expressions for transmission and reflection

We saw in section 9.5 that the displacement wave $y(x, t)$ is not isomorphic to any of the quantities that characterize other wave types. Instead, the velocity wave $\dot{y}$ is isomorphic to, for example, B/μ_0 for an electromagnetic wave in vacuum, and F_L is isomorphic to, for example, the electric field E of an electromagnetic wave in vacuum. We will see that all these isomorphisms remain faithful when we consider the boundary conditions for each of the various types of waves. Therefore, it is worthwhile for us to consider the reflection and transmission for the $\dot{y}$ wave and the F_L wave. However, for the most general case we must allow the tension T to be different in the two parts of the rope. To accomplish this, we imagine that the two halves are connected by a massless ring which slides frictionlessly on a pole, so that the pole can supply the force needed for the tensions to be different. To the left of the pole we have rope 1, and to the right we have rope 2.

What is the boundary condition on F_L? Consider the forces exerted on the massless ring. The force exerted on the ring by the section of rope to its left is, of course, F_{L1}, where "1" in the subscript indicates the force is exerted on the left side of the interface. The ring exerts a force F_{L2} on the rest of the rope to its right, where "2" indicates the force is exerted on the right side of the interface. By Newton's third law, this is opposite to the force exerted by the rope to its right on the ring, so that the net force (in the y-direction) exerted on the ring is $F_{L1} - F_{L2}$. Since the ring is massless, the net force must be zero (otherwise its acceleration would be infinite), so that

$$\left[F_{L1} = F_{L2}\right]_{x=0}. \tag{10.4.1}$$

Boundary condition for F_L

In other words, the boundary condition for F_L is that it must be continuous at the interface.

As we did when considering the displacement y of the rope, we can express F_L in rope 1 (to the left of the interface) as the sum of the right-traveling incident wave $F_{L,R1}$ and the left-traveling reflected wave $F_{L,L1}$:

$$F_{L1} = F_{L,R1} + F_{L,L1}.$$

On the right side of the interface, there is only the transmitted wave:

$$F_{L2} = F_{L,R2}.$$

Therefore, the boundary condition becomes

$$\left[F_{L,R1} + F_{L,L1} = F_{L,R2}\right]_{x=0}. \tag{10.4.2}$$

What is the boundary condition for $\dot{y}$? We know that the rope itself must be continuous, that is, $\left[y_1 = y_2\right]_{x=0}$. Taking the time derivative of this gives

$$\left[\dot{y}_1 = \dot{y}_2\right]_{x=0}. \tag{10.4.3}$$

Boundary condition for $\dot{y}$

Expressing this in the R1, L1, R2 notation, we have

$$\left[\dot{y}_{R1} + \dot{y}_{L1} = \dot{y}_{R2}\right]_{x=0}. \tag{10.4.4}$$

We will begin by finding the reflected and transmitted $\dot{y}$ waves. For notational convenience, in what follows we omit the explicit reminders that we are considering the behavior at the interface $x = 0$. Dividing equation (10.4.2) by (10.4.4) gives

$$\frac{F_{L,R1} + F_{L,L1}}{\dot{y}_{R1} + \dot{y}_{L1}} = \frac{F_{L,R2}}{\dot{y}{R2}}.$$

From equation (10.3.2), $Z_2 = \dfrac{F_{L,R2}}{\dot{y}_{R2}}$, so that above becomes

$$F_{L,R1} + F_{L,L1} = Z_2 \left(\dot{y}_{R1} + \dot{y}_{L1} \right). \tag{10.4.5}$$

From equation (10.3.4), we have that, for the reflected wave (which travels left)

$$Z_1 = -\frac{F_{L,L1}}{\dot{y}_{L1}} \Leftrightarrow F_{L,L1} = -\dot{y}_{L1} Z_1.$$

Substituting this into equation (10.4.5) gives

$$F_{L,R1} - \dot{y}_{L1} Z_1 = Z_2 \left(\dot{y}_{R1} + \dot{y}_{L1} \right).$$

Dividing by $\dot{y}_{R1}$ gives

$$\underbrace{\frac{F_{L,R1}}{\dot{y}_{R1}}}_{Z_1} - \frac{\dot{y}_{L1}}{\dot{y}_{R1}} Z_1 = Z_2 \left(1 + \frac{\dot{y}_{L1}}{\dot{y}_{R1}} \right) \Rightarrow Z_1 - Z_2 = \frac{\dot{y}_{L1}}{\dot{y}_{R1}} \left(Z_1 + Z_2 \right) \Rightarrow$$

$$\dot{y}_{L1} = \dot{y}_{R1} \frac{Z_1 - Z_2}{Z_1 + Z_2}, \text{ or} \tag{10.4.6}$$

$$\boxed{\left[\dot{y}_{\text{reflected}} = \dot{y}_{\text{incident}} \frac{Z_i - Z_t}{Z_i + Z_t} \right]_{x=0}}, \tag{10.4.7}$$

where $Z_i = Z_1$ is the impedance for the rope on the incident side and $Z_t = Z_2$ is the impedance for the rope on the transmitted side.

Your turn: Use equation (10.4.6) to substitute for $\dot{y}_{L1}$ in equation (10.4.4) and show that

$$\boxed{\left[\dot{y}_{\text{transmitted}} = \dot{y}_{\text{incident}} \frac{2Z_i}{Z_i + Z_t} \right]_{x=0}}, \tag{10.4.8}$$

Concept test (answer below[8]): A right-traveling pulse gets to the end of a rope, which is attached to a massless ring which slides frictionlessly on a pole. Is the reflected $\dot{y}$ pulse inverted or not with respect to the incident $\dot{y}$ pulse? Use the results of this section to get your answer.

For F_L, the expressions for reflection and transmission are different. The reflected wave moves left in rope 1, so

$$Z_1 = -\frac{F_{L,L1}}{\dot{y}_{L1}} \Leftrightarrow F_{L,L1} = -\dot{y}_{L1} Z_1 = -\dot{y}_{L1} \frac{F_{L,R1}}{\dot{y}_{R1}}.$$

---

8. The massless ring has zero impedance (since an infinitesimal force in the y-direction is all that's needed to produce a velocity and $Z = F_L/\dot{y}$). Plugging $Z_2 = 0$ into equation (10.4.6) gives $|\dot{y}_{\text{reflected}} = \dot{y}_{\text{incident}}|_{x=0}$, so the reflected velocity pulse is uninverted.

Using equation (10.4.6) to substitute for $\dot{y}_{L1}$, we then have that, at the interface,

$$F_{L,L1} = -F_{L,R1}\frac{Z_1 - Z_2}{Z_1 + Z_2} \text{ or}$$

$$\boxed{\left[F_{L,\text{reflected}} = F_{L,\text{incident}}\frac{Z_t - Z_i}{Z_i + Z_t}\right]_{x=0}.} \qquad (10.4.9)$$

The transmitted wave moves right in rope 2, so

$$Z_2 = \frac{F_{L,R2}}{\dot{y}_{R2}} \Leftrightarrow F_{L,R2} = \dot{y}_{R2}Z_2.$$

Using (10.4.8) to substitute for $\dot{y}_{R2}$, we get that, at the interface

$$F_{L,R2} = \left(\dot{y}_{R1}\frac{2Z_1}{Z_1 + Z_2}\right)Z_2.$$

Using $Z_1 = \dfrac{F_{L,R1}}{\dot{y}_{R1}} \Leftrightarrow \dot{y}_{R1} = \dfrac{F_{L,R1}}{Z_1}$, we obtain

$$F_{L,R2} = \left(\frac{F_{L,R1}}{Z_1}\frac{2Z_1}{Z_1 + Z_2}\right)Z_2 = F_{L,R1}\frac{2Z_2}{Z_1 + Z_2}, \text{ so that}$$

$$\boxed{\left[F_{L,\text{transmitted}} = F_{L,\text{incident}}\frac{2Z_t}{Z_i + Z_t}\right]_{x=0}.} \qquad (10.4.10)$$

Concept test (answer below[9]): As in the previous concept test, a right-traveling pulse gets to the end of a rope, which is attached to a massless ring which slides frictionlessly on a pole. Is the reflected F_L pulse inverted or not with respect to the incident F_L pulse? Again, use the results of this section to get your answer.

10.5 Reflected and transmitted waves for transmission lines

Now, let's consider what happens when a wave pulse on a transmission line encounters an interface where the properties of the line (such as the inductance per length L_0 or the capacitance per length C_0) change suddenly, as suggested in figure 10.5.1. As for

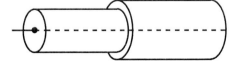

Figure 10.5.1 An interface between two types of coaxial cable.

---

9. Plugging $Z_t = 0$ into equation (10.4.9), we see that $\left[F_{L,\text{reflected}} = -F_{L,\text{incident}}\right]_{x=0}$, so the reflected F_L pulse is inverted, and thus is opposite in phase to the reflected $\dot{y}$ pulse.

a rope, we can anticipate that there will be a reflected pulse and a transmitted pulse. We will be able to use our isomorphisms to find the expressions for the reflected and transmitted pulses with no additional work, after checking to make sure the boundary conditions are the same. We have previously made isomorphisms between rope waves and electromagnetic waves in vacuum, and between electromagnetic waves in vacuum and waves on transmission lines:

Table 10.5.1. Isomorphism between rope waves, em waves in vacuum, and waves on a transmission line

Rope waves	Electromagnetic waves in vacuum	Waves on a transmission line
Transverse velocity $\dot{y}$	B/μ_0	I
y-component of force from the left $F_L = -T\dfrac{\partial y}{\partial x}$	Electric field E	V
Mass per length μ	Permeability of free space μ_0	Inductance per length L_0
Tension T	Inverse of permittivity of free space $1/\varepsilon_0$	Inverse of capacitance per length $1/C_0$

Therefore, $\dot{y}$ for the rope wave is isomorphic to I for the transmission line wave, and so on.

We have two boundary conditions:
1 The voltage at the left side of the interface must equal the voltage on the right side, that is,

$$\left[V_i = V_t\right]_{x=0}.$$ (10.5.1)

From equation (10.4.1), this is the same as the boundary condition for the isomorphic quantity F_L: $\left[F_{Li} = F_{Lt}\right]_{x=0}$.
2 The current at the left side of the interface must equal the current at the right side of the interface, that is,

$$\left[I_i = I_t\right]_{x=0}.$$

From equation (10.4.3), this is the same as the boundary condition for the isomorphic quantity $\dot{y}$, $\left[\dot{y}_i = \dot{y}_t\right]_{x=0}$. Therefore, the isomorphism is complete even for reflections and transmissions, and we can immediately make use of the isomorphic expressions for transmission and reflection for a rope wave. To do so, we need the expression for the impedance of the transmission line. For the rope $Z = \sqrt{T\mu}$, so, reading off the isomorphism, the impedance for the transmission line is

$$\boxed{Z = \sqrt{L_0/C_0}.}$$ (10.5.2)

By equation (10.3.2), for the rope wave $Z = \dfrac{F_L}{\dot{y}}$ for a right-moving wave, so for waves on transmission lines

$$\boxed{Z = \frac{V}{I},}$$
(10.5.3)

Impedance for right-moving waves on transmission lines

in accordance with our expectations for the impedance of electrical circuits.

The voltage wave on the transmission line is isomorphic to F_L. So, translating equations (10.4.9) and (10.4.10), we have

$$\boxed{\left[V_{\text{reflected}} = V_{\text{incident}} \frac{Z_t - Z_i}{Z_i + Z_t} \right]_{x=0}.}$$
(10.5.4)

Reflection for the V of a transmission line wave.

and

$$\boxed{\left[V_{\text{transmitted}} = V_{\text{incident}} \frac{2Z_t}{Z_i + Z_t} \right]_{x=0}.}$$
(10.5.5)

Transmission for the V of a transmission line wave.

The I wave on the transmission line is isomorphic to $\dot{y}$. So, translating equations (10.4.7) and (10.4.8) we have

$$\boxed{\left[I_{\text{reflected}} = I_{\text{incident}} \frac{Z_i - Z_t}{Z_i + Z_t} \right]_{x=0},}$$
(10.5.6)

Transmission for the I of a transmission line wave

$$\boxed{\left[I_{\text{transmitted}} = I_{\text{incident}} \frac{2Z_i}{Z_i + Z_t} \right]_{x=0},}$$
(10.5.7)

Reflection for the I of a transmission line wave

Concept test (answer below[10]): A right-travelling wave gets to the end of a coaxial cable, at which point a wire has been attached connecting the inner conductor to the outer conductor. Is the reflected current pulse inverted or not with respect to the incident current pulse?

---

10. The wire connecting the inner to the outer conductor has zero impedance, so $Z_t = 0$. Plugging this into equation (10.5.6), we see that $[I_{\text{reflected}} = I_{\text{incident}}]_{x=0}$, so the reflected current pulse is uninverted.

a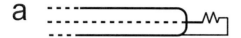

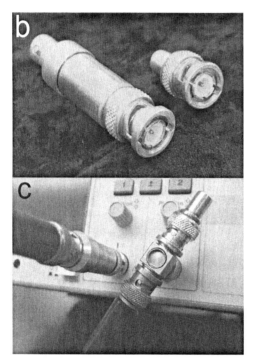

Figure 10.5.2 a: Schematic picture of a terminator for a transmission line. A resistance equal to the characteristic impedance of the transmission line is connected between the two conductors. This "feels" like an infinite length of cable, and so generates no reflections b: Two types of terminators for coaxial cables. On the left is an inline terminator, and on the right is a stub terminator. Both have 50 Ω between inner and outer conductors. c: Typical arrangements for suppressing reflections at the inputs of an oscilloscope, using an inline terminator on the left, and a coax T and stub terminator on the right.

Concept test (answer below[11]): As in the previous concept test, a right-traveling wave gets to the end of a coaxial cable, at which point a wire has been attached connecting the inner conductor to the outer conductor. Is the reflected voltage pulse inverted or not with respect to the incident current pulse?

Finally, to suppress reflections, at the end of the cable we simply connect a resistor with resistance equal to the characteristic impedance of the cable, as illustrated in figure 10.5.2a. Standard RG58 coaxial cable (the type used in laboratories, and the type that has a BNC coaxial connector at the end) has $C_0 = 100 \, \text{pF/m}$ and $L_0 = 250 \, \text{nH/m}$; plugging in these numbers gives $Z = \sqrt{\dfrac{L_0}{C_0}} = 50 \, \Omega$. Thus, a 50 Ω resistor connected from the inner to the outer conductor feels (electrically) the same as an infinite length of cable. In practice, one either connects the end of the cable to a target device that already has an input impedance of 50 Ω, or instead one uses a "terminator." There are

---

11. Plugging $Z_t = 0$ into equation (10.5.4), we see that $\left[V_{\text{reflected}} = -V_{\text{incident}}\right]_{x=0}$, so the reflected voltage pulse is inverted, and thus is opposite in phase to the reflected current pulse. This makes sense, since we know that for left-traveling waves the voltage is opposite in phase to the current.

two common styles, as shown in figure 10.5.2b and c, either of which can be used to make a nonreflecting connection to a device with high input impedance such as an oscilloscope. Either of the styles of terminators has a 50 Ω resistor connecting the inner and outer conductors of the coax.

We are now equipped to really understand what the term "characteristic impedance of a transmission line" means:

> **The characteristic impedance of a transmission line is (1) The resistance one could connect between the two wires at the end of the line to suppress reflections and (2) The resistance one would measure at the left end of an infinitely-long transmission line.**

Meaning 2 may be difficult to understand at first. One way to measure resistance is to apply a known voltage and measure the resulting current. When you first apply the voltage to the transmission line, you launch a voltage wavefront down the line, accompanied of course by a current wavefront, as shown in figure 10.5.3. The equipment you're using is supplying current to charge up the capacitance between the two conductors of the transmission line; as time goes on, and the voltage wavefront propagates out, more and more length must be charged up. Since the capacitance per unit length is constant, this means that charge must be delivered at a constant rate to the inner conductor (and removed from the outer conductor). Thus, you would measure a constant current going into the inner conductor, and coming out of the outer conductor. While the wavefronts are propagating out along the infinite cable, the current you measure would be exactly the same DC current as you'd get from a 50 Ω resistor.

If you take a 1-m length of standard coaxial cable, with nothing connected to either end, apply a fixed voltage between the inner and outer conductors at the left end, and measure the resulting current, you'll get zero current, instead of the DC current mentioned above. This is because it takes a very short time for the propagating

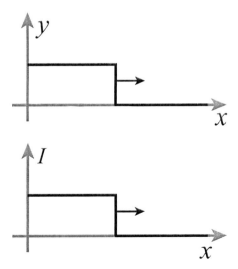

Figure 10.5.3 When you first apply a voltage to the left end of a transmission line, you generating a wavefront in the voltage which propagates down the line, accompanied by a wavefront in the current.

wavefronts in figure 10.5.3 to reach the end of this cable. Because it is not terminated with a 50 Ω resistor, a reflection is generated, which comes back to the left end of the cable, generating a new reflection which goes back to the right end, and so on. Very quickly, the whole thing settles down into a state with no propagating waves, but rather a steady DC voltage difference and no current flowing.

> **Self-test (answer below[12]):** A voltage difference of 50 V is suddenly applied to the inner conductor of a standard coax cable, relative to the outer conductor. The cable is 100-m long. For a very short time, the observed current going into the inner conductor (and coming out of the outer conductor) has a well-defined constant value. It then starts changing, and settling toward zero current. What is the well-defined constant value, and for how long is it observed?

10.6   Reflection and transmission for electromagnetic waves in matter: Normal incidence

Next, we consider what happens when a light wave encounters the boundary between two linear materials, assuming the wave propagates in a direction perpendicular to the boundary. We will be able to use our isomorphisms to find the expressions for the reflected and transmitted pulses with no additional work, after checking to make sure the boundary conditions are the same. We have previously made isomorphisms between rope waves and electromagnetic waves in vacuum, and between electromagnetic waves in vacuum and in linear materials:

Table 10.6.1. Isomorphism between rope waves, em waves in vacuum, and em waves in linear materials

Rope waves	Electromagnetic waves in vacuum	Electromagnetic waves in linear materials
Transverse velocity $\dot{y}$ y-component of force from the left $F_{\mathrm{L}} = -T\dfrac{\partial y}{\partial x}$	B/μ_0 Electric field E	$H = B/\mu$ Electric field E
Mass per length μ Tension T	Permeability of free space μ_0 Inverse of permittivity of free space $1/\varepsilon_0$	Permeability μ Inverse of permittivity $1/\varepsilon$

So, for example, H is isomorphic to $\dot{y}$.

---

12. Until the wavefront propagates to the end of the cable <u>and back</u>, the information that the cable is not infinitely long is unknown. As we saw in section 9.7, the speed of waves on standard coax cable is 2/3 c, so the constant current condition lasts for a time $\dfrac{2 \times (100\ \text{m})}{\frac{2}{3}c} =$ 1μ s. The amount of current is $I = V/Z = 1$A.

What is the boundary condition for E? At the interface between two linear materials, there can be surface charge, which affects the component of E perpendicular to the interface but not the component parallel to the interface. Therefore, for an electromagnetic wave which propagates perpendicular to the interface (so that **E** is parallel to the interface), E must be continuous at the interface. This is the same boundary condition as for the isomorphic quantity for a rope wave, F_L, which must also be continuous at the interface as discussed in section 10.4.

What is the boundary condition for H? Recall from section 9.6 that in linear material, we can create an alternate version of Maxwell's equations by replacing μ_0 with μ, ε_0 with ε, and all references to charge with references to free charge. Therefore, the alternate version of Ampère's law in an isotropic linear medium is

$$\oint \mathbf{B} \cdot \vec{\mathrm{d}\ell} = \mu\, I_{\substack{\text{free} \\ \text{net} \\ \text{threading}}} + \mu\varepsilon \frac{\mathrm{d}}{\mathrm{d}t} \int \mathbf{E} \cdot \hat{\mathbf{n}}\, \mathrm{d}A.$$

If μ and ε vary in space, then we must bring them inside the integrals:

$$\oint \frac{1}{\mu} \mathbf{B} \cdot \vec{\mathrm{d}\ell} = I_{\substack{\text{free} \\ \text{net} \\ \text{threading}}} + \frac{\mathrm{d}}{\mathrm{d}t} \int \varepsilon\, \mathbf{E} \cdot \hat{\mathbf{n}}\, \mathrm{d}A. \qquad (10.6.1)$$

Consider the Ampèrian loop shown in figure 10.6.1, which straddles the interface. If we make the height of the loop infinitesimal, then terms that depend on the area of the loop, such as $\int \mathbf{E} \cdot \hat{\mathbf{n}}\, \mathrm{d}A$, and any ordinary current flowing through the loop become negligible. *Surface* currents can still be important, since their current density is infinite. However, we will assume that the materials in question are nonconducting, so the surface current is zero. Then, since $\mathbf{H} = \dfrac{1}{\mu}\mathbf{B}$ for a linear material, equation (10.6.1)

simplifies to $\oint \mathbf{H} \cdot \vec{\mathrm{d}\ell} = 0$. The contributions to this integral from the vertical arms of the loop are negligible, since we have made the height of the loop infinitesimal. For a plane wave propagating perpendicular to the interface, we align the top and bottom arms of the loop along the axis of **H**, so that we get $H_1\ell - H_2\ell = 0$. (The direction of $\vec{\mathrm{d}\ell}$ for the bottom arm of the loop is opposite to that for the top arm, leading to the minus sign for the $H_2\ell$ term.) Therefore, $H_1 = H_2$, meaning that H is continuous across the interface. This is the same boundary condition as for the isomorphic quantity in a rope wave, $\dot{y}$, which is continuous across the interface.

Therefore, the isomorphism is complete even for reflections and transmissions, and we can immediately make use of the isomorphic expressions for transmission and reflection for a rope wave. To do so, we need the expression for the impedance of the transmission line. For the rope $Z = \sqrt{T\mu}$, so, reading off the isomorphism, the

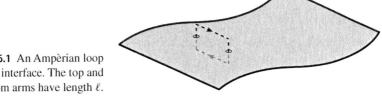

Figure 10.6.1 An Ampèrian loop at the interface. The top and bottom arms have length ℓ.

impedance for electromagnetic waves in linear materials is

$$Z = \sqrt{\mu/\varepsilon}. \tag{10.6.2}$$

By equation (10.3.2), for the rope wave $Z = \dfrac{F_L}{\dot{y}}$ for a right-moving wave, so for waves in linear materials

$$Z = \frac{E}{H}. \tag{10.6.3}$$

Impedance for right-moving electromagnetic waves in linear materials

The E in an electromagnetic wave is isomorphic to F_L. So, translating equation (10.4.9) we have

$$\left[E_{\text{reflected}} = E_{\text{incident}} \frac{Z_t - Z_i}{Z_i + Z_t} \right]_{x=0}. \tag{10.6.4}$$

For problems in optics, it is conventional to work in terms of the index of refraction

$$n \equiv \frac{c}{v_p} = \sqrt{\frac{\mu\varepsilon}{\mu_0\,\varepsilon_0}}. \tag{10.6.5}$$

For most materials, $\mu \cong \mu_0$. Making use of this as we plug equation (10.6.2) into (10.6.4), we obtain at $x = 0$

$$E_{\text{reflected}} \cong E_{\text{incident}} \frac{\sqrt{\dfrac{\mu_0}{\varepsilon_t}} - \sqrt{\dfrac{\mu_0}{\varepsilon_i}}}{\sqrt{\dfrac{\mu_0}{\varepsilon_i}} + \sqrt{\dfrac{\mu_0}{\varepsilon_t}}} = E_{\text{incident}} \frac{\sqrt{\varepsilon_i} - \sqrt{\varepsilon_t}}{\sqrt{\varepsilon_t} + \sqrt{\varepsilon_i}} \Rightarrow$$

$$\left[E_{\text{reflected}} \cong E_{\text{incident}} \frac{n_i - n_t}{n_t + n_i} \right]_{x=0}. \tag{10.6.6}$$

Concept test (answer below[13]): If $n_i > n_t$, is the reflection inverted relative to the incident wave or not? If $n_i < n_t$, is the reflection inverted or not?

---

13. From equation (10.6.6), we can see that if $n_i > n_t$ (e.g., a wave travelling from glass to air), then the coefficient for the reflected amplitude is positive, so that the reflected wave is uninverted (albeit smaller than the incident wave). On the other hand, if $n_i < n_t$ (e.g., a wave travelling from air to glass), then the coefficient is negative, so the reflection is inverted. This distinction is important for various interference phenomena (see, for example, problem 10.16), and so it is worth remembering. Here's a mnemonic that may be helpful: "Air to glass, switch head and ass. Glass to air, the phase don't care."

Often in optics we are more interested in the intensity than the amplitude. From equation (9.6.8), $S = \sqrt{\dfrac{\varepsilon}{\mu}}\, E^2$, so

$$S_{\text{reflected}} \cong S_{\text{incident}} \left(\frac{n_i - n_t}{n_t + n_i} \right)^2. \tag{10.6.7}$$

Your turn: Use the isomorphism to show that

$$\left[E_{\text{transmitted}} = E_{\text{incident}} \frac{2Z_t}{Z_i + Z_t} \right]_{x=0}. \tag{10.6.8}$$

and (assuming $\mu \cong \mu_0$ for both materials)

$$\boxed{\left[E_{\text{transmitted}} = E_{\text{incident}} \frac{2n_i}{n_t + n_i} \right]_{x=0}.} \tag{10.6.9}$$

In problem 10.13, you can show that if $\mu \cong \mu_0$ then

$$S_t = S_i \frac{4 n_i n_t}{\left(n_i + n_t \right)^2}. \tag{10.6.10}$$

10.7 Reflection and transmission for sound waves, and summary of isomorphisms

Finally, we consider what happens when a sound wave encounters the boundary between two media (e.g., air and water), assuming the wave propagates in a direction perpendicular to the boundary. In section 9.8, we showed that the velocity wave $v \equiv \dfrac{d\delta}{dt}$ (where δ is the longitudinal displacement) is isomorphic to B/μ_0 for an electromagnetic wave in vacuum, which is in turn isomorphic to the transverse velocity $\dot{y}$ for a rope wave. We also showed that the pressure variation P' is isomorphic to E an electromagnetic wave in vacuum, which is in turn isomorphic to F_L for a rope wave.

What is the boundary condition for the pressure variation P'? Consider a volume element of infinitesimal thickness centered on the interface. The net force on this element must be zero (to avoid infinite acceleration). Therefore, the pressure must be the same on either side of the element, so that P' must be continuous at the interface. This is the same boundary condition as for the isomorphic quantity for a rope wave, F_L, which must also be continuous at the interface as discussed in section 10.4.

What is the boundary condition for v? The two materials at the interface must remain in contact with each other, so that the displacement δ must be continuous at the interface. Therefore, its derivative $v \equiv \dfrac{d\delta}{dt}$ is continuous across the interface. This is the same boundary condition as for the isomorphic quantity in a rope wave, $\dot{y}$, which is continuous across the interface. Therefore, the isomorphism is again complete.

We are now ready to summarize all the isomorphisms we have discussed:

Table 10.7.1. Summary of Isomorphisms for waves

	Rope waves	Electromagnetic waves in linear matter	Waves on transmission lines	Sound waves
	Transverse velocity $\dot{y}$	$H = B/\mu$ $(B/\mu_0$ in vacuum$)$	I	Velocity $v \equiv \partial\delta/\partial t$
	y-component of force from the left $F_L = -T\dfrac{\partial y}{\partial x}$	Electric field E	V	Pressure variation P'
	Mass per length μ	Permeability μ	Inductance per length L_0	Equilibrium density ρ_0
	Tension T	Inverse permittivity $1/\varepsilon$	Inverse of capacitance per length: $1/C_0$	γP_0
Phase velocity	$\sqrt{T/\mu}$	$1/\sqrt{\varepsilon\mu}$	$1/\sqrt{C_0 L_0}$	$\sqrt{\gamma P_0/\rho_0}$ (for a gas)
Impedance Z	$\sqrt{\mu T}$	$\sqrt{\mu/\varepsilon}$	$\sqrt{L_0/C_0}$	$\sqrt{\rho_0 \gamma P_0}$ (for a gas)
Impedance Z for right-moving waves	$F_L/\dot{y}$	E/H	V/I	P'/v
Amplitude reflection coefficients	$F_{Lr} = F_{Li}\dfrac{Z_t - Z_i}{Z_i + Z_t}$ $\dot{y}_r = \dot{y}_i\dfrac{Z_i - Z_t}{Z_i + Z_t}$	$E_r = E_i\dfrac{Z_t - Z_i}{Z_i + Z_t}$ $H_r = H_i\dfrac{Z_i - Z_t}{Z_i + Z_t}$	$V_r = V_i\dfrac{Z_t - Z_i}{Z_i + Z_t}$ $I_r = I_i\dfrac{Z_i - Z_t}{Z_i + Z_t}$	$P' = P'_i\dfrac{Z_t - Z_i}{Z_i + Z_t}$ $v_r = v_i\dfrac{Z_i - Z_t}{Z_i + Z_t}$
Amplitude transmission coefficients	$F_{Lt} = F_{Li}\dfrac{2Z_t}{Z_i + Z_t}$ $\dot{y}_t = \dot{y}_i\dfrac{2Z_i}{Z_i + Z_t}$	$E_t = E_i\dfrac{2Z_t}{Z_i + Z_t}$ $H_t = H_i\dfrac{2Z_i}{Z_i + Z_t}$	$V_t = V_i\dfrac{2Z_t}{Z_i + Z_t}$ $I_t = I_i\dfrac{2Z_i}{Z_i + Z_t}$	$P'_t = P'_i\dfrac{2Z_t}{Z_i + Z_t}$ $v_t = v_i\dfrac{2Z_i}{Z_i + Z_t}$

10.8 Snell's Law

What happens when waves of any type in a two- or three-dimensional system encounter an interface at an angle, rather than encountering it at normal incidence? We define the "wavefront" to be a line drawn along the crest of the wave. (For waves in three dimensions, the wavefront is a plane.) Usually, the wavefront is perpendicular to the direction of propagation; this direction is shown as a "ray." Experimentally, we find that if a wave has straight wavefronts in medium 1, the transmitted wavefronts in medium 2 are also straight, as shown in figure 10.8.1. We found in our study of one-dimensional waves that the frequency in medium 2 is the same as in medium

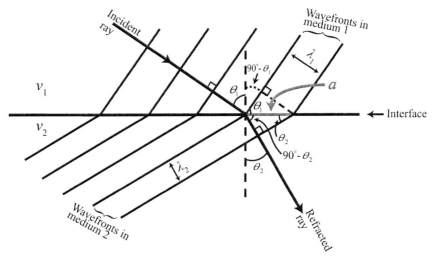

Figure 10.8.1 Straight wavefronts incident from medium 1 lead to straight wavefronts in medium 2, with the same frequency.

1; this is a consequence of causality, since each crest in the incident wave launches a crest in the transmitted wave.

Therefore, $v = \dfrac{\lambda}{T} = \lambda f \Rightarrow \dfrac{v_1}{v_2} = \dfrac{\lambda_1 f}{\lambda_2 f} = \dfrac{\lambda_1}{\lambda_2}$.

From the figure, $\lambda_1 = a \sin \theta_1$ and $\lambda_2 = a \sin \theta_2$. Therefore, $\dfrac{v_1}{v_2} = \dfrac{\sin \theta_1}{\sin \theta_2}$.

Your turn: Using the definition of the index of refraction, $n \equiv \dfrac{c}{v}$ and the above equation, show that

$$\boxed{n_1 \sin \theta_1 = n_2 \sin \theta_2} , \qquad (10.8.1)$$

Snell's Law

where n_1 is the index of refraction in medium 1 and n_2 is the index in medium 2.

Snell's Law describes refraction. For example, if medium 1 is air (for which $n_1 \cong 1$) and medium 2 is glass (for which n_2 is in the range 1.5–1.9), so that $n_2 > n_1$, then θ_2 must be smaller than θ_1, meaning that the ray "refracts toward the normal," as shown in figure 10.8.1. On the other hand, if $n_2 < n_1$ (as would be the case for a glass-to-air interface), then $\theta_2 > \theta_1$, meaning that the ray refracts away from the normal.

Example: A plano-convex lens. All of geometric optics can be derived from the simple idea that there must be some curved surface which refracts parallel incoming rays so that they meet a distance f behind the lens, where f is called the focal length. For manufacturing convenience, most lenses are made with spherical surfaces,

continued

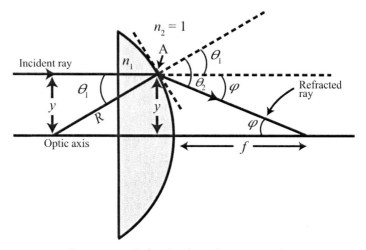

Figure 10.8.2 Refraction for a plano-convex lens.

even though this only approximately produces the desired convergence to a point. (The lack of ideal performance for such lenses is called "spherical aberration.") We consider a plano-convex lens, as shown in figure 10.8.2, made from glass of refractive index n_1. The lens is immersed in air, which we take to have index of refraction $n_2 \cong 1$. The "optic axis" is the horizontal line drawn through the center of the lens. Consider a ray of light parallel to the optic axis, a distance y above it. Because it strikes the planar surface of the lens at normal incidence, it is not refracted there, however it is refracted at the point where it encounters the curved surface of the lens, at point A. This surface has radius of curvature R; because the surface is spherical, a radius line drawn from A to the center of curvature is perpendicular to the lens surface. Therefore, the angle of incidence θ_1 is the angle between this radius and the incident ray, as shown. We will consider only rays traveling close to the optic axis; this is called the "paraxial approximation." Therefore, all the angles involved are small; for example, the smaller y is, the smaller θ_1 and θ_2 are. At point A, Snell's Law gives $n_1 \sin \theta_1 = n_2 \sin \theta_2$. Since the angles are small, we have $\sin \theta \cong \theta$, so that Snell's Law becomes

$$n_1 \theta_1 \cong n_2 \theta_2 = \theta_2.$$

From the figure, the "bending angle" is given by

$$\varphi = \theta_2 - \theta_1 \cong n_1 \theta_1 - \theta_1 \Rightarrow$$
$$\varphi \cong \theta_1 (n_1 - 1) \qquad (10.8.2)$$

and

$$\tan \varphi = \frac{y}{f} \Rightarrow$$
$$\varphi \cong \frac{y}{f}, \qquad (10.8.3)$$

where, in the paraxial approximation φ is small, so $\sin \varphi \cong \varphi$, $\cos \varphi \cong 1$, and $\tan \varphi \cong \varphi$. Combining equations (10.8.2) and (10.8.3) gives

$$\frac{y}{f} \cong \theta_1 (n_1 - 1).$$

From the figure, y is approximately equal to the arclength $\theta_1 R$, so that

$$\frac{\theta_1 R}{f} \cong \theta_1 (n_1 - 1) \Rightarrow$$

$$\frac{1}{f} \cong \frac{1}{R} (n_1 - 1). \tag{10.8.4}$$

So, a more strongly curved lens (one with small R) has a shorter focal length f, as is reasonable.

10.9 Total internal reflection and evanescent waves

Consider what happens when the incident medium has a higher index of refraction than the transmitted medium. Snell's law, equation (10.8.1), gives,

$$n_i \sin \theta_i = n_t \sin \theta_t \Leftrightarrow \sin \theta_t = \frac{n_i}{n_t} \sin \theta_i.$$

Since $n_i > n_t$, the refracted angle θ_t is greater than the incident angle θ_i, as shown in figure 10.9.1a. As θ_i increases, θ_t also increases, reaching $90°$ at a critical value of θ_i determined by

$$\sin 90° = \frac{n_i}{n_t} \sin \theta_i \Rightarrow$$

$$\theta_i \equiv \theta_c = \sin^{-1} \frac{n_t}{n_i}. \tag{10.9.1}$$

For values of θ_i greater than θ_c, there is no real angle θ_t which satisfies Snell's law, so there is no transmitted ray. Therefore, all the incident energy goes into the reflected ray; this is called total internal reflection.

If you've ever gone under water and looked up, you have seen total internal reflection. If you look straight up, you can see whatever is above the water (usually the sky). But, if you look off to the side, you'll notice that the underside of the water acts like a mirror, as shown in figure 10.9.2a.

Perhaps the most important application of total internal reflection is for fiber optics. As shown in figure 10.9.2b, if you shine a laser into the end of a long piece of glass or plastic, at an angle reasonably close to parallel to the long axis (i.e., at an angle such that $\theta_i > \theta_c$), then it reflects back and forth between the sides, as shown in the figure. By encoding information into pulses of a laser beam, this can be used to transmit information (e.g., telephone and internet signals) over long distances. In fact, essentially all long-range information transmission is now done with either fiber optics or satellite transmission.

The plastic used for the figure scatters the laser beam quite a bit (which allows us to visualize the beam), resulting in considerable energy loss, as you can see. Commercial fiber optics are made using ultra high purity glass, so that a signal can be transmitted for very long distances. There is still the need to regenerate the signal periodically, using a "repeater" which includes a receiver, elements that sharpen the pulses,

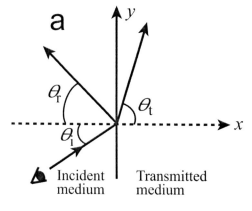

a

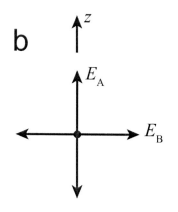

b

Figure 10.9.1 a: Angles of reflection and refraction for a ray incident from medium 1. For the case shown, $n_i > n_t$, so that the refracted ray is bent away from the normal. If θ_i is large enough, then $\theta_t = 90°$. For even larger values of θ_i, there is no real value of θ_t that satisfies Snell's Law, so there is no transmitted propagating wave. The incident, transmitted, and reflected rays all lie in the x–y plane. b: View looking along the incident ray (the viewpoint shown by the eye symbol in part (a)). The electric field vector for the incident wave can oscillate along any axis perpendicular to the direction of propagation. For example, it can oscillate in the $\pm z$-direction (which of course is perpendicular to the direction of propagation); this polarization is labeled E_A. Alternatively, the electric field can oscillate in the x–y plane (but perpendicular to the direction of propagation); this polarization is labeled E_B. Of course, the polarization could be along any axis in between those of E_A and E_B.

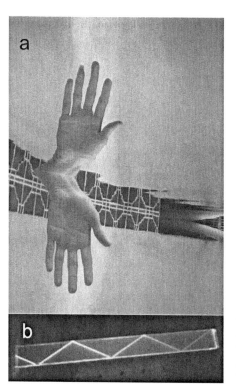

Figure 10.9.2 a: The hand's image is reflected by the underside of the water, due to total internal reflection. (Image courtesy of and © Jacob J. Loman) b: A laser beam bounces inside a block of the plastic PMMA, due to total internal reflection. (Image from Wikimedia Commons.)

and a transmitter. However, because of the very low absorption, the repeater stations can be 100 km apart, roughly 100 times further than for conventional copper wire cable. In fact, the maximum distance between repeaters for fiber optics is not limited by absorption, but rather by effects of dispersion (see section 9.12) which tends to smear out the pulses as they propagate. To avoid such dispersion, engineers must design the fiber optic system to use wavelengths of light over which the dispersion relation is close to linear. There are several "windows" of such wavelengths for properly processed glass, and fortuitously some of these coincide with wavelengths at which other critical components, such as optical amplifiers, can be made to work efficiently. Most fiber optic cables are operated at wavelengths near 1,500 nm, corresponding to the infrared. The frequency of this infrared light is 2×10^{14} Hz, so that, in principle, the length of each laser pulse can be extremely short, allowing a vast amount of information to be transmitted quickly over a single fiber. There are practical limitations associated with dispersion and with the equipment connected to the fibers that limit the data rate to much less than the theoretical maximum, although recent tests have achieved transmission rates of up to 1.4×10^{13} bits per second, over distances of up to 160 km between repeaters.[14] By contrast, a typical copper cable has a maximum transmission rate of 4×10^{9} bits per second, with a much shorter run between repeaters. Thus, a coaxial cable assembly containing 144 fibers, which is only 1.2 cm in diameter, can carry the same information as three copper cable assemblies, each 7.5 cm in diameter, each of which contains hundreds of twisted copper wire pairs. The weight and cost of the fiber is *far* less. Additionally, the different twisted wire pairs within a copper cable assembly exhibit significant "crosstalk" (the signal from one pair leaks onto another); this is completely eliminated for fiber optics. Copper wiring is usually used for local connections, because it is much simpler to connect the wires together, and the equipment costs are lower.

Fiber optics are made with a central core that has a high index of refraction, surrounded by a cladding layer with a lower index. As we have discussed earlier, by using total internal reflection, one can eliminate the refracted beam. However, as we will now show, the oscillating electric and magnetic fields actually do extend beyond the interface, in the form of an "evanescent wave." As we'll see, this has important implications for the design of fiber optics, as well as for the operation of certain scientific instruments.

In sections 9.5 and 9.6, we considered a wave propagating in the x-direction, with the electric field oscillating in the $\pm y$-direction, and the H field (or the B field for a wave in vacuum) oscillating in the $\pm z$-direction. However, we could also make a wave propagating in the x-direction for which the electric field oscillates in the $\pm z$-direction (and the H field in the $\mp y$-direction), or in fact any direction perpendicular to the direction of propagation. The axis along which the electric field points determines the "polarization" of the light. For the geometry of figure 10.9.1a, there are two distinct polarizations: (A) The electric field can oscillate parallel to the z-axis; this possible polarization is labeled E_A in part b of the figure. For this polarization, the electric field

---

14. NTT press release, Sept. 29, 2006.

is parallel to the interface. (B) The electric field can oscillate in the x–y plane, but in a direction perpendicular to the propagation; this possible polarization is labeled E_B. The x–y plane contains the incident, refracted, and reflected rays, and is sometimes called the "plane of incidence." Of course, the electric field could oscillate along any axis between these two extremes, but these other angles can be formed as a superposition of E_A and E_B. We will consider E_A and E_B separately, and apply boundary conditions to find the electric and magnetic fields in the transmitted medium for $\theta_i > \theta_c$, that is, for angles of incidence such that there is no transmitted ray. We will find that the results are similar for both cases.

Case A: Electric field polarized parallel to interface. We begin by considering the case $\theta_i < \theta_c$, so that there is a transmitted ray. A wave traveling in the x-direction can be written $f(kx - \omega t)$, where $k \equiv 2\pi / \lambda$ is the wavenumber. In problem 9.8, you can show that $f(\mathbf{k} \cdot \mathbf{r} - \omega t)$ represents a wave traveling in the direction of $\mathbf{k}$, with $|\mathbf{k}| = k = 2\pi / \lambda$, where $\mathbf{k}$ is called the wavevector. Thus, the incident electric field can be written

$$\mathbf{E}_i = \hat{\mathbf{k}} E_{i0} \cos\left(\mathbf{k}_i \cdot \mathbf{r} - \omega t\right) = \text{Re } \tilde{\mathbf{E}}_i, \quad \text{where } \tilde{\mathbf{E}}_i = \hat{\mathbf{k}} E_{i0}\, e^{i(\mathbf{k}_i \cdot \mathbf{r} - \omega t)}, \quad (10.9.2)$$

where $\mathbf{k}_i$ is the incident wavevector and $\hat{\mathbf{k}}$ is the unit vector in the z-direction. From figure 10.9.1a, we have $\mathbf{k}_i = k_i \left(\hat{\mathbf{i}} \cos\theta_i + \hat{\mathbf{j}} \sin\theta_i\right)$. So, we can rewrite the magnitude of the incident electric field as

$$E_i = \text{Re } \tilde{E}_i, \quad \text{where } \tilde{E}_i = E_{i0}\, e^{i(xk_i \cos\theta_i + yk_i \sin\theta_i - \omega t)}.$$

At the interface at $t = 0$, this reduces to

$$\tilde{E}_i\,(x = 0, t = 0) = E_{i0}\, e^{i\, yk_i \sin\theta_i}. \quad (10.9.3)$$

Now, for the reflected beam. Recall from the concept test at the end of section 10.6 that when light reflects off a glass-to-air interface there is no inversion of the phase ("glass to air, the phase don't care"). Therefore, the reflected electric field is simply

$$\mathbf{E}_r = \text{Re } \tilde{\mathbf{E}}_r, \quad \text{where } \tilde{\mathbf{E}}_r = \hat{\mathbf{k}} E_{r0}\, e^{i(\mathbf{k}_r \cdot \mathbf{r} - \omega t)}.$$

From figure 10.9.1a, we have $\mathbf{k}_r = k_r \left(-\hat{\mathbf{i}} \cos\theta_r + \hat{\mathbf{j}} \sin\theta_r\right)$. Plugging this into the above gives

$$\tilde{E}_r = E_{r0}\, e^{i(-xk_r \cos\theta_r + yk_r \sin\theta_r - \omega t)} \Rightarrow$$

$$\tilde{E}_r\,(x = 0,\, t = 0) = E_{r0}\, e^{i\, yk_r \sin\theta_r}. \quad (10.9.4)$$

Finally, the transmitted electric field is

$$\mathbf{E}_t = \text{Re}\tilde{\mathbf{E}}_t, \quad \text{where } \tilde{\mathbf{E}}_t = \hat{\mathbf{k}} E_{t0}\, e^{i(\mathbf{k}_t \cdot \mathbf{r} - \omega t)} \Rightarrow$$

$$\tilde{E}_t = E_{t0} e^{i(xk_t \cos\theta_t + yk_t \sin\theta_t - \omega t)} \Rightarrow \quad (10.9.5)$$

$$\tilde{E}_t\,(x = 0,\, t = 0) = E_{t0}\, e^{i\, yk_t \sin\theta_t}. \quad (10.9.6)$$

Recall from section 10.6 that one of the boundary conditions is that the component of the electric field parallel to the interface is continuous across the interface. For the polarization we are considering in this case, the entire electric field is parallel to the interface. Therefore, we must have

$$\tilde{E}_i\,(x = 0, t = 0) + \tilde{E}_r\,(x = 0, t = 0) = \tilde{E}_t\,(x = 0, t = 0)\,.$$

Substituting from equations (10.9.3), (10.9.4), and (10.9.6), we get

$$E_{i0}\,e^{i\,yk_i\sin\theta_i} + E_{r0}\,e^{i\,yk_r\sin\theta_r} = E_{t0}\,e^{i\,yk_t\sin\theta_t}\,. \tag{10.9.7}$$

This equation can only hold for all values of y if the factors that appear in the exponentials are equal.[15] Thus,

$$iyk_i\,\sin\theta_i = iyk_r\,\sin\theta_r = iyk_t\,\sin\theta_t, \tag{10.9.8}$$

and therefore

$$E_{i0} + E_{r0} = E_{t0}. \tag{10.9.9}$$

From equation (10.9.8) we get two significant results. First, since the speed is the same for the incident and reflected waves (as is the frequency), and since $v = \omega/k \Leftrightarrow k = \omega/v$, we have that $k_i = k_r$. So, using the first equality in equation (10.9.8), we have

$$\boxed{\theta_i = \theta_r,} \tag{10.9.10}$$

or "the angle of incidence equals the angle of reflection," as you probably already knew.

From equation (10.9.8) we also have

$$\sin\theta_t = \frac{k_i}{k_t}\sin\theta_i = \frac{\omega/v_i}{\omega/v_t}\sin\theta_i.$$

Since $n \equiv c/v$, this gives

$$\sin\theta_t = \frac{n_i}{n_t}\sin\theta_i, \tag{10.9.11}$$

which is Snell's Law arrived at by a somewhat different route from that in section 10.8.

Even for total internal reflection, where $\theta_i > \theta_c$, the boundary conditions still hold, and so the above arguments still hold. Therefore, there is still a transmitted wave, although we will see that it has a different mathematical form. (You can see

---

15. This may not be immediately obvious; the following may help to clarify. Define a new variable $\alpha \equiv e^{iy}$. Then, equation (10.9.7) becomes $E_{i0}\alpha^{k_i\,\sin\theta_i} + E_{r0}\alpha^{k_r\,\sin\theta_r} = E_{t0}\alpha^{k_t\,\sin\theta_t}$. You know that in any equation, the coefficients of like powers of the variable must be equal. In this equation, we have three different-looking exponents of α. If the two exponents on the left side of the equation ($k_i\sin\theta_i$ and $k_r\sin\theta_r$) were different, then the exponent on the right side ($k_r\sin\theta_r$) could only equal one of them, so the equation could not hold. Therefore, all three exponents must be equal to each other. (Also, we must have $E_{i0} + E_{r0} = E_{t0}$.)

from equation (10.9.9) that there must be a transmitted wave, since we have defined the amplitudes E_{i0}, E_{r0}, and E_{t0} to be positive.) We still have $\sin \theta_t = \dfrac{n_i}{n_t} \sin \theta_i$, but for $\theta_i > \theta_c$ we have $\dfrac{n_i}{n_t} \sin \theta_i > 1$, so that θ_t is an imaginary angle.

Your turn: Using $\sin \theta_t = \dfrac{e^{i\theta_t} - e^{-i\theta_t}}{2}$, show that we can have $\sin \theta_t > 1$ if θ_t is imaginary. Must θ_t be a positive imaginary number (of the form $+iA$) or a negative imaginary number (of the form $-iA$)?

We can write

$$\cos \theta_t = \sqrt{1 - \sin^2 \theta_t} = \sqrt{1 - \frac{n_i^2}{n_t^2} \sin^2 \theta_i} \Rightarrow$$

$$\cos \theta_t = i \sqrt{\frac{n_i^2}{n_t^2} \sin^2 \theta_i - 1} \Rightarrow$$

$$\cos \theta_t = \frac{i}{n_t} \sqrt{n_i^2 \sin^2 \theta_i - n_t^2}, \tag{10.9.12}$$

where for $\theta_i > \theta_c$ the quantity inside the square root of equation (10.9.12) is positive. The transmitted electric field is still given by equation (10.9.5):

$$\tilde{E}_t = E_{t0} e^{i(xk_t \cos \theta_t + yk_t \sin \theta_t - \omega t)}.$$

Substituting for $\cos \theta_t$ using equation (10.9.12), and for $\sin \theta_t$ using equation (10.9.11) gives

$$\tilde{E}_t = E_{t0}\, e^{-x\frac{k_t}{n_t}\sqrt{n_i^2 \sin^2 \theta_i - n_t^2}}\, e^{i\left(yk_t \frac{n_i}{n_t} \sin \theta_i - \omega t\right)}.$$

Your turn: Since $k = \dfrac{\omega}{v}$ and $n = \dfrac{c}{v}$, we have that $k_t = \dfrac{\omega n_t}{c}$. Use this to show that

$$\tilde{E}_t = E_{t0}\, e^{-\kappa x}\, e^{i(k_e y - \omega t)}, \tag{10.9.13}$$

where

$$\kappa \equiv \frac{\omega}{c} \sqrt{n_i^2 \sin^2 \theta_i - n_t^2}, \tag{10.9.14}$$

and the wavenumber for the "evanescent" wave in the transmitted medium is

$$k_e \equiv \frac{\omega n_i}{c} \sin \theta_i. \tag{10.9.15}$$

Equation (10.9.13) describes a wave that travels in the $+y$-direction, parallel to the interface. (The polarization is still along the z-axis.) As described by the $e^{-\kappa x}$ term, the wave dies off exponentially with distance into the transmitted medium. We will explore the implications of this "evanescent wave" after we consider the other polarization.

Case B: electric field polarized in the plane of incidence. If the electric field lies in the x–y plane, then $\mathbf{H}$ is parallel to the interface, that is, parallel to the z-axis. Recall from section 10.6 that the component of $\mathbf{H}$ that is parallel to the interface is continuous across the interface. Since this is the same condition as we had for $\mathbf{E}$ for the case A polarization, the entire rest of the argument is identical, with the substitution of H for E. Thus, we obtain

$$\tilde{H}_t = H_{t0}\, e^{-\kappa x}\, e^{i(k_e y - \omega t)}, \qquad (10.9.16)$$

with the same values for κ and k_e. Again, this describes a wave propagating in the $+y$-direction, and exponentially decaying in the x-direction. The connections between E and H are more complicated in this circumstance than in an infinite expanse of linear matter (we don't simply have $E = \sqrt{\dfrac{\mu}{\varepsilon}}H$), so the waves in the transmitted medium are different for the two polarizations (in ways beyond their polarization), but they are qualitatively very similar.

Because the evanescent wave propagates parallel to the interface, there is ordinarily no transmission of power into the transmitted medium, so that the power of the reflected beam equals the power of the incident beam. However, the oscillating electromagnetic field of the evanescent wave *can* cause local excitation of atoms which absorb at that frequency. For this reason, the cladding which provides the low index of refraction around the high-index core of a fiber optic must be very pure, to avoid absorption by impurity atoms.

Connection to current research: Total Internal Reflection Fluorescence Microscopy. This type of absorption is used to advantage in Total Internal Reflection Fluorescence Microscopy (TIRFM). In fluorescence, an electron in an atom is excited to a high energy state using ultraviolet light. As the electron returns to its original state through a series of steps, it emits visible light of a characteristic wavelength. In TIRFM, a sample of interest (usually biological) is applied to the surface of a quartz substrate. Ultraviolet light is transmitted within the substrate, contained by total internal reflection. The associated evanescent wave extends at most 100 nm into the biological sample, so only atoms within this thickness can be excited by the wave and fluoresce. Thus, the image of the fluorescent light comes from the very thin layer of sample in contact with the quartz; this localization can be very important for various experiments. For example, Gunnarson and co-authors used a sequence of tethers to attach individual DNA molecules to a quartz substrate, as shown schematically in figure 10.9.3a. The arrows at the bottom of the figure schematically represent the rays of UV light bouncing off the top surface of the substrate because of total internal reflection. The bound DNA molecules then bind to complementary DNA strands which are attached to fluorescent self-assembled nanospheres in the solution above. Only the bound nanospheres fluoresce, due to the small extent of the evanescent wave. Thus, each bright spot in the image (figure 10.9.3b) represents the binding of a nanosphere to a single DNA molecule. The simplest way to use this technique is to test for the presence of the complementary DNA strand in a target solution. However, it can also be used in other ways. For example, by measuring how much time each nanosphere stays bound, the scientists can detect whether there are any mismatches in the base sequences of the two complementary DNA strands.

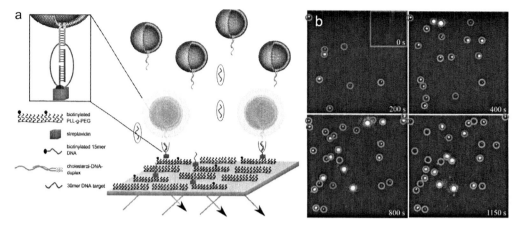

Figure 10.9.3 a: Schematic diagram showing how DNA molecules that are 30 bases long ("30-mers") are attached via a series of tethering molecules to a quartz substrate. Self-assembled fluorescent nanospheres in the solution then attach to the DNA. b: Sequence of TIRFM images. Each spot (highlighted by a grey circle) is a single nanosphere attached to a single DNA molecule. Reprinted with permission from A. Gunnarson et al., *Nano Letters* **8**, 183–8 (2008). Copyright 2008 American Chemical Society.

Concept and skill inventory for chapter 10

After reading this chapter, you should fully understand the following terms:

Boundary conditions (10.1)
Characteristic impedance for mechanical systems (10.3)
F_L (10.3)
Terminator (10.5)
Characteristic impedance of a transmission line (10.5)
Index of refraction (10.6)
Snell's Law (10.8)
Refraction (10.8)
Total internal reflection (10.9)
Evanescent wave (10.9)

You should understand the following connections:

$\dot{y}$, B, H, I, & v (10.5–10.7)
F_L, E, V, & P' (10.5–10.7)
μ, μ_0, μ, L_0, & ρ_0 (10.5–10.7)
T, $1/\varepsilon_0$, $1/\varepsilon$, $1/C_0$, & γP_0 (10.5–10.7)

You should understand the difference between:

The phase of an em wave reflected off a glass-to-air interface & the phase of an em wave reflected off an air-to-glass interface (10.6)

You should be familiar with the following additional concepts:

For a right-traveling wave, the motion of the left end of the rope determines the wave shape. (10.2)

You should be able to:

Calculate the characteristic impedance of a mechanical system (10.3)

Find the shape of the transmitted and reflected pulses on ropes given the shape of the incident pulse and the speeds on each side of the interface (10.1–10.2)

Find the shape of the transmitted and reflected pulses for rope waves, transmission line waves, em waves in vacuum or linear media, and sound waves, given the shape of the incident pulse and the information needed to calculate the impedance (10.4–10.7)

Explain how to suppress reflections with a compact terminator for rope waves (10.3) or transmission line waves (10.5)

Calculate the angle of refraction (10.8)

Calculate the critical angle for total internal reflection (10.8)

Calculate the amplitude of an evanescent wave in the transmitted medium (10.8)

In addition to all of the above, you should be able to combine the concepts you've learned to address new situations.

Problems

Note: Additional problems are available on the website for this text.

Instructor: Ratings of problem difficulty, full solutions, and important additional support materials are available on the website.

10.1 **"The Wave"** Spectators in large, round stadiums often perform a maneuver called "the wave," in which a "pulse" of people standing and waving their arms propagates around the stadium while the rest of the people remain seated. In this problem, you'll model this phenomenon with a "peer pressure" model, in which the motion of a particular spectator is completely determined by what the people to his left and right are doing. (In other words, there is no free will in this model!) Here's the mathematical form for our model:

$$\ddot{y}_p = -s\left(y_p - y_{p-1}\right)^{\beta} - s\left(y_p - y_{p+1}\right)^{\beta},$$

where β and s are constants, y_p is the position of the pth spectator, y_{p-1} is the position of the spectator to his left, and y_{p+1} is the position of the spectator to his right. Assume that negative values of y are allowed (perhaps by crouching down in the seat), as well as positive values. The normal sitting position corresponds to $y = 0$. We will focus only on spectators in the front row, and assume that this row extends without a break around the entire circular stadium. Note that, for pulses shorter than the circumference of the stadium, this is like an infinitely long chain of people. We'll assume that

Figure 10.P.1 Waveform for "the wave."

the big changes in y occur on length scales much longer than the spacing between spectators, but still much shorter than the stadium circumference.

(a) A pulse with a triangular shape (as shown in figure 10.P.1) propagates without changing shape, and an identical pulse travels in the opposite direction. When the peaks of the two pulses reach the same point, the instantaneous value of y is twice the peak height of either pulse alone. The pulses continue on, without having affected each other by their "collision." What must be the value of β? Explain.

(b) Now there is a single triangular pulse, as shown in the figure. Queen Elizabeth II is in the front row. It is an experimental fact that she will not participate in the wave. Mathematically, we could say that $y = 0$ for the queen, no matter what the people near her are doing. Describe what happens when the pulse gets to the queen's position.

(c) The wave has now died away, and everyone in the stadium is sitting (i.e., $y = 0$ for everyone). Now, the queen decides to stand up, and continues standing. Describe qualitatively what happens.

10.2 Reflection and transmission of two pulses. Two ropes under tension are joined together, with $v_1 = \frac{1}{2}v_2$, where v_1 is the velocity of waves in the left rope and v_2 is the velocity in the right rope. As shown figure 10.P.2, pulses approach the interface from both sides. The pulse on the right has half the height of that on the left, but the same width.

(a) Sketch the shape of the rope for the instant when the halfway point of the left pulse reaches the interface.

(b) Sketch the shape of the rope well after both pulses have reached the interface.

10.3 For sinusoidal waves on a rope that encounter an interface with a different kind of rope, show that energy is conserved, that is, that the power of the

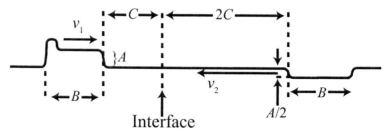

Figure 10.P.2 Pulses incident from left and right on an interface.

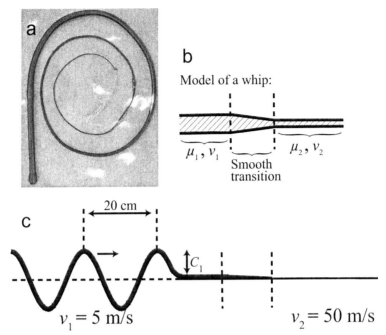

Figure 10.P.3 a: Photo of an Australian bull whip. b: Model for a whip. c: A sinusoidal wave is incident from the left. Image credit for part a: © 1998 by C. Goodwin (license at artlibre.org/licence/lal/en/).

incident wave equals the sum of the powers for the reflected and transmitted waves.

10.4 Modeling a whip. Instead of an abrupt interface from one material to another, one can create a "gradual" interface, in which the speed of wave propagation changes smoothly from v_1 to v_2. If such a gradual interface is made just right, reflections can be almost eliminated. This has important practical consequences for coupling high-frequency circuits to antennas, and also for designing a whip. As shown in figure 10.P.3, a real whip has a gradually tapering diameter. We will use a simplified model, as shown in part (b). At the left, we assume a uniform diameter, with wave speed v_1. In the middle, there is a tapering region. At the right, there is a uniform, smaller diameter, with wave speed v_2. The tension T is the same throughout. Assume the tapering region is made "perfectly," so that there is no reflection for waves coming in from the left.

A sinusoidal wave train propagates to the right, along the thick portion, as shown in part (c). To make the whip "crack," the transverse velocity $\dot{y}$ must exceed the speed of sound in air ($v_s = 330$m/s), **once the wave train propagates into the thinner part of the rope on the right.** What is the minimum value of C_1 required for this to happen, given the other information in the figure? You should give a numerical answer. *Hint: Because of the gradual interface, you cannot use the formula for an abrupt interface to find the amplitude of the transmitted wave. Instead, use conservation of energy to find this amplitude.*

10.5 **(a)** Write an expression for the mechanical impedance of a damped, driven harmonic oscillator as a function of frequency in terms of k, b, m, F_0, and ω_d. To keep things neat, you may write your expression in terms of other quantities, so long as you define these other quantities in terms of k, b, m, F_0, and ω_d. Assume the oscillator is in steady state. **(b)** You should find that your expression for Z is complex. What does that imply?

10.6 A weasel holds the left end of a long rope which is under tension and initially at rest. The weasel then starts to move the end up and down sinusoidally. Show that power delivered by the weasel to the rope, averaged over a cycle, is equal to the power in the wave.

10.7 In section 10.5, I claimed that if you had an infinitely long coaxial cable and applied a constant voltage V to the left end of it, you would need to supply a constant current $I = V/Z$, where Z is the characteristic impedance of the coax, in order to charge up the capacitance between the inner and outer conductors, as the wavefront of V propagates to the right. Although we have already made arguments that show this is correct in section 10.5, sometimes it is helpful to see things from more than one point of view. Starting from $Q = CV$, and taking into account the time dependence of the effective capacitance (as the wavefront propagates to the right), show that, indeed, $I = V/Z$.

10.8 The characteristic impedance of standard coaxial cable is 50 Ω. **(a)** Explain what this means. **(b)** Why do we not measure a resistance of 50 Ω when we connect a normal ohmmeter between the inner and outer conductors of a coaxial cable found in any of our labs?

10.9 Standard coaxial cable has $C_0 = 100$ pF/m and $L_0 = 250$ nH/m. The right end of a length of standard coax is connected to a length of a different kind of coax which also has $C_0 = 100$ pF/m but has $L_0 = 500$ nH/m. A right moving current pulse is sent down this cable. Describe quantitatively what happens when the pulse encounters the junction. Include comments about the relative height and width of the incident current pulse, reflected current pulse, and transmitted current pulse. Also include comments about the relative height and width of the incident voltage pulse, reflected voltage pulse, and transmitted voltage pulse.

10.10 **Disappearing beads**. If colorless, transparent plastic beads are put into a liquid that has the same refractive index as the beads, they become invisible. Explain.

10.11 **RFID tags**. A radio frequency identification (RFID) tag is a small device attached to an item that needs to be tracked. To locate or identify the item, one uses a "reader" unit to broadcast a radio frequency (RF) electromagnetic wave (which we'll call the interrogation signal) into the area containing the item with the attached RFID tag. The tag then sends back an RF response signal indicating both its presence and a tracking number. The lowest cost versions of these tags are powered by the energy in the interrogation signal. There are several different methods for generating the response wave. We can model one of these technologies by an antenna which can be connected to ground by an electrically controlled switch. When the switch is closed, current can flow from the antenna to ground. When the switch is open, no current flows. When the circuit within the RFID tag detects the interrogation signal, it opens and closes the switch in a specific pattern determined by the tracking number associated with that particular tag. By controlling the flow of current in the antenna, the circuit controls the strength of the reflected electromagnetic wave (as you'll explore in this problem); this reflected wave is used as the response signal. **(a)** Explain qualitatively why the reflected signal is stronger when the switch is closed than when it is open. **(b)** Assume the electric field of the interrogation signal at the position of the RFID tag is $E_i \cos \omega t$, where the incident amplitude E_i is positive. Is the reflected wave given by $E_r \cos \omega t$ or instead by $-E_r \cos \omega t$, where E_r is positive. Explain your reasoning. (Note: you are not expected to determine the magnitude of E_r.)

10.12 Impedance matching. Suppose you want to maximize the transmission of a wave from medium 1 to medium 2, but they have different impedances Z_1 and Z_2. You might think that inserting a slab of a third medium, call it medium A, in between would lower the transmission, because now there are two interfaces (one between 1 and A, and another between A and 2), with reflections occurring at both interfaces. However, if the impedance of A, Z_A, is chosen correctly, the net transmission from 1 to 2 can be significantly enhanced. **(a)** Assuming that all the impedances are real (and positive), what is the best choice for Z_A to maximize net transmission? (Note: you should ignore second-order reflection effects. For example, you should not worry about the fact that the wave which is reflected to the left off of the A-2 interface will be partly reflected back to the right off of the 1-A interface. Such effects would only increase the overall enhancement achieved by adding the slab of A.) **(b)** For the case $Z_2 = 3Z_1$, by what factor can the overall amplitude transmission be enhanced?

10.13 For electromagnetic plane waves that propagate in a direction perpendicular to the interface between two linear materials, both of which have $\mu \cong \mu_0$, show that

$$S_t \cong S_i \frac{4n_i n_t}{\left(n_i + n_t\right)^2}.$$

10.14 Show that, for electromagnetic waves in linear materials with $\mu \cong \mu_0$

$$\left[H_{\text{reflected}} \cong H_{\text{incident}} \frac{n_t - n_i}{n_i + n_t}\right]_{x=0} \quad \text{and}$$

$$\left[H_{\text{transmitted}} \cong H_{\text{incident}} \frac{2n_t}{n_i + n_t}\right]_{x=0}.$$

10.15 A plane wave of light with intensity 1.25 kW/m² travels through the air, and strikes a thick pane of glass at normal incidence. (The intensity is averaged over one cycle of the wave; intensity is always quoted in this way.) The glass has $\mu \cong \mu_0$, and dielectric constant $\kappa = 5.6$. Assume that the electric and magnetic fields are mutually perpendicular in the glass and the air. **(a)** What are the rms amplitudes of the electric and magnetic fields for the incident wave? **(b)** What are the rms amplitudes of the electric and magnetic fields for the reflected wave? **(c)** What are the rms amplitudes of the electric and magnetic fields for the transmitted wave?

10.16 **Thin film interference.** Extraordinarily beautiful patterns can be made by reflection of light off thin films. You have probably seen the colorful patterns of light reflecting from a thin film of oil or gasoline as it spreads on a water surface. By suspending a soap film vertically, one can observe spectacular patterns as the soapy fluid flows toward the bottom. (For examples, simply do an internet image search for "soap film.") In this problem, you will investigate this phenomenon. The index of refraction for soapy liquid is approximately 1.4. (Take this as exact for this problem.) Consider a film in the y–z plane of soapy liquid with thickness d. The film was created by dipping a hoop into a bucket of soapy liquid, and then holding the hoop vertically. Over time, the soapy liquid flows from the top to the bottom, so that the thickness of the film at the top gets smaller. We aim a beam of light at this top part. **(a)** Take the incident light to be traveling in the $+x$-direction, so that it is at normal incidence to left side of the film (hereafter referred to as the "front side"). Show that only 2.78% of the incident intensity is reflected. **(b)** The result of part (a) means that the beam transmitted into the soapy liquid is essentially as strong as the incident beam. Explain briefly why the intensity reflected off the back surface of the film (i.e., the right surface, where there is a liquid-to-air interface) is 2.78% of the intensity of the beam that was transmitted through the front surface.

Almost all the light that is reflected off the back surface will get through the front surface. (Again, only 2.78% gets re-reflected off the front surface back to the right.) Therefore, there are two beams reflected to the $-x$-direction, one from the front surface and one from the back, and these beams are of almost equal intensity. Experimentally, one observes that as the soap film gets thinner, eventually the top part (which is thinnest) "disappears," meaning that it reflects essentially no light. At this point, the thickness of the film at the top part is less than 10 nm. **(c)** Explain why this reflects essentially no light. (Bear in mind that the wavelength of visible light in air is 380–750 nm.) **(d)** At a point lower down in the film, where it is thicker, the film appears blue (corresponding to a wavelength in air of 480 nm), meaning that the light reflected back to the left is primarily blue. What is the approximate thickness of the soap film at this point?

10.17 **Anti-reflection coatings.** (You should do problems 10.12 and 10.16 before this.) In problem 10.12, you showed how to maximize transmission of a wave from one medium to another by inserting a slab of appropriate impedance between them. This, of course, also minimizes reflection. This method for minimizing reflection works for all wavelengths of incident waves, and for all types of waves. In this problem, we consider light waves in particular. Frequently, optical components such as lenses have "anti-reflection" coatings applied, to minimize the reflection. **(a)** For a lens made from glass with index of refraction 1.52, which is to be used in air, what index of refraction should the coating have? (Assume the light impinges on the lens at normal incidence.) **(b)** The impedance matching condition tells you the index of refraction needed for the coating, but says nothing about the optimum thickness. For the soap film treated in problem 10.16, the reflection was minimized at all wavelengths for a film thickness less than 10 nm. Why not use this same strategy, and just make the anti-reflection coating less than 10-nm thick? **(c)** In fact, one must know the wavelength of the light being used to choose the best thickness for the coating. If the lens is to be used for light from an Argon ion laser (wavelength of 488 nm in air), what is the best choice for the thickness of the anti-reflection coating?

10.18 Prisms. Figure 10.P.4 shows a schematic picture of white light entering a prism (from the left), and being split into its component colors. Based on this, make a qualitative sketch of the dispersion relation for light in glass. Your sketch should show whether the curve slopes up or down, and whether it is concave up, concave down, straight, or some other shape. *Hint: The wavelength of red light is longer than that of blue light.*

10.19 In a particular setup for Total Internal Reflection Fluorescence Microscopy (TIRFM), the ultraviolet light is provided by a laser, so that the angle of the light relative to the surface of the quartz is well-defined, but can be varied by turning a knob on the apparatus. Explain how this would allow the experimenter to control the thickness of the sample that is being probed.

10.20 Dispersion in a fiber optic cable. A fiber optic cable is used to transmit a series of laser pulses that represent computer data. The range of wavenumbers used to Fourier synthesize the pulses spans from k_1 to k_2, with an average value of k_{av}. Over this range, the dispersion relation for the glass used for the fiber optic can be approximated by $\omega = \alpha k + \alpha \beta\, k^2/k_{av}$, where β is a small dimensionless number. For the pulses to remain readable, the wave at k_1 must stay in almost exactly the same alignment relative to the wave at k_2 as the pulse propagates; the maximum allowable relative shift of these two waves is a distance of $0.01/k_{av}$. What is the approximate maximum distance between repeater stations, assuming that this dispersion is the limiting factor?

10.21 One-way viewing? You have probably seen "one-way mirrors" in television shows. Typically, a criminal suspect is being interrogated in one room, while several police officers watch through a one-way mirror from another

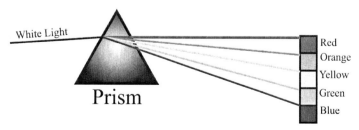

Figure 10.P.4 A prism is used to separate white light into beams of different wavelengths.

room. To the criminal, the mirror looks like a mirror, so that he only sees his own reflection, but for the policemen it works like a window, so that they can see the criminal. In fact, this one-way effect is completely due to the difference in lighting between the two rooms. The "one-way mirror" is coated with a thinner layer of reflective metal than a normal mirror, so that perhaps 80% of the light is reflected and 20% is transmitted, instead of having 100% transmission. The fraction transmitted and reflected is the same in both directions. However, the room with the criminal is kept brightly lit, whereas the other room is kept fairly dark. Thus, almost all of the light the criminal sees comes from his own room, whereas most of the light that the policemen see comes from the criminal's room.

You can easily demonstrate this for yourself. Find a CD-R (i.e., a CD that you might write data onto using your computer) that is silvery on both sides. (Some CD-R's are painted or have a label on one side; these won't work for our purpose.) If you hold the CD-R up to a bright light, you should be able to see the light dimly through the CD-R, because the metal coating is not very thick. Now stand in front of a mirror and hold the CD-R tight against your right eye. Looking at yourself in the mirror with your left eye, you can't see your right eye. But, looking with your right eye, you can still see the reflection of yourself in the mirror. (It's a bit hazy, because of the grooves in the CD.)

Having recently learned about total internal reflection, an inventor proposes a scheme for a true one-way viewing system that would work even if both rooms are equally illuminated. The inventor wishes to patent his idea. As a patent office clerk, you are called on to decide whether the idea has merit. In the drawing that accompanies the patent application (figure 10.P.5), the two rooms are separated by a large piece of glass with a triangular cross section. The application reads, "Light from the police officer on the left undergoes total internal reflection at the glass-to-air interface, as shown by the black ray, and so does not reach the criminal. However, since the index of refraction of air is lower than that of glass, there

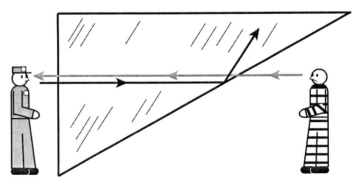

Figure 10.P.5 A proposed scheme for one-way viewing. According to the scheme, light from the prisoner reaches the policeman, but light from the policeman undergoes total internal reflection, and so does not reach the prisoner. What is wrong with this reasoning?

is no such total internal reflection for the light coming from the criminal (as shown by the gray ray), so the police officer can see him." What is wrong with this argument?

10.22 Your head is 1-m below the surface of a swimming pool. Looking straight up, you can see a circle of the sky, but beyond this circle (i.e., at larger angles relative to straight up) the bottom surface of the water looks silvery. The index of refraction for water is 1.33. What is the radius of the circle on the surface of the water outside of which the surface is silvery?

A Group Velocity for an Arbitrary Envelope Function

In this appendix, we wish to show that the group velocity $v_g = \dfrac{d\omega}{dk}\bigg|_{k_c}$ is the velocity of the envelope of a wavepacket that is composed of Fourier components with wavenumbers centered on k_c. First, we will show that we can easily construct such a wave packet by multiplying a sinusoidal oscillation with wavenumber k_c by an envelope function.

We consider an envelope function $f(x)$ that is composed of only low-wavenumber (long wavelength) Fourier components, that is, the Fourier components are only nonzero for $|k| < k_m$, where k_m is the maximum wavenumber needed to Fourier synthesize $f(x)$. We can write $f(x)$ as a sum of its Fourier components:

$$f(x) = \frac{1}{\sqrt{2\pi}} \int_{-\infty}^{\infty} F(k)\, e^{ikx} dk, \tag{A1}$$

where we write the Fourier transform as $F(k)$ rather than $Y(k)$, and $F(k)$ is nonzero only from $-k_m$ to k_m, that is, only for $|k| < k_m$. We place no other restrictions on $F(k)$, so that our arguments are valid for a very general envelope function $f(x)$, restricted only by the range of frequencies of the sinusoids summed to create it. We use $f(x)$ as an envelope function for a carrier wave:

$$y(x) = f(x)\cos k_c x = \mathrm{Re}\left[z(x)\right], \tag{A2}$$

where

$$z(x) = f(x)\, e^{ik_c x}.$$

An example of the function $y(x)$ is shown in figure A1. Using (A1), we can write

$$z(x) = \left[\frac{1}{\sqrt{2\pi}} \int_{-\infty}^{\infty} F(k)\, e^{ikx} dk\right] e^{ik_c x}.$$

This presentation is based in part on that in *The Physics of Waves*, by Howard Georgi, Prentice-Hall, Englewood Cliffs, NJ, 1993, pp. 235–6.

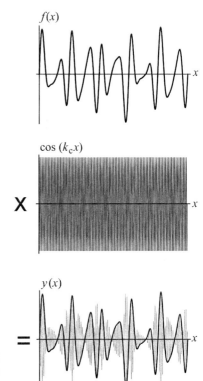

Figure A1 Multiplying an envelope function by a high-wavenumber "carrier wave" creates the modulated function shown at the bottom.

Because the integral is over k, we can take the factor $e^{ik_c x}$ inside the integral, giving

$$z(x) = \frac{1}{\sqrt{2\pi}} \int_{-\infty}^{\infty} F(k)\, e^{i(k+k_c)x} dk.$$

We define $k' = k + k_c \Leftrightarrow k = k' - k_c$, so that

$$z(x) = \frac{1}{\sqrt{2\pi}} \int_{-\infty}^{\infty} F(k' - k_c)\, e^{ik'x} dk'.$$

Since k' is merely a variable of integration, we can drop the prime, so that

$$z(x) = \frac{1}{\sqrt{2\pi}} \int_{-\infty}^{\infty} F(k - k_c)\, e^{ikx} dk. \tag{A3}$$

Note that this has the same form as the Fourier expansion of $z(x)$:

$$z(x) = \frac{1}{\sqrt{2\pi}} \int_{-\infty}^{\infty} Z(k)\, e^{ikx} dk.$$

Thus, the Fourier transform of $z(x)$ is $Z(k) = F(k - k_c)$.

The only contributions to the integral in (A3) are from wavenumbers for which $F\left(k-k_c\right)$ is nonzero. Recall that $F(k)$ is nonzero only for $|k| < k_m$, so that $F\left(k-k_c\right)$ is only non-zero for $\left|k-k_c\right| < k_m$, as shown in figure A2a and b. Thus, by using an envelope function to modulate a carrier wave of wavenumber k_c, we have indeed constructed a wavepacket that is composed of Fourier components with wavenumbers centered on k_c.

Now, we want to make this function move as a traveling wave. To make a simple function such as Ae^{ikx} move to the right as a traveling wave, we replace the e^{ikx} by the traveling wave $e^{i(kx-\omega t)}$, giving $Ae^{i(kx-\omega t)}$. We apply this same operation to each of the factors e^{ikx} (A3) giving

$$z(x,t) = \frac{1}{\sqrt{2\pi}} \int_{-\infty}^{\infty} F\left(k-k_c\right) e^{i(kx-\omega t)}\, dk, \qquad \text{(A4)}$$

where ω is a function of k, as described by the dispersion relation for the particular system.

Again, the only contributions to the integral in (A4) are from the wavenumbers $\left|k-k_c\right| < k_m$ for which $F\left(k-k_c\right)$ is nonzero. We assume that, over this range of

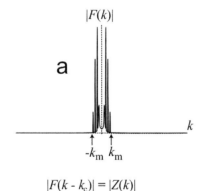

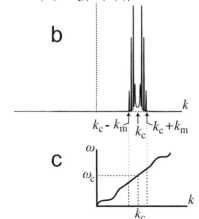

Figure A2 a: Magnitude of $F(k)$, the Fourier transform of the $f(x)$ shown in fig A1. $F(k)$ is nonzero only for the range $-k_m$ to k_m. b: Multiplying $f(x)$ by a sinusoidal carrier wave with wavenumber k_c shifts the Fourier spectrum so that it is now centered on k_c. c: If the dispersion relation is linear over the range of nonzero $Z(k)$, then we can show that the envelope of the wave moves at v_g without changing shape.

wavenumbers we can write the dispersion relation as

$$\omega = \omega_c + (k - k_c) \left. \frac{d\omega}{dk} \right|_{k_c}, \tag{A5}$$

where $\omega_c = \omega(k_c)$ is the angular frequency corresponding to the wavenumber k_c of the carrier wave. In words, our assumption (A5) says that ω depends linearly on k, though it need not be directly proportional. An example of such a dispersion relation is shown in figure A2c. (In fact, if k_m is small enough, then any dispersion relation will satisfy this restriction, since any function is linear over a small enough range.) Defining $v_g = \left. \frac{d\omega}{dk} \right|_{k_c}$, we can rewrite equation (A5) as

$$\omega = \omega_c + (k - k_c) v_g. \tag{A6}$$

Substituting this into equation (A4) gives us

$$z(x, t) = \frac{1}{\sqrt{2\pi}} \int_{-\infty}^{\infty} F(k - k_c) \, e^{ikx} e^{-i\omega_c t} e^{-i\, k v_g t} e^{ik_c v_g t} dk.$$

Now, we define $k'' \equiv k - k_c \Leftrightarrow k = k'' + k_c$, so that

$$z(x, t) = \frac{1}{\sqrt{2\pi}} \int_{-\infty}^{\infty} F(k'') \, e^{i(k'' + k_c)x} e^{-i\omega_c t} e^{-i(k'' + k_c)v_g t} e^{ik_c v_g t} dk.$$

$$= \frac{1}{\sqrt{2\pi}} \int_{-\infty}^{\infty} F(k'') \, e^{i(k_c x - \omega_c t)} e^{ik''(x - v_g t)} dk''.$$

Again, k'' is just a variable of integration, so we can rewrite it as k:

$$z(x, t) = \frac{1}{\sqrt{2\pi}} \int_{-\infty}^{\infty} F(k) \, e^{i(k_c x - \omega_c t)} e^{ik(x - v_g t)} dk \Rightarrow$$

$$z(x, t) = e^{i(k_c x - \omega_c t)} \frac{1}{\sqrt{2\pi}} \int_{-\infty}^{\infty} F(k) \, e^{ik(x - v_g t)} dk. \tag{A7}$$

Since equation (A1): $f(x) = \frac{1}{\sqrt{2\pi}} \int_{-\infty}^{\infty} F(k) \, e^{ikx} dk$, we have that

$$f(x - v_g t) = \frac{1}{\sqrt{2\pi}} \int_{-\infty}^{\infty} F(k) \, e^{ik(x - v_g t)} dk,$$

so that equation (A7) becomes

$$z(x, t) = e^{i(k_c x - \omega_c t)} f(x - v_g t).$$

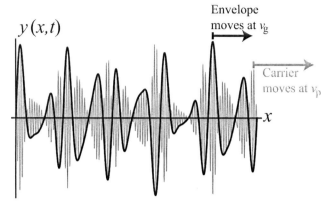

Figure A3 The traveling wave $y(x, t)$.

Note that $f\left(x - v_g t\right)$ is just the shape $f(x)$ moving to the right at speed v_g. The actual wave is found by taking the real part:

$$y(x, t) = \text{Re}\left[z(x, t)\right] = \underbrace{\cos\left(k_c x - \omega_c t\right)}_{\substack{\text{carrier wave} \\ \text{travels at} \\ v_p = \dfrac{\omega_c}{k_c}}} \underbrace{f\left(x - v_g t\right)}_{\substack{\text{envelope} \\ \text{travels at} \\ v_g}}$$

This is what we set out to show, and is illustrated in figure A3.

Index